Remote Sensing of Environmental Changes in Cold Regions

Remote Sensing of Environmental Changes in Cold Regions

Special Issue Editors

Jinyang Du
Jennifer D. Watts
Hui Lu
Lingmei Jiang
Paolo Tarolli

MDPI • Basel • Beijing • Wuhan • Barcelona • Belgrade

MDPI

Special Issue Editors

Jinyang Du
University of Montana
USA

Jennifer D. Watts
Woods Hole Research Center
USA

Hui Lu
Tsinghua University
China

Hui Lu
Tsinghua University
China

Lingmei Jiang
Beijing Normal University
China

Paolo Tarolli
University of Padova
Italy

Editorial Office
MDPI
St. Alban-Anlage 66 4052
Basel, Switzerland

This is a reprint of articles from the Special Issue published online in the open access journal *Remote Sensing* (ISSN 2072-4292) from 2018 to 2019 (available at: https://www.mdpi.com/journal/remotesensing/special_issues/cold_rs)

For citation purposes, cite each article independently as indicated on the article page online and as indicated below:

LastName, A.A.; LastName, B.B.; LastName, C.C. Article Title. *Journal Name* **Year**, *Article Number*, Page Range.

ISBN 978-3-03921-570-6 (Pbk)
ISBN 978-3-03921-571-3 (PDF)

Cover image courtesy of Jennifer Watts

Contents

About the Special Issue Editors

Jinyang Du is a research scientist at the University of Montana, Missoula, Montana, USA. His research interests include quantitative remote sensing of landscape freeze/thaw state, vegetation water content, soil moisture and snow water equivalent for global environmental change studies. He was awarded IEEE Geoscience and Remote Sensing Society's Highest Impact Paper Award for 2015.

Jennifer D. Watts is an assistant scientist at the Woods Hole Research Center, Falmouth, Massachusetts. She is also an affiliate assistant professor in the Department of Land Resources and Environmental Sciences (LRES) at Montana State University, Bozeman, Montana. Her research interests include using satellite remote sensing for the detection of ecosystem change across high latitude regions and the application of Earth observation data in carbon cycle and climate change studies.

Hui Lu is now an associate professor in the Department of Earth System Science, Tsinghua University, Beijing, China. He received his B.Eng. and M.Eng. degrees from Tsinghua University and his Ph.D. degree in hydrology from the University of Tokyo, Tokyo, Japan, in 2006. His current research interests include the development of hydrologic models, land surface models, and data assimilation systems; microwave remote sensing of land surface parameters; and application of Earth observation data in water cycle and global change studies. He has published more than 100 papers in journals and conference proceedings. He is a senior member of IEEE, an editorial board member of *Remote Sensing of Environment*, and a recipient of Publons' 2018 Peer Review Award in Geoscience and Multidisciplinary Fields.

Lingmei Jiang is currently an associate professor in the Faculty of Geographical Science, Beijing Normal University. Her research interests include microwave emission/scattering modeling of land surface, remote sensing of snow cover and snow water equivalent, and remote sensing data assimilated into land surface models. She has authored/co-authored over 150 scientific publications and has been awarded the Shi Yafeng Prize for Young Scientists in Cryosphere and Environment in 2018.

Paolo Tarolli is currently an associate professor and head of the Earth Surface Processes and Society research group at the University of Padova (Italy). He is Deputy President of the Natural Hazards (NH) Division of the European Geosciences Union (EGU) and Deputy President of sub-division VII (Information and Communication Technologies) of the Italian Society of Agricultural Engineering. He is an expert in digital terrain analysis, earth surface processes analysis, natural hazards, geomorphology, hydro-geomorphology, LIDAR, and structure-from-motion photogrammetry. His new research directions include the analysis of topographic signatures of human activities from the local to regional scale. Tarolli is also Executive Editor of Natural Hazards and Earth System Sciences (Copernicus) and an Associate Editor of Remote Sensing (MDPI) and Land Degradation & Development (Wiley). He is the author of more than 100 papers published in international peer-reviewed journals. He has given more than 20 invited talks in international research institutions and foreign Academies (i.e., Princeton University, Ecole Polytechnique Fédérale de Lausanne, AgroParisTech, National Cheng Kung University, China University of Geosciences, Dalian University and Technology, Chinese Academy of Sciences) and at international meetings (IGC, AAG, ISPRS, RGS-IBG, AOGS-AGU, Soil Science Society of China).

remote sensing

MDPI

Editorial

Editorial for Special Issue: "Remote Sensing of Environmental Changes in Cold Regions"

Jinyang Du [1,*], Jennifer D. Watts [2], Hui Lu [3], Lingmei Jiang [4] and Paolo Tarolli [5]

[1] Numerical Terradynamic Simulation Group, W.A. Franke College of Forestry and Conservation, The University of Montana, Missoula, MT 59812, USA
[2] Woods Hole Research Center, Falmouth, MA 02540, USA; jwatts@whrc.org
[3] Department of Earth System Science, Tsinghua University, Beijing 100084, China; luhui@tsinghua.edu.cn
[4] State Key Laboratory of Remote Sensing Science, Jointly Sponsored by Beijing Normal University and Institute of Remote Sensing and Digital Earth of Chinese Academy of Sciences, Faculty of Geographical Science, Beijing Normal University, Beijing 100875, China; jiang@bnu.edu.cn
[5] Department of Land, Environment, Agriculture and Forestry, University of Padova, viale dell'Università 16, 35020 Legnaro (PD), Italy; paolo.tarolli@unipd.it
* Correspondence: jinyang.du@ntsg.umt.edu

Received: 9 September 2019; Accepted: 16 September 2019; Published: 18 September 2019

Cold regions, characterized by the presence of permafrost and extensive snow and ice cover, are significantly affected by changing climate. Of great importance is the ability to track abrupt and longer term changes to ice, snow, hydrology and terrestrial ecosystems that are occurring within these regions. Remote sensing allows for measurement of environmental variables at multiple spatial and temporal scales, providing key support for monitoring and interpreting the environmental changes occurring in cold regions. The recent advances in the application of remote sensing for the analysis of environmental changes in cold regions are documented in this Special Issue.

Theoretical modeling—For improving the current understanding of L-band microwave emissions from snow-covered soil, the Wave Approach for LOw-frequency MIcrowave emission in Snow (WALOMIS) model, initially developed for semi-infinite snow-firn conditions, was adapted and parameterized for seasonal snow. Evaluations of the model simulations against ground-based radiometer measurements show that the WALOMIS model can well reproduce the observed brightness temperature (Tb) with overall root-mean-square error (RMSE) between 7.2 and 10.5 K and have higher performance over larger incidence angles and H-polarization. The wave approach of WALOMIS also enables better quantification of the effects of interference and snow layering [1].

Ice—Satellite-based sea ice concentration (SIC) products have been widely used in monitoring global warming and navigating ships but are difficult to validate over the remote Arctic regions. For assessing the performance of satellite products and algorithms, SIC data sets were derived from ship-borne photographic observations acquired along cruise paths and compared with six passive microwave remote sensing products. The comparisons suggest that satellite products likely over/underestimate SIC under low/high SIC conditions mainly due to the presence of melt ponds; and the Special Sensor Microwave Imager Sounder (SSMIS) NASA Team algorithm has the overall best accuracy [2].

Ice-jam flood is one of the major hazards threatening riverine communities in the sub-arctic regions. Early forecasting of ice-jam flood can benefit from accurate locating and discriminating different types of ice. A novel method of differentiating ice runs from intact ice covers was developed using spaceborne synthetic-aperture radar (SAR) observations and the Freeman–Durden decomposition technique. The method was demonstrated using RADARSAT-2 imagery acquired along the Athabasca River for the spring of 2018, showing the distinct scattering signatures of ice runs and intact ice and its potentials in flood monitoring [3].

Snow—Snow properties including snow cover area and snow water equivalent (SWE) are vital inputs for numerical weather predictions and hydrologic model simulations. The quantification of global snow depth (SD) and SWE distributions generally relies on the observations from multi-frequency satellite microwave radiometers such as the Advanced Microwave Scanning Radiometer for Earth Observing System (AMSR-E) and China's FengYun-3D (FY-3D) Microwave Radiometer Image (MWRI). For developing FY-3D SD algorithm for regions of China, five operational algorithms were first evaluated using in-situ measurements. Considerable underestimate for deep snowpack (>20 cm) or persistent overestimate of SD by these algorithm outputs are mainly caused by inaccurate representation of snowpack characteristics in China. The FY-3D SD algorithm was then built using an empirical retrieval formula calibrated by weather station measurements. The refined algorithm shows improved retrieval accuracy over the baseline products with a RMSE of 6.6 cm and bias of 0.2 cm [4].

Frozen soil—One of the key issues in satellite microwave sensing of frozen soil is the determination of microwave radiation response depth (MRRD). A parameterized model to estimate MRRD was developed using the combination of theoretical model simulations and field measurements. According to the model, MRRD can be accurately determined from soil temperature, soil texture and microwave frequency. The estimated errors of MRRD of frozen loam soil at 6.9 GHz, 10.65 GHz, 18.7 GHz and 36.5 GHz were about 0.537 cm [5].

Surface water—Near-nadir interferometric imaging SAR techniques are well suited for measuring terrestrial water body extent and surface height at relatively fine spatial and temporal resolutions. The concept of near-nadir interferometric measurements was implemented in the experimental Interferometric Imaging Radar Altimeters (InIRA) mounted on Chinese Tian Gong 2 (TG-2) space laboratory. Both theoretical simulations and InIRA imagery showed that water and surrounding land pixels can be well distinguished by near-nadir SAR and the intensity of radar signals is determined by surface dielectric properties, roughness and incidence angles. A dynamic threshold approach was developed for InIRA and tested over Tibetan lakes where in-situ observations are sparse. Validations using a 30-m LandSat water mask suggest that high accuracy (>90%) of water and land classification can be achieved by InIRA [6].

Alternatively, optical remote sensing enables surface water mapping at sub-meter to meter scales. For mitigating the risks of glacier lake outburst flood, multi-resolution satellite imageries from LandSat (30-m resolution), Sentinel-2 (10-m resolution), WorldView and GeoEye (0.5–2 m resolution) were synergistically used to analyze the dynamics of supraglacial ponds in the Himalayan region. The analyses showed a continuous increase in the area and number of supraglacial ponds from 1989–2017, consistent seasonal patterns and a great diversity of pond features. The satellite images also revealed high persistency and density of the ponds (>0.005 km^2) near the glacier terminuses; and a fast expanding of spillway lakes on the Ngozompa, Bhote Koshi, Khumbu and Lumsamba glaciers [7].

Landsat imageries (1985–2015) and higher resolution aerial photographs were used to quantify surface water changes in the high Arctic pond complexes of western Banks Island, Northwest Territories. Analysis based on remote sensing, field sampling and geostatistic approaches showed an overall drying trend of high Arctic lakes mainly driven by climate factors and also affected by intensive occupation by lesser snow geese [8].

Vegetation—Multi-year Landsat and MODIS (Moderate Resolution Imaging Spectroradiometer) data sets were examined to reconstruct vegetation recovery from wildfire disturbances in Alaska. Breakpoint analysis using the BFAST (Breaks for Additive Seasonal and Trend) approach was able to capture the wildfire-related structural change in the MODIS normalized difference vegetation index (NDVI) time series. Further analysis of the change detection results suggested that vegetation cover density in the Alaskan wetlands likely recovers to pre-fire levels in less than 10 years [9].

In summary, continuous warming has altered the hydrologic and ecologic conditions across the cold regions, resulting in a myriad of changes including glacier melting, active layer deepening, permafrost degradation, snow and ice phenology changes, water body shrink and expansion, and regional greening and browning. Remote sensing is essential in tracking and understanding the environmental

Remote Sens. **2019**, *11*, 2165

changes and revealing the underlying physical mechanisms. Multi-source data fusion approaches, emerging techniques such as microsatellites and artificial intelligence, light detection and ranging (LIDAR) and structure from motion photogrammetry, and next generation satellite missions will enable unprecedented remote sensing performance in cold land studies [10].

Conflicts of Interest: The authors declare no conflict of interest.

References

1. Roy, A.; Leduc-Leballeur, M.; Picard, G.; Royer, A.; Toose, P.; Derksen, C.; Lemmetyinen, J.; Berg, A.; Rowlandson, T.; Schwank, M. Modelling the L-band snow-covered surface emission in a winter canadian prairie environment. *Remote Sens.* **2018**, *10*, 1451. [CrossRef]
2. Wang, Q.; Lu, P.; Zu, Y.; Li, Z.; Leppäranta, M.; Zhang, G. Comparison of Passive Microwave Data with Shipborne Photographic Observations of Summer Sea Ice Concentration along an Arctic Cruise Path. *Remote Sens.* **2019**, *11*, 2009. [CrossRef]
3. Lindenschmidt, K.E.; Li, Z. Radar Scatter Decomposition to Differentiate between Running Ice Accumulations and Intact Ice Covers along Rivers. *Remote Sens.* **2019**, *11*, 307. [CrossRef]
4. Yang, J.; Jiang, L.; Wu, S.; Wang, G.; Wang, J.; Liu, X. Development of a Snow Depth Estimation Algorithm over China for the FY-3D/MWRI. *Remote Sens.* **2019**, *11*, 977. [CrossRef]
5. Zhang, T.; Jiang, L.; Zhao, S.; Chai, L.; Li, Y.; Pan, Y. Development of a Parameterized Model to Estimate Microwave Radiation Response Depth of Frozen Soil. *Remote Sens.* **2019**, *11*, 2028. [CrossRef]
6. Li, S.; Tan, H.; Liu, Z.; Zhou, Z.; Liu, Y.; Zhang, W.; Liu, K.; Qin, B. Mapping high mountain lakes using space-borne near-nadir SAR observations. *Remote Sens.* **2018**, *10*, 1418. [CrossRef]
7. Chand, M.B.; Watanabe, T. Development of Supraglacial Ponds in the Everest Region, Nepal, between 1989 and 2018. *Remote Sens.* **2019**, *11*, 1058. [CrossRef]
8. Campbell, T.; Lantz, T.; Fraser, R. Impacts of Climate Change and Intensive Lesser Snow Goose (Chen caerulescens caerulescens) Activity on Surface Water in High Arctic Pond Complexes. *Remote Sens.* **2018**, *10*, 1892. [CrossRef]
9. Potter, C. Recovery rates of wetland vegetation greenness in severely burned ecosystems of Alaska derived from satellite image analysis. *Remote Sens.* **2018**, *10*, 1456. [CrossRef]
10. Du, J.; Watts, J.D.; Jiang, L.; Lu, H.; Cheng, X.; Duguay, C.; Farina, M.; Qiu, Y.; Kim, Y.; Kimball, J.S.; et al. Remote Sensing of Environmental Changes in Cold Regions: Methods, Achievements and Challenges. *Remote Sens.* **2019**, *11*, 1952. [CrossRef]

remote sensing

MDPI

Article

Modelling the L-Band Snow-Covered Surface Emission in a Winter Canadian Prairie Environment

Alexandre Roy [1,2,3,*], Marion Leduc-Leballeur [4], Ghislain Picard [5], Alain Royer [1,3], Peter Toose [6], Chris Derksen [6], Juha Lemmetyinen [7], Aaron Berg [8], Tracy Rowlandson [8] and Mike Schwank [9,10]

[1] Université de Sherbrooke, 2500 boul. Université, Sherbrooke, QC J1K 2R1, Canada; alain.royer@usherbrooke.ca
[2] Département des Sciences de l'Environnement, Université du Québec à Trois-Rivières, 3351 Boulevard des Forges, Trois-Rivières, QC G9A 5H7, Canada
[3] Centre d'études Nordiques, Université Laval, Québec, QC G1V 0A6, Canada
[4] Institute of Applied Physics—Ational Research Council, 50019 Sesto Fiorentino, Italy; m.leduc@ifac.cnr.it
[5] Université Grenoble Alpes, CNRS, IGE, F-38000 Grenoble, France; Ghislain.picard@univ-grenoble-alpes.fr
[6] Climate Research Division, Environment and Climate Change Canada, Toronto, ON M3H 5T4, Canada; peter.toose@canada.ca (P.T.); chris.derksen@canada.ca (C.D.)
[7] Finnish Meteorological Institute, FI-00101 Helsinki, Finland; Juha.Lemmetyinen@fmi.fi
[8] Department of Geography, Environment, and Geomatics University of Guelph, Guelph, ON N1G 2W1, Canada; aberg@uoguelph.ca (A.B.); trowland@uoguelph.ca (T.R.)
[9] Gamma Remote Sensing AG, CH-3073 Gümligen, Switzerland; schwank@gamma-rs.ch
[10] Swiss Federal Research Institute WSL, CH-8903 Birmensdorf, Switzerland
*   Correspondence: Alexandre.Roy@UQTR.ca; Tel.: +1-819-376-5011 (ext. 3680)

Received: 10 August 2018; Accepted: 5 September 2018; Published: 11 September 2018

Abstract: Detailed angular ground-based L-band brightness temperature (T_B) measurements over snow covered frozen soil in a prairie environment were used to parameterize and evaluate an electromagnetic model, the Wave Approach for LOw-frequency MIcrowave emission in Snow (WALOMIS), for seasonal snow. WALOMIS, initially developed for Antarctic applications, was extended with a soil interface model. A Gaussian noise on snow layer thickness was implemented to account for natural variability and thus improve the T_B simulations compared to observations. The model performance was compared with two radiative transfer models, the Dense Media Radiative Transfer-Multi Layer incoherent model (DMRT-ML) and a version of the Microwave Emission Model for Layered Snowpacks (MEMLS) adapted specifically for use at L-band in the original one-layer configuration (LS-MEMLS-1L). Angular radiometer measurements (30°, 40°, 50°, and 60°) were acquired at six snow pits. The root-mean-square error (RMSE) between simulated and measured T_B at vertical and horizontal polarizations were similar for the three models, with overall RMSE between 7.2 and 10.5 K. However, WALOMIS and DMRT-ML were able to better reproduce the observed T_B at higher incidence angles (50° and 60°) and at horizontal polarization. The similar results obtained between WALOMIS and DMRT-ML suggests that the interference phenomena are weak in the case of shallow seasonal snow despite the presence of visible layers with thicknesses smaller than the wavelength, and the radiative transfer model can thus be used to compute L-band brightness temperature.

Keywords: L-band emission; snow; WALOMIS; Frozen soil; ground-based radiometer

1. Introduction

Three spaceborne L-band passive microwave radiometer missions were successfully launched in recent years for global monitoring of soil moisture and sea surface salinity. The European Space Agency (ESA) Soil Moisture and Ocean Salinity (SMOS) mission [1] was launched in November

2009 and continues to operate. The NASA Aquarius instrument on board the Aquarius/Satélite de Aplicaciones Científicas (SAC-D) mission, developed collaboratively between the U.S. National Aeronautics and Space Administration (NASA) and Argentina's space agency, Comisión Nacional de Actividades Espaciales (CONAE) acquired L-band observations between September 2012 and July 2015 [2], and the NASA Soil Moisture Active Passive (SMAP) satellite was launched in January 2015 [3]. These missions also provide useful information for cryosphere applications including monitoring the freeze/thaw (F/T) state of the land surface [4–7], estimating snow density and ground permittivity [8,9], and retrieving the thickness of thin sea ice [10].

Many studies have improved the modelling of L-band brightness temperature (T_B) for non-frozen surfaces [11], while numerous snow emission models exist for higher frequencies [12–15]. Comparatively few studies have calibrated and validated a snow emission model over frozen soil at L-band, as conventionally snow cover has been thought to have little relevance for L-band emissions because of inherent low scattering and absorption in dry snow. Recent studies, however, show the non-negligible impact of dry snow on L-band emission [16–18]. On the basis of the Microwave Emission Model for Layered Snowpacks (MEMLS, [12]) and the L-band microwave emission of the biosphere (L-MEB) model [11], Naderpour et al. [19] developed a simplified emission model specifically for L-band (called LS-MEMLS hereafter). The model neglects volume scattering in the snow layer, which is a plausible approximation for L-band. In cases of wet snow, absorption is considered by LS-MEMLS, whereas dry snow is assumed to be fully transparent, which is reasonable for seasonal snowpacks with thicknesses much smaller than L-band emission depth in dry snow (>300 m [20]). However, sensitivity to dry snow is retained though impedance matching and changes in the refraction angle at the snow-soil interface due to variable snow permittivity, which is in turn controlled by the dry snow density. The study also introduces a dual parameter retrieval approach for dry snow density and ground permittivity. The model and retrieval methods were evaluated with experimental data in boreal forest [9] and Canadian prairie environments [17]. However, the LS-MEMLS model approach does not take into account wave coherence effects [21], which potentially induces multiple reflections within a thin layer of snow or ice and associated interferences. Coherence effects may arise when the thickness of the layer is less than about a quarter of the wavelength (λ; about 5 cm at L-band; [22]), and when layers are sufficiently homogeneous and parallel in the horizontal direction within the radiometer field of view. This can lead to significant variation in T_B especially at the horizontal polarization [23,24].

In this study, we focus on modelling the snow contribution to L-band emission to better understand the effect of snow layering and interference for improved F/T monitoring and snow density retrieval. Three electromagnetic models were compared: the Dense Media Radiative Transfer-Multi Layer incoherent model (DMRT-ML) [14], the LS-MEMLS model [19], and the Wave Approach for LOw-frequency MIcrowave emission in Snow (WALOMIS) model [18,25]. The latter is a coherent model successfully used at L-band in the case of semi-infinite snow-firn over Antarctica. It was not previously applied to seasonal snow cover, so some improvements are introduced in this study.

Ground-based L-band radiometer measurements acquired in a Canadian prairie environment are used to first implement the WALOMIS model for seasonal snow, followed by comparisons with the LS-MEMLS and the DMRT-ML. In the following sections, we first present the ground-based radiometer observations and in situ measurements as well as the three snow microwave emission models and the soil emission model. The model parameterizations are presented, after which we present results and a comparison of the model performance.

2. Site and Data

During the 2014–2015 winter, a ground-based L-band radiometer measurement campaign was conducted at the Kernen Crop Research Farm (KCRF; 52.149°N; 106.545°W), a 380 ha property within the city of Saskatoon owned and operated by the University of Saskatchewan, Canada. L-band radiometer measurements and coincident snow pit and meteorological observations were performed. The study area and the campaign and datasets are described in detail in Reference [17].

At KCRF, tree scenes were located adjacent to each other within the same field to ensure similarity in background emission, with only the overlying snow conditions altered. The three scenes include: Scene 1—Undisturbed snow: A scene of naturally accumulating snow-covered ground; Scene 2—Snow free: snow was removed on a weekly basis to maintain bare ground; Scene 3—Artificially compacted snow: a scene with deep and dense snow. Additional snow was manually added to Scene 3 and compacted on December 10th, 19th, 2014 and January 11th 2015 and then left to evolve naturally for the rest of the season. As this study focuses on snow emission modelling, Scene 2 was not used in this study. The scenes were characterized by silt-loam bare soil conditions. Wheat residue was noted and not disturbed during the study. Surface roughness was derived using a terrestrial Light Detection and Ranging (LiDAR) system using a surface roughness tool called Roughness from Point Cloud Profiles (RPCP) [26] implemented in Whitebox Geospatial Analysis Tools (GAT) software [27]. Surface roughness had a root-mean-square height (RMSH) of 1.78 cm and 1.64 cm within Scene 1 and 3, respectively, at the beginning of the study and remained almost unchanged throughout the study (RMSH = 1.79 cm and 1.75 cm).

L-band measurements were acquired by a surface-based hyperspectral dual polarization L-band Fourier transform radio-frequency interference (RFI) detecting radiometer with 385 channels designed for a frequency range from 1400 MHz to ≈1550 MHz. The radiometer antenna is a 19-element air loaded conformal muffin tin design that has a 30° half-power (−3 dB) beamwidth. A method was developed for separating out the thermal spectrum from RFI-contaminated channels to get unique RFI-free T_B from the measured spectrum [28]. Only the protected radio-astronomy frequency spectrum of 1400–1427 MHz was used to calculate the T_B. The radiometer was set 2.75 m above the surface, and measurements at the angles 30°, 40°, 50°, and 60° relative to nadir were taken of the three scenes on a weekly basis. On 9 November 2014, radiometric measurements were taken while the soil was frozen and snow-free. From December 2014 to March 2015, six radiometric measurements were taken, coincident with manual snow pit measurements in the vicinity of Scene 1 and 3. The snow pits included documenting the snow stratigraphy, including the presence of ice lenses. Profiles of snow temperature and snow density were taken for the observed snow layers. Mass density was measured using a 100 cm^3 density cutter, and samples were weighed with a digital scale with an accuracy of ±0.1 g. The snow and soil temperature at 2.5 cm intervals were measured with a digital temperature probe (±0.1 °C). Soil was frozen at each visit.

Figure 1 shows snow pits performed close to Scene 1 and 3 during each visit. Note that on 7 December 2014, only a single snow pit is available and refers to both scenes. Snow pits in the vicinity of Scene 1 were generally shallow (Table 1) and composed of a depth hoar layer at the bottom and a high-density rounded grain winds slab snow layer at the surface. One or two high-density melt/ice crust layers and/or ice lenses were present within the snowpack, resulting from mid-winter melt events (see in Reference [17] the Figure 4 and details). Note that this strong stratification between the top and bottom of the snowpack made the snow density measurements a challenge because of the hardness (surface wind slab and melt/ice crust) and the instability (depth hoar) of the snow layers. Because of the artificial compaction of snow, the snow density of the bottom layer is higher in Scene 3. There was still a high snow density observed in February showing that the artificial high-density snow/ice crust made up a large proportion of the lower 10 cm of the snowpack in Scene 3. However, as the season progressed and metamorphism continued within the snowpack, there was a decrease in the density of the snow/ice crust layers found within the bottom layers. Note that all air temperature measurements below −6 °C ensure that the snow was dry during each visit (Table 1).

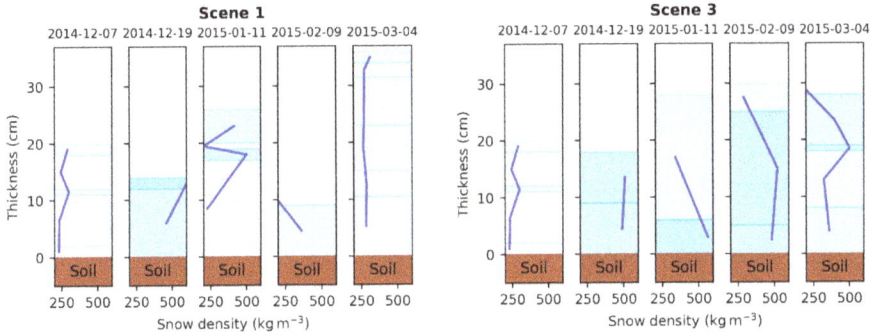

Figure 1. Snow pits measurements in the vicinity of Scene 1 (**left**) and Scene 3 (**right**) performed at each visit. Dark blue lines are the snow layer density. Scene 2 was not considered in this study.

Table 1. Snow and air temperature measurements during each visit. "-" means no data. Note that missing data are related to technical issues with instruments.

Dates	Snow Depth (cm)		Snow Bulk Density (kg m^{-3})	Air Temperature (°C)	Magna Probe Mean Snow Depth and Standard Deviation (cm)
	Scene 1	Scene 3	Scene 1	Scene 3	
2014-12-7	20	20	230	230	−6.6	12 ± 5
2014-12-19	14	18	460	490	−9.0	-
2015-1-11	26	28	226	563	−20.2	18 ± 10
2015-2-9	12	30	360	480	−13.9	18 ± 7
2015-3-4	36	31	283	358	−16.8	25 ± 7

3. Emission Models

All three snow microwave emission models and the soil emission model used in this study are already well described in detail (see previously provided references). Accordingly, we only recall here the principal components of each model, the model inputs and adjustments made for this study.

3.1. WALOMIS

The WALOMIS [18] coherent snow emission model is based on a wave approach, i.e., solving Maxwell's equation for a multi-layered medium [29,30]. Each layer is characterized by thickness, temperature, and density. The most important simplification in this model is to neglect scattering by snow grains. This assumption is invalid for high microwave frequencies, however, in the case of L-band, scattering by grains is insignificant in comparison with absorption and reflection at the interfaces between layers due to the L-band wavelength being several orders of magnitude larger than snow grain size. Under these assumptions, the vertically and horizontally polarized T_B of a given snowpack is calculated with the propagation-matrix derived from Reference [31].

WALOMIS was initially implemented to investigate the microwave emission at L-band for semi-infinite snow-firn in Antarctica [18,25]. In the case of the Antarctic ice-sheet, the soil emission can be ignored because of the high ice thickness (>1000 m). Thus, the lowest layer of the model is considered a semi-infinite ice layer. In contrast, in the case of seasonal snowpack, the soil emission is not negligible for the total emission of the snow-covered ground. Therefore, for the present study WALOMIS was adapted to take into account the soil emission from below the snowpack replacing the semi-infinite bottom ice layer by a soil layer characterized by the observed temperature and permittivity.

Because of the high sensitivity of the interference phenomena to the layer thickness with the wave approach (which is not the case with the non-coherent radiative transfer approach), the result obtained for a specific snowpack configuration (i.e., a given set of inputs) may differ considerably from those obtained with a slightly different snowpack. To account for the variable nature and the imperfect layering of the snowpack within the footprint of the radiometer, it is essential to average a large number of simulations using inputs that represent natural snow variability. As thousands of simulations are required, it would be impossible to obtain the input profiles from direct measurements. Several studies suggested stochastic methods to generate such profiles from measurements in Antarctica (e.g., Reference [18,29,32]). A similar procedure used here is described in Section 4.2. The output of the model is the average T_B from all the generated profiles.

3.2. DMRT-ML

The DMRT-MultiLayer (DMRT-ML) is an incoherent model that describes the snowpack as a multilayer medium, where each snow layer is characterized by its thickness, temperature, density, grain optical radius, stickiness parameter, and liquid water content. The model is available from http://gp.snow-physics.science/dmrtml. It is based on the DMRT theory [30]. In this study, stickiness is not investigated because scattering by grains is negligible, and this parameter has no effect at L-band (typically less than 0.1 K). Because all the measurements were made in cold conditions with dry snow, the liquid water content was considered to be zero. For each layer, the effective dielectric constant is represented using the first order quasi-crystalline approximation and the Percus–Yevik approximation for spherical grains. The absorption and scattering coefficients are calculated assuming a medium of "ice spheres in air background" and the emission and propagation of radiation through the snowpack are computed using the Discrete Ordinate Method (DISORT: [33]) with 64 streams, which takes multiple scattering between the layers into account, but not the interferences.

3.3. LS-MEMLS-1L

The LS-MEMLS [19] model estimates L-band microwave emission from a ground surface covered by a layer of dry snow. This emission model is based on parts of MEMLS, [12] with the assumptions of no absorption and no volume scattering in dry snow, which are applicable to the L-band frequencies in dry snow. Once the interface reflectivities are known, the Kirchhoff coefficients associated with a single (snow) layer above an infinite half-space (ground) are computed to derive T_B. Snow is characterized only by its permittivity, controlled by the dry snow density. Schwank et al. [8] assumed a single snow layer with a homogeneous density distribution, which allowed a numerical inversion of the model with minimal a priori information, for purposes of retrieval of snow and ground parameters. Although e.g., Reference [34] applied the model also in a configuration exhibiting a vertical distribution of snow densities, in this study LS-MEMLS is applied in the original one-layer configuration (LS-MEMLS-1L) to evaluate its applicability for snow density retrievals [8,9].

3.4. Soil Emission Model

At L-band, soil emission has a significant contribution to the signal emerging from the surface in environments with seasonal snow [35]. Hence, a soil reflectivity model is a prominent component of seasonal snow microwave-emission models. In this study, the soil reflectivity is calculated from the Fresnel equations and the roughness is considered as negligible. A specular soil reflectivity model is used in this study because WALOMIS needs electric field reflectivity between layers, while known rough soil emission models (i.e., Reference [36]) provide only the power reflectivity without phase information. Because the main purpose of this study is to evaluate the performance of snow emission models, it is important that the same soil emission model is used for the three snow emission models in order to avoid any bias in simulations that come from soil emission modelling. The same specular soil reflectivity model is thus used with each of the three snow emission models. The hypothesis of a specular soil is plausible in our case because the root-mean-square height (RMSH: 1.79 cm and 1.75 cm; see Section 2) of the soil measured with the LiDAR is much smaller than the L-band wavelength (RMSH < $\lambda/12$).

Fresnel equations calculate the soil reflectivity from the permittivity of the frozen soil and the permittivity of the layer on top (air or snow). Snow permittivity is calculated from snow emission models, but frozen soil permittivity (ε_g) was not measured and remains an unknown. Hence, ε_g was inferred from frozen ground snow-free radiometric measurements taken on 9 November (see Table 1). An iterative process with an increment of 0.1 was used to calculate the frozen soil permittivity that minimized the root-mean-square error (RMSE) between the measured T_B ($T_{B\,mes}$), and simulated T_B ($T_{B\,sim}$) at vertical (V-pol) and horizontal polarization (H-pol) at the four measured incidence angles such that:

$$RMSE_{\varepsilon g} = \sqrt{\frac{\sum_{i=1}^{N} \left(T_{B\sim,i}^{V-pol} - T_{Bmes,i}^{V-pol} \right)^2 + \left(T_{B\sim,i}^{H-pol} - T_{Bmes,i}^{H-pol} \right)^2}{2N}} \tag{1}$$

The optimization of ε_g was done on Scene 1 and 3 separately. In this case, ε_g should be considered as an effective parameter that allows representation of frozen soil emission for the three models, but can also partially compensate for the specular assumption.

4. Results

4.1. Frozen Soil Permittivity

The optimized ε_g was calculated from T_B simulations for ε_g ranging from 1 to 15 and using November snow-free frozen soil parameters. $RMSE_{\varepsilon g}$ was computed from angular radiometer measurements performed on 1st November 2015 on Scene 1 and 3 at vertical and horizontal polarization (Figure 2). The optimal value of ε_g was 4.6 and 4.9 for Scene 1 and 3, respectively. $RMSE_{\varepsilon g}$ of 10.7 K and 8.2 K, respectively, was observed for Scene 1 and 3, but an important component of this error was the poor simulation performance at 60° V-pol (discussed in more detail later). $RMSE_{\varepsilon g}$ computed without the T_B at 60° are 6.9 K and 5.7 K for an optimized ε_g of 4.3 and 4.8 for Scene 1 and 3, respectively. The small differences between both sites are not significant, and could be related to differences in soil RMSH, which is not considered in the soil model. These optimized ε_g were used in the following simulations.

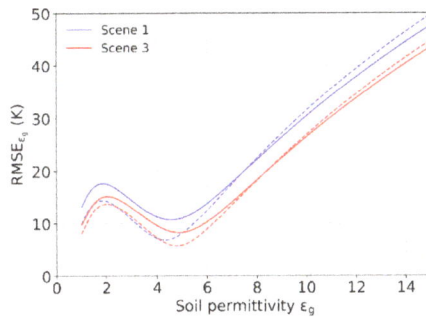

Figure 2. $RMSE_{\varepsilon g}$) obtained from Dense Media Radiative Transfer-Multi Layer (DMRT-ML) simulations and angular radiometer measurements performed in November 2015 on snow-free frozen soil on Scene 1 (blue) and 3 (red) for different values of soil permittivity (ε_g) with a specular soil reflectivity model. $RMSE_{\varepsilon g}$ including all incidence angles (solid lines) and without 60° (dashed lines) are represented.

4.2. WALOMIS Gaussian Noise Parameterization

As is true of any model based on the wave approach, the result obtained for a specific snowpack configuration (i.e., a given set of inputs) may differ considerably from those obtained with a slightly different snowpack, which is not the case with the radiative transfer approach. This is due to the high sensitivity of interference phenomena to layer characteristics. Hence, on the basis of Reference [18],

10,000 snow density profiles were generated by adding a Gaussian noise (σ_d) to the measured density profile and T_B was obtained from the average of 10,000 WALOMIS simulations performed with these profiles.

Figure 3 shows the effect of the Gaussian noise for an example of simulations for Scene 1 on 4 March 2015 with fixed layer thickness. Even with a very high Gaussian noise of $\sigma_d = 90$ kg m^{-3}, the simulations show a wavy angular pattern owing to the high sensitivity of interference phenomena, very different to the measured angular spectrum. These results suggest that for shallow seasonal snow, adding Gaussian noise to density profiles does not reproduce the variability of snow cover within the radiometer field of view.

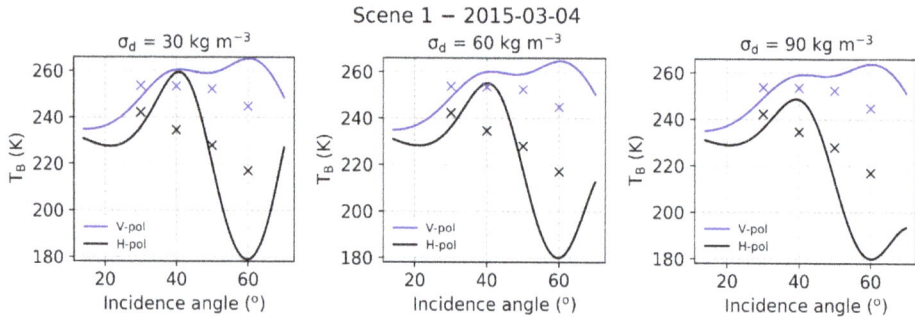

Figure 3. T_BV-pol (blue) and T_BH-pol (black) on 4 March 2015 at Scene 1 measured (symbols) and simulated (lines) with the Wave Approach for LOw-frequency MIcrowave emission in Snow (WALOMIS) with added Gaussian noise (σ_d) to the measured density of 30 kg m^{-3} (**left**), 60 kg m^{-3} (**center**) and 90 kg m^{-3} (**right**).

In situ measurements performed during the campaign revealed strong variability in the thickness of internal layers within the snowpack (see Reference [17]), which could better represent interference phenomena in the model. WALOMIS simulations were also performed from 10,000 profiles of layers thickness by adding a Gaussian noise (σ_h) to the measured layer thickness, keeping the measured snow density. Figure 4 shows that with increasing σ_h, the angular pattern of the simulation gets closer to the T_B measurements. Increasing the variability of layer thickness to 2–4 cm results in agreement with the T_B observations (Figure 4). With a $\sigma_h = 2$ cm and 4 cm, the simulated H-pol also capture the T_B decrease with incidence angle. However, contrary to observations which slowly decrease with incidence angle, simulations at V-pol tend to increase with incidence angle, before decreasing at 60°. As expected, because interference phenomena are highly sensitive to optical path-length across layers, a Gaussian noise of $\sigma_h = 2$ cm, was found to give the best agreement with measurements and was used for the subsequent simulations.

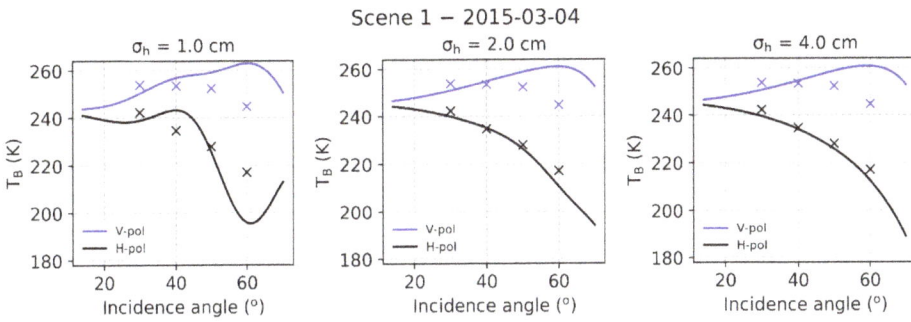

Figure 4. T_BV-pol (blue) and T_BH-pol (black) for 4 March 2015 at Scene 1 measured (symbols) and simulated (lines) with WALOMIS with added Gaussian noise (σ_h) to the measured layer thickness of 1 cm (**left**), 2 cm (**center**) and 4 cm (**right**).

4.3. Footprint Integration

Because of the large antenna full beamwidth ($30°$) and the footprint geometry of the surface-based radiometer, the simulated T_B of a single directional incidence angle (θ) might not be representative of the measured T_B over a large footprint, especially at higher incidence angles [37]. Hence, in this study, a weighting function was computed to estimate the integrated T_B within the footprint for a range of incidence angles. For this estimation, the area included in $\theta \pm 15°$ was considered. A Gaussian weighting was applied to θ with a standard deviation of $7.5°$ in order to represent the antenna directional power sensitivity pattern. Then, a factor $1/r^2$, with r the distance from the radiometer, was applied to the obtained coefficients to attenuate the contribution as a function of the location within the footprint. The coefficients used for the weighting are illustrated in Figure 5, normalized by the maximum.

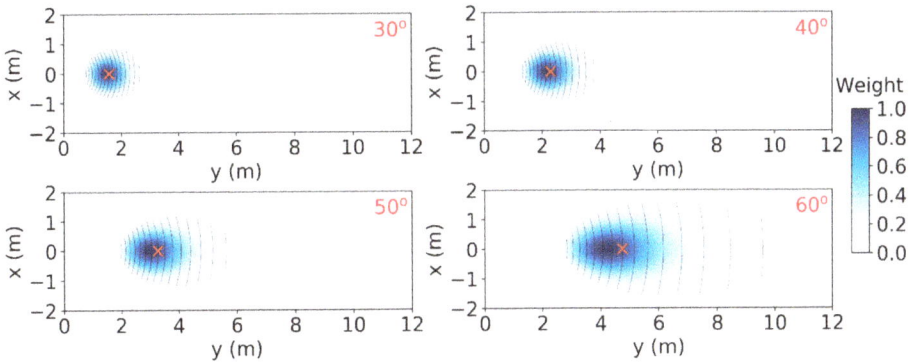

Figure 5. Weight relative to maximum diagram where X and Y are coordinates (in meters) on the field from the radiometer $(0,0)$. Red crosses are the incidence angles of the antenna footprint center. Grey lines for incidence angles inside the footprint by $2°$ step.

When the weighting function was applied to the simulations of the three snow emission models at Scene 3 on 4 March 2015, there is a decrease of T_B at both H-pol and V-pol mostly at incidence angles higher than $55°$ (Figure 6). This decrease in T_B slightly improved the results for all three models at high incidence angles, thus it was used in the following simulations.

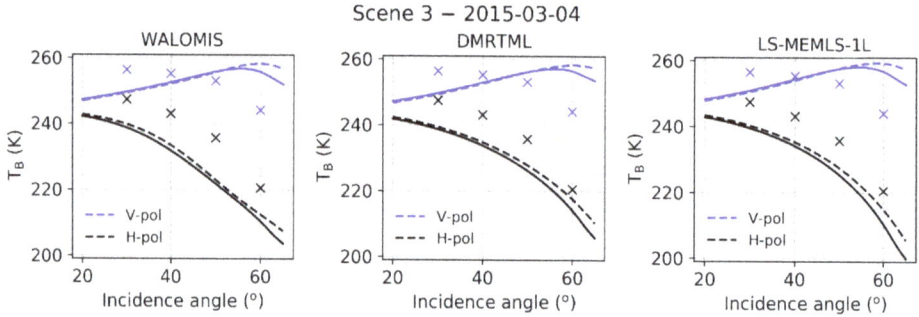

Figure 6. T_BV-pol (blue) and T_BH-pol (black) measured (symbols) and simulated using non-weighted (dashed) and weighted (lines) signal within the footprint; with WALOMIS (**left**), DMRT-ML (**center**) and LS-MEMLS-1L (**right**) at Scene 3 for 4 March 2015.

4.4. Snow Emission Model Intercomparison

The effect of dry snow at L-band is mostly related to refraction and impedance matching [16]. Impedance matching by dry snow reduces dielectric gradients and consequently increases thermal emission of the scene, while refraction caused by the snow layer in contact with the ground surface leads to a steeper incidence angle at the ground surface in comparison with the observation angle [16]. The impact is in general higher at H-pol, since at V-pol these effects are partly compensatory, and even fully compensatory around an incidence angle of 51°. At H-pol, the effect of snow typically increases with incidence angle. At the plot scale, the presence of snow can change T_B at H-pol from 5 K at 30° up to 20 K at 60° [17]. The specular soil reflectivity model with optimized permittivity and the optimized weighting function were used to simulate the T_B at the Scene 1 and 3 and for the six sampling periods with the three snow emission models (Figure 7). The three models show very similar overall RMSE at V-pol and H-pol with values ranging between 7.2 K and 10.5 K. LS-MEMLS-1L gives lower RMSE at H-pol while DMRT-ML gives slightly better results at V-pol. Note that for the three models and both polarizations, the RMSE increases with increasing incidence angle (Table 2). The worst results are obtained at 60°, which has a strong impact on the overall RMSE (noted "All" in Table 2). It is thus not possible to state that a specific snow emission model gives better results overall. Nevertheless, at 50° and 60°, LS-MEMLS-1L clearly underestimates the variability of T_B H-pol, with a standard deviation of the simulations much lower than the standard deviation of observations. Simulations at 60° ranged between 203.9 K and 213.9 K, while the measured T_B H-pol ranged between 192.8 K and 225.8 K. On the other hand, the standard deviations of WALOMIS and DMRT-ML simulations at T_B H-pol are much closer to the measurements (Table 3). These results suggest that snow layers significantly impact the T_B H-pol. The multi-layer model configurations are able to better capture this effect, but the complex interaction of the radiation within the layers and the difficulty to precisely measure the snow layer characteristics in the field at the meter scale (footprint) leads to an overall RMSE comparable to LS-MEMLS-1L applied in a 1-layer configuration. We thus face the problem where more complex and more sensitive radiometric models require precise in situ information for comprehensive evaluation, a condition that will provide limitations for more general use, especially at the satellite scale.

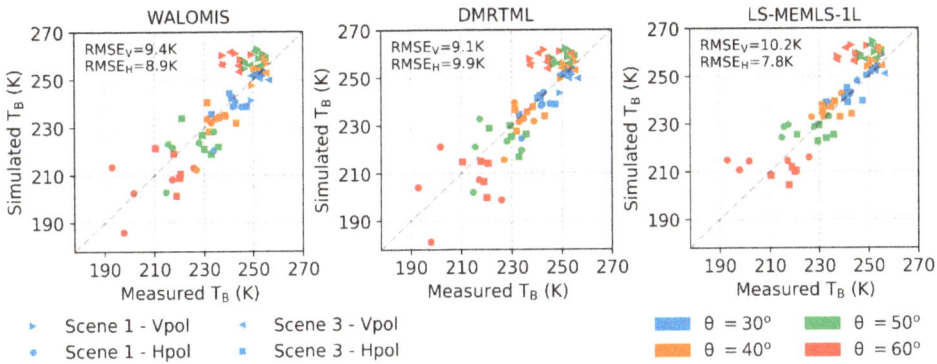

Figure 7. T_B simulated from WALOMIS (**left**), DMRT-ML (**center**) and LS-MEMLS-1L (**right**) for Scene 1 and 3 at V-pol and H-pol (symbols) and at four incidence angles (colors).

Table 2. Root-mean-square height (RMSE) of all the dates and Scene 1 and 3 together. "All" is the RMSE calculated with incidence angles of 30°, 40°, 50° and 60°.

	RMSE for T_BV-pol (K)					RMSE for T_BH-pol (K)				
	30°	40°	50°	60°	All	30°	40°	50°	60°	All
WALOMIS	4.6	2.9	7.8	15.4	9.1	6.9	7.8	10.2	12.9	9.7
DMRT-ML	4.1	2.6	7.7	14.8	8.8	5.9	7.0	11.0	15.7	10.5
LS-MEMLS-1L	2.5	3.8	9.2	16.1	9.6	3.7	4.3	7.3	11.2	7.2

Table 3. Standard deviation (Std) of all the dates and, Scene 1 and 3 together. "All" is the Std calculated with incidence angles of 30°, 40°, 50° and 60°.

	Std for T_BV-pol (K)				Std for T_BH-pol (K)			
	30°	40°	50°	60°	30°	40°	50°	60°
WALOMIS	3.8	3.1	2.6	2.5	6.7	7.5	8.1	10.1
DMRT-ML	3.1	2.5	2.3	2.4	5.6	6.8	8.6	11.1
LS-MEMLS-1L	2.6	2.5	2.4	2.4	2.9	2.9	2.8	2.8
Measures	2.3	2.7	2.9	4.2	5.3	7.8	11.3	13.8

5. Discussion

5.1. Soil Permittivity Parameterization

The optimized values of frozen soil permittivity are close to other studies [16,17,35,38,39]. Rautiainen et al. [35] obtained L-band real part soil permittivity values between 3.3 to 3.8 in boreal forest frozen soils; Schwank et al. [16] showed that for L-band, the real part of the frozen soil permittivity varies from 3.5 to 4.5, while Hallikainen et al. [38] showed that it varies from 5 to 8 in the 10 to 18 GHz frequency range. At the same frequency range, Mironov et al. [39] developed a temperature dependent permittivity model and showed that the permittivity could vary from 3 to 4.5 for a frozen soil at -25 °C. The optimized values are similar to the ones obtain at the same site ([17]; $\varepsilon_g = 5.1$), but using a two parameter (ε_g and snow density) retrieval method [8]. The plausible optimized ε_g values thus show that the assumption of a flat soil at L-band was reasonable for our site and the impact of soil roughness is reasonably included in the ε_g. However, in this study, soil emission was empirically parameterized to minimize its impact on the simulation of snow covered ground, but it remains that the snow-free frozen soil RMSE are similar to the snow simulations. In particular, Figure 6 shows that at V-pol, there is a clear underestimation of the T_B at an incidence angle of 30°, while there is an overestimation of T_B at higher incidence angle (50° and 60°). There is thus an opposite trend in the angular spectrum between the measurements and the observations, an effect most likely related to the soil model. It should be note that those error impact substantially the results with snow-covered surface. Simulations and observations at V-pol will have to be investigated further.

5.2. WALOMIS Gaussian Noise Parameterization

This study shows that an approach developed by Reference [18], which applies Gaussian noise to in situ snow density measurements to estimate T_B variability is insufficient to smooth the snow layer interference phenomena in WALOMIS. However, Gaussian noise applied to snow depth layer thickness of σ_h = 2 cm leads to an angular spectrum comparable to the observations. A value of σ_h = 2 cm is plausible considering the high standard deviation of snow depth in the area and the variable internal layering of the snowpack. Snow depth surveys around the area suggest that despite the surface appearing relatively homogeneous, the standard deviation of the snow depth is in the range of 28 to 56% (Table 1). The high variability in snow depth is related to wind redistribution, a key process in this prairie environment [40]. Strong winds also lead to a high snow density layer observed at top of the snowpack due to compaction processes [41], thus resulting in the optimized snow depth layer thickness variability σ_h of 2 cm at the plot scale.

5.3. Footprint Integration

Roy et al. [37] showed that spatially distributed surface-based radiometer observations within a SMOS pixel gives constantly lower T_B (between 8.5 and 22.9 K) than SMOS observations at 60° V-pol (not apparent at lower incidence angles). It was hypothesized that these discrepancies were due to the large beamwidth (30°) of the surface-based radiometer, which produced an incidence angle in the far range of up to 75°. This may exaggerate the influence of high incidence angles on reducing the magnitude of V-pol T_B. The calculation of the weighted footprint integration in this paper shows that the effective incidence angle can explain a small part of the ground-based radiometer bias to lower T_B at 60° V-pol (Figure 6). For the three models, RMSE was reduced 2.4 K at 60° V-pol because of the footprint integration, but no clear improvement was observed at H-pol. The footprint integration cannot fully account for the entire difference between simulations and measurements at higher incidence angles. Thus, part of the difference may be the result of the sky contribution (low T_B) captured by antenna side lobes.

5.4. Snow Emission Model Intercomparison

Results show that all models give similar overall RMSE values between 7.2 K and 10.5 K. It is thus difficult to identify a preferred model for this environment. Those snow-covered RMSE are very similar to snow-free RMSE, which suggest that a large part of the errors come from the soil parameterization (see Section 5.1). However, our results suggest that the multi-layer models, DMRT-ML and WALOMIS, are able to better simulate the range of observed T_B compared to the one-layer model. DMRT-ML and WALOMIS simulations are similar, and suggest that the interference phenomena, simulated by WALOMIS, have a marginal impact on the simulations. Leduc-Leballeur et al. [14] showed that the coherence effect considered in WALOMIS significantly improve the simulation for the semi-infinite snow layered medium in Antarctica, which is not the case for shallow seasonal snowpack. It seems that DMRT-ML, by considering the refraction between layers incorporate the main features needed to reproduce variations in the T_B similar to the observations. However, the snowpack during the winter was shallow with very strong meter scale spatial variability and a stratigraphy that made the snow density measurements challenging to document within the field of view of the radiometer (Table 1). Furthermore, because of the several melt/refreeze events that occurred during the winter several ice layers were present in the snowpack resulting in substantial spatial variability. Because of this high spatial variability in the presence of ice layers and snow density within the snowpack, it is unlikely that individual snow pit measurements are representative of the radiometer footprint, and thus a further source of model uncertainty. Therefore, it is difficult to precisely quantify the effect of coherence in an ice layer, even in a relatively controlled experiment such as that described. A similar experiment but with deeper and more homogeneous snow could help to improve our understanding of snow layer effects on L-band emission. The higher complexity of simulations with WALOMIS may not be necessary, unless coherence and layering effects are very strong.

Although the one-layer model does not capture stratigraphy and coherence effects captured in the more detailed multi-layered models, the overall performance was similar. Hence, in a snow density inversion scheme such as that proposed by Reference [8], the one-layer model is a practical solution because it includes only a small number of free parameters, which is almost a prerequisite to achieve unique values of respective retrievals.

6. Summary and Conclusions

Recent studies show the non-negligible impact of snow on L-band emission [8,17,19] even when the snowpack is shallow. In this study, the Wave Approach for LOw-frequency MIcrowave emission in Snow was adapted and parameterized for seasonal snow using ground-based L-band radiometer observations in a prairie environment. A specular soil reflectivity model was added to WALOMIS, and frozen soil permittivity (ε_g) of 4.3 and 4.8 used for Scene 1 and 3, respectively, was estimated using an optimisation scheme that compared, observed and simulated T_B (DMRT-ML) from snow free frozen ground measurements.

Gaussian noise of snow depth layers of σ_h = 2 cm obtained a comparable multi-angular T_B response but still underestimated the observations at H-pol. The same simulations also underestimated at lower incidence angles and overestimated at higher incidence angles compared to V-pol observations. The calculation of a weighted footprint integration shows that the effective incidence angle can explain part, but not all, of the ground-based radiometer bias to lower T_B at 60° V-pol, but no clear improvement was observed at H-pol. The WALOMIS simulations were then compared to two other radiative transfer models, the DMRT-ML and a simplified adaptation of MEMLS for L-band (LS-MEMLS-1L), used in a single-layer configuration. RMSE between simulated and measured T_B were similar for the three models with overall RMSE between 7.2 and 10.5 K. However, WALOMIS and DMRT-ML were able to better reproduce the range of T_B of the observations at higher V-pol incidence angles (50° and 60°) and across all incidence angles at H-pol.

The impact of snow on surface emission at L-band is relatively small compared to higher frequencies, where the wavelength is similar to snow grain size, inducing scattering. However, there is still valuable and complementary snow information in L-band observations, such as sensitivity to snow density [9] and ice layers [6]. Hence, it is important to develop and assess existing emission models that will help to better quantify these different effects for applications such as the observation of the soil freeze/thaw status. In this study, an effort was made to develop and parameterize WALOMIS for seasonal snow. While the results of all three models are similar, the wave approach of WALOMIS suggests that it is an appropriate new tool to better understand and model L-band emission when interference and snow layering is present.

Author Contributions: A.R. (Alexandre Roy), P.T., C.D., A.B., A.R. (Alain Royer) and T.R. designed the study. A.R. (Alexandre Roy), P.T., C.D., J.L., A.B., and T.R. collected the data. M.L.-L. and A.R. (Alexandre Roy) performed the simulation and the analysis. G.P., J.L. and M.S. developed the softwares. All authors contributed in editing the manuscript.

Funding: The authors would like to thank the Canadian Space Agency (CSA), the Natural Sciences and Engineering Research Council of Canada (NSERC), the French Space Agency (CNES–TOSCA SMOS program), the European Space Agency (ESA) and the Conseil Franco-Québécois de Coopération Universitaire (CFQCU) for their financial support.

Acknowledgments: Thanks to Laurent Bergeron (Université de Sherbrooke) for performing some of the simulations. We would like to thank Erika Tetlock, Craig Smith, Lauren Arnold (Environment and Climate Change Canada), Matthew Williamson (University of Guelph) and the staff at the University of Saskatchewan's Kernen Crop Research Farm for their contributions to the field work.

Conflicts of Interest: The authors declare no conflict of interest.

References

1. Kerr, Y.H.; Waldteufel, P.; Wigneron, J.P.; Delwart, S.; Cabot, F.; Boutin, J.; Escorihuela, M.-J.; Font, J.; Reul, N.; Gruhier, C.; et al. The SMOS mission: New tool for monitoring key elements of the global water cycle. *Proc. IEEE* **2010**, *98*, 666–687. [CrossRef]

2. Lagerloef, G.; deCharon, A.; Lindstrom, E. Ocean salinity and the Aquarius/SAC-D mission: A new frontier in ocean remote sensing. *Mar. Technol. Soc. J.* **2013**, *47*, 26–30. [CrossRef]

3. SMAP Handbook, Mapping Soil Moisture and Freeze/Thaw from Space. Available online: https://smap.jpl.nasa.gov/system/internal_resources/details/original/178_SMAP_Handbook_FINAL_1_JULY_2014_Web.pdf (accessed on 5 September 2018).

4. Rautiainen, K.; Parkkinen, T.; Lemmetyinen, J.; Schwank, M.; Wiesmann, A.; Ikonen, J.; Derksen, C.; Davydov, S.; Davydova, A.; Boike, J.; et al. SMOS prototype algorithm for detecting autumn soil freezing. *Remote Sens. Environ.* **2016**, *180*, 346–360. [CrossRef]

5. Rautiainen, K.; Lemmetyinen, J.; Schwank, M.; Kontu, A.; Ménard, C.B.; Mätzler, C.; Drusch, M.; Wiesmann, A.; Ikonen, J.; Pulliainen, J. Detection of soil freezing from L-band passive microwave observations. *Remote Sens. Environ.* **2014**, *147*, 206–218. [CrossRef]

6. Roy, A.; Royer, A.; Derksen, C.; Brucker, L.; Langlois, A.; Mialon, A.; Kerr, Y.H. Evaluation of Spaceborne L-Band Radiometer Measurements for Terrestrial Freeze/Thaw Retrievals in Canada. *IEEE J. Sel. Top. Appl. Earth Obs. Remote Sens.* **2015**, *8*, 4442–4459. [CrossRef]

7. Derksen, C.; Xu, X.; Dunbar, S.R.; Colliander, A.; Kim, Y.; Kimball, J.S.; Black, T.A.; Euskirchen, E.; Langlois, A.; Loranty, M.M.; et al. Retrieving landscape freeze/thaw state from Soil Moisture Active Passive (SMAP) radar and radiometer measurements. *Remote Sens. Environ.* **2017**, *194*, 48–62. [CrossRef]

8. Schwank, M.; Mätzler, C.; Wiesmann, A.; Wegmüller, U.; Pulliainen, J.; Lemmetyinen, J.; Drusch, M. Snow density and ground permittivity retrieved from L-band radiometry: A synthetic analysis. *IEEE J. Sel. Top. Appl. Earth Observ. Remote Sens.* **2015**, *8*, 3833–3845. [CrossRef]

9. Lemmetyinen, J.; Schwank, M.; Rautiainen, K.; Kontu, A.; Parkkinen, T.; Mätzler, C.; Wiesmann, A.; Wegmüller, U.; Derksen, C.; Toose, P.; et al. Snow density and ground permittivity retrieved from L-Band radiometry: Application to experimental data. *Remote Sens. Environ.* **2016**, *180*, 377–391. [CrossRef]

10. Kaleschke, L.; Tian-Kunze, X.; Maaß, N.; Mäkynen, M.; Drusch, M. Sea ice thickness retrieval from SMOS brightness temperatures during the Arctic freeze-up period. *Geophys. Res. Lett.* **2012**, *39*, L05501. [CrossRef]

11. Wigneron, J.-P.; Kerr, Y.; Waldteufel, P.; Saleh, K.; Escorihuela, M.-J.; Richaume, P.; Schwank, M. L-band microwave emission of the biosphere (L-MEB) model: Description and calibration against experimental data sets over crop fields. *Remote Sens. Environ.* **2007**, *107*, 639–655. [CrossRef]

12. Wiesmann, A.; Mätzler, C. Microwave emission model of layered snowpacks. *Remote Sens. Environ.* **1999**, *70*, 307–316. [CrossRef]

13. Tsang, L.; Kong, J.A.; Ding, K.-H. *Scattering of Electromagnetic Waves, Vol. 1: Theories and Applications*; Wiley-Interscience: New York, NY, USA, 2000.

14. Picard, G.; Brucker, L.; Roy, A.; Dupont, F.; Fily, M.; Royer, A.; Harlow, C. Simulation of the microwave emission of multilayered snowpacks using the Dense Media Radiative transfer theory: The DMRT-ML model. *Geosci. Model Dev.* **2013**, *6*, 1061–1078. [CrossRef]

15. Pulliainen, J.T.; Grandell, J.; Hallikainen, M.T. HUT snow emission model and its applicability to snow water equivalent retrieval. *IEEE Trans. Geosci. Remote Sens.* **1999**, *37*, 1378–1390. [CrossRef]

16. Schwank, M.; Rautiainen, K.; Mätzler, C.; Stähli, M.; Lemmetyinen, J.; Pulliainen, J.; Vehviläinen, J.; Kontu, A.; Ikonen, J.; Ménard, C.B.; et al. Model for microwave emission of a snow-covered ground with focus on L band. *Remote Sens. Environ.* **2014**, *154*, 180–191. [CrossRef]

17. Roy, A.; Toose, P.; Williamson, M.; Rowlandson, T.; Derksen, C.; Royer, A.; Berg, A.A.; Lemmetyinen, J.; Arnold, L. Response of L-Band brightness temperatures to freeze/thaw and snow dynamics in a prairie environment from ground-based radiometer measurements. *Remote Sens. Environ.* **2017**, *191*, 67–80. [CrossRef]

18. Leduc-Leballeur, M.; Picard, G.; Milaon, A.; Arnaud, L.; Lefebvre, E.; Possenti, P.; Kerr, Y.H. Modeling L-band brightness temperature at dome C in Antarctica and comparison with SMOS observations. *IEEE Trans. Geosci. Remote Sens.* **2015**, *53*, 4022–4032. [CrossRef]

19. Naderpour, R.; Schwank, M.; Mätzler, C. Davos-Laret Remote Sensing Field Laboratory: 2016/2017 Winter Season L-Band Measurements Data-Processing and Analysis. *Remote Sens.* **2017**, *9*, 1185. [CrossRef]

20. Hofer, R.; Mätzler, C. Investigations on snow parameters by radiometry in the 3- to 60-mm wavelength Region. *J. Geophys. Res.* **1980**, *85*, 453–460. [CrossRef]

21. Schwank, M.; Stähli, M.; Wydler, H.; Leuenberger, J.; Mätzler, C.; Member, S.; Flühler, H. Microwave L-Band Emission of Freezing Soil. *IEEE Trans. Geosci. Remote Sens.* **2004**, *42*, 1252–1261. [CrossRef]

22. Mätzler, C. Applications of the interactions of micowaves with natural snow cover. *Remote Sens. Rev.* **1987**, *2*, 259–392. [CrossRef]

23. Montpetit, B.; Royer, A.; Roy, A.; Langlois, L.; Derksen, D. Snow microwave emission modeling of ice lenses within a snowpack using the microwave emission model for layered snowpacks. *IEEE Trans. Geosci. Remote Sens.* **2013**, *51*, 4705–4717. [CrossRef]

24. Rees, A.; Lemmetyinen, J.; Derksen, C.; Pulliainen, J.; English, M. Observed and modelled effects of ice lens formation on passive microwave brightness temperatures over snow covered tundra. *Remote Sens. Environ.* **2010**, *114*, 116–126. [CrossRef]

25. Leduc-Leballeur, M.; Picard, G.; Macelloni, G.; Arnaud, L.; Brogioni, M.; Mialon, A.; Kerr, Y.H. Influence of snow surface properties on L-band brightness temperature at Dome C, Antarctica. *Remote Sens. Environ.* **2017**, *199*, 427–436. [CrossRef]

26. Chabot, M.; Lindsay, J.; Rowlandson, T.; Berg, A.A. Comparing the Use of Terrestrial LiDAR Scanners and Pin Profilers for Deriving Agricultural Roughness Statistics. *Can. J. Remote Sens.* **2018**, 1–16. [CrossRef]

27. Lindsay, J.B. Whitebox GAT: A case study in geomorphometric analysis. *Comput. Geosci.* **2016**, *95*, 75–84. [CrossRef]

28. Toose, P.; Roy, A.; Solheim, F.; Derksen, C.; Royer, A.; Walker, A. Radio frequency interference mitigating hyperspectral L-band radiometer. *Geosci. Instrum. Method Data Syst.* **2017**, *6*, 39–51. [CrossRef]

29. West, R.D.; Winebrenner, D.P.; Tsang, L.; Rott, H. Microwave emission from density-stratified Antarctic firn at 6 cm wavelength. *J. Glaciol.* **1996**, *42*, 63–76. [CrossRef]

30. Tsang, L.; Kong, J.A. *Scattering of Electromagnetic Waves, Vol. 3: Advanced Topics*; Wiley-Interscience: Paris, France, 2001.

31. Tsang, L.; Kong, J.; Shin, R. *Theory of Microwave Remote Sensing*; Wiley-Interscience: New York, NY, USA, 1987.

32. Tan, S.; Aksoy, M.; Brogioni, M.; Macelloni, G.; Durand, M.; Jezek, K.C.; Wang, T.-L.; Tsang, L.; Johnson, J.T.; Drinkwater, M.R.; et al. Physical models of layered polar firn brightness temperatures from 0.5 GHz to 2 GHz. *IEEE J. Sel. Top. Appl. Earth Obs. Remote Sens.* **2015**, *8*, 3681–3691. [CrossRef]

33. Jin, Y. *Electromagnetic Scattering Modelling for Quantitative Remote Sensing*; World Scientific: Singapore, 1994.

34. Naderpour, R.; Schwank, M.; Mätzler, C.; Lemmetyinen, J.; Steffen, K. Snow Density and Ground Permittivity Retrieved From L-Band Radiometry: A Retrieval Sensitivity Analysis. *IEEE J. Sel. Top. Appl. Earth Obs. Remote Sens.* **2017**, *10*, 3148–3161. [CrossRef]

35. Rautiainen, K.; Lemmetyinen, J.; Pulliainen, J.; Vehvilainen, J.; Drusch, M.; Kontu, A.; Kainulainen, J.; Seppänen, J. L-band radiometer observations of soil processes in boreal and subarctic environments. *IEEE Trans. Geosci. Remote Sens.* **2012**, *50*, 1483–1497. [CrossRef]

36. Wegmüller, U.; Mätzler, C. Rough bare soil reflectivity model. *IEEE Trans. Geosci. Remote* **1999**, *37*, 1391–1395. [CrossRef]

37. Roy, A.; Toose, P.; Derksen, C.; Rowlandson, T.; Berg, A.; Lemmetyinen, J.; Royer, A.; Tetlock, E.; Helgason, W.; Sonnentag, O. Spatial Variability of L-Band Brightness Temperature during Freeze/Thaw Events over a Prairie Environment. *Remote Sens.* **2017**, *9*, 894. [CrossRef]

38. Hallikainen, M.T.; Ulaby, F.T.; Dobson, M.C.; El-Rayes, M.A.; Wu, L.-K. Microwave dielectric behavior of wet soil—Part I: Empirical models and experimental observations. *IEEE Trans. Geosci. Remote Sens.* **1985**, *GE-23*, 25–34. [CrossRef]

39. Mironov, V.L.; De Roo, R.D.; Savin, I.V. Temperature-dependable microwave dielectric model for an Arctic soil. *IEEE Trans. Geosci. Remote Sens.* **2010**, *48*, 2544–2556. [CrossRef]

40. Fang, X.; Pomeroy, J.W. Modelling blowing snow redistribution to prairie wetlands. *Hydrol. Process.* **2009**, *23*, 2557–2569. [CrossRef]

41. Li, L.; Pomeroy, J. Estimates of threshold wind speeds for snow transport using meteorological data. *J. Appl. Meteorol.* **1999**, *36*, 205–213. [CrossRef]

remote sensing

MDPI

Article

Comparison of Passive Microwave Data with Shipborne Photographic Observations of Summer Sea Ice Concentration along an Arctic Cruise Path

Qingkai Wang [1], Peng Lu [1], Yongheng Zu [1], Zhijun Li [1,*], Matti Leppäranta [2] and Guiyong Zhang [3]

1   State Key Laboratory of Coastal and Offshore Engineering, Dalian University of Technology, Dalian 116024, China
2   Institute of Atmospheric and Earth Sciences, University of Helsinki, 00014 Helsinki, Finland
3   State Key Laboratory of Structural Analysis for Industrial Equipment, School of Naval Architecture, Dalian University of Technology, Dalian 116024, China
*   Correspondence: lizhijun@dlut.edu.cn; Tel.: +86-0411-84708271

Received: 29 June 2019; Accepted: 22 August 2019; Published: 26 August 2019

Abstract: Arctic sea ice concentration (SIC) has been studied extensively using passive microwave (PM) remote sensing. This technology could be used to improve navigation along vessel cruise paths; however, investigations on this topic have been limited. In this study, shipborne photographic observation (P-OBS) of sea ice was conducted using oblique-oriented cameras during the Chinese National Arctic Research Expedition in the summer of 2016. SIC and the areal fractions of open water, melt ponds, and sea ice (A_w, A_p, and A_i, respectively) were determined along the cruise path. The distribution of SIC along the cruise path was U-shaped, and open water accounted for a large proportion of the path. The SIC derived from the commonly used PM algorithms was compared with the moving average (MA) P-OBS SIC, including Bootstrap and NASA Team (NT) algorithms based on Special Sensor Microwave Imager/Sounder (SSMIS) data; and ARTIST sea ice, Bootstrap, Sea Ice Climate Change Initiative, and NASA Team 2 (NT2) algorithms based on Advanced Microwave Scanning Radiometer 2 (AMSR2) data. P-OBS performed better than PM remote sensing at detecting low SIC (< 10%). Our results indicate that PM SIC overestimates MA P-OBS SIC at low SIC, but underestimates it when SIC exceeds a turnover point (TP). The presence of melt ponds affected the accuracy of the PM SIC; the PM SIC shifted from an overestimate to an underestimate with increasing A_p, compared with MA P-OBS SIC below the TP, while the underestimation increased above the TP. The PM algorithms were then ranked; SSMIS-NT and AMSR2-NT2 are the best and worst choices for Arctic navigation, respectively.

Keywords: sea ice concentration; passive microwave; shipborne observation; Arctic navigation

1. Introduction

Arctic sea ice cover has undergone substantial changes in recent decades, such as reductions in sea ice thickness [1,2] and extent [3,4], loss of sea ice volume [5] and multiyear ice coverage [6,7], and a rapid decline in sea ice concentration (SIC) in summer and early autumn [8,9]. These changes have made the Arctic more accessible, which has led to numerous studies of sea ice along the Arctic Passage in the fields of remote sensing, engineering, and geoscience. SIC is one of the main parameters, not only for Arctic sea surface albedo [10] and climate change [11], but also for ice resistance and the safety of ships during Arctic navigation [12].

Passive microwave (PM) remote sensing is a powerful approach to detect SIC at a pan-Arctic scale. This method is widely used because of its relatively low sensitivity to atmospheric conditions, such as clouds and humidity, and because it is independent of daylight hours. In addition, PM has also been adopted to investigate the snow properties such as snow water equivalent [13,14]. PM Arctic sea ice information has been available since the launch of the Scanning Multichannel Microwave Radiometer (SMMR) in 1978, followed by the Special Sensor Microwave/Imager (SSM/I) and the Special Sensor Microwave Imager/Sounder (SSMIS) in 1987. Later, more useful PM information on Arctic sea ice has been provided by the Advanced Microwave Scanning Radiometer for the Earth Observing System (AMSR-E) and its successor, the Advanced Microwave Scanning Radiometer 2 (AMSR2), launched in 2002 and 2012, respectively. Several algorithms have been developed to estimate Arctic SIC based on PM data, such as NASA Team (NT) and its enhanced version (NASA Team 2, NT2), Bootstrap, Arctic radiation and turbulence interaction study (ARTIST) sea ice (ASI), and the European Space Agency (ESA) Sea Ice Climate Change Initiative (SICCI) algorithms. Based on these algorithms, more variability and unknown aspects of the SIC in the Northern Polar Region have been revealed [15–17].

However, the accuracy of the SIC derived from PM algorithms is limited. Each algorithm uses a set of brightness temperatures for ice-free ocean (SIC = 0) and closed ice cover (SIC = 100%) to retrieve SIC [18], which represents the radiometric characteristics of different polar sea surface types [19]. The brightness temperature of sea ice depends on the real temperature and emissivity, which in turn depends on the phase (ice/water) and salinity [20]. In summer, the real temperature is at the melting point, but the phase and salinity experience high variability. Therefore, the brightness temperature of summer sea ice varies considerably in time and space, and the PM algorithms have large uncertainties [21]. Inter-comparison of SIC algorithms has shown that there are differences between the SIC retrieved from different PM algorithms [21–24]. Ivanova et al. [25] also showed that the differences between the PM SIC algorithms are greater in summer than in winter.

Due to the potential errors in the PM SIC, field measurements are essential to obtain SIC at higher resolution and to validate PM SIC measurements. Several methods are available for shipborne observations. The visual observation (V-OBS) aboard a vessel is a direct way to record SIC along the cruise transect, according to the protocol of the Antarctic Sea Ice Processes and Climate program (ASPeCt) or the Arctic Shipborne Sea Ice Standardization Tool (ASSIST) standard, and V-OBS has been carried out extensively on expeditions in the Polar Regions [26–30]. Ozsoy-Cicek et al. [31] compared NT2 and Bootstrap SIC based on AMSR-E data with V-OBS SIC in Antarctica, and found that in general the NT2 algorithm produces slightly higher SIC measurements than the Bootstrap algorithm. Beitsch et al. [19] found that Bootstrap and ASI SIC based on AMSR-E data were in better agreement with V-OBS SIC than the SIC based on SSM/I data using the same algorithms. Pang et al. [32] found that AMSR2-ASI SIC matched better with V-OBS SIC at the Arctic ice edge than AMSR2-Bootstrap SIC.

However, because human subjectivity during V-OBS cannot be avoided [33], instruments, such as digital cameras, have been applied. Hall et al. [34] used an oblique-oriented camera to capture the ice conditions on a scientific cruise to the Greenland Sea. When retrieving SIC, they only used a sub-scene close to the ship without a geometric correction of the images. To promote the use of photographs and retrieval accuracy, algorithms have been developed to retrieve SIC from oblique-captured photographs that have been processed with image partitioning and geometric orthorectification. Band thresholding, the simplest classification method for pixel extraction, has been used extensively for image partitioning [27,35–37]. Other methods, such as the K-means technique, has also been employed [38,39]. For geometric orthorectification, Weissling et al. [38] used a Delauney Triangulation method, which needs a considerable number of ground control points, stating that this method was just an ad hoc solution. A more general method of geometric correction for oblique-view photographs was developed by Lu and Li [40] based on photogrammetric theory, in which the actual image pixel size can be calculated using the camera system parameters. Using the shipborne photographic observation (P-OBS) method enables the recording of small-scale ice features that cannot be detected by PM.

With the development of commercial activities in the Arctic, ice management requires more small-scale SIC information along the ship routes to guide navigation. In fact, vessels are always directed to the areas with wide leads [41]. However, as the most widely used method to detect sea ice condition, PM observations are deficient in detecting SIC at small scales due to its coarse grid. Therefore, the performances of PM SIC algorithms need to be evaluated to validate their agreements with the SIC along the vessel routes from the perspective of navigation. Furthermore, ranking of the PM SIC algorithms is necessary for choosing an optimal option to guide ice navigation. Although a few studies have reported the differences between PM and shipborne SIC along a cruise path in the Arctic [28–30,42], very few ones have adopted an objective method, such as P-OBS.

In order to achieve the present objective, a shipborne sea ice P-OBS program was carried out on the icebreaker R/V *Xuelong* as it sailed in the sea ice zone within the Pacific Arctic sector during the Chinese National Arctic Research Expedition (CHINARE) in the summer of 2016. Based on the P-OBS SIC, the performances of several most commonly used PM SIC algorithms (including Bootstrap and NT SIC based on SSMIS data and ASI, Bootstrap, SICCI, and NT2 SIC based on AMSR2 data) were evaluated. The text is divided as follows: Section 2 provides details of the P-OBS method and image processing. Section 3 presents the distribution of SIC along the cruise path by P-OBS and the comparisons between PM SIC and P-OBS SIC. In Section 4, we discuss the potential factors influencing the differences in the SIC derived from the two sources, the effects of melt ponds, and an inter-comparison of SIC from 2010 to 2016. Finally, these PM algorithms are ranked in Section 5.

2. Data and Methods

2.1. Overview of Ice Navigation

The cruise path of R/V *Xuelong* in the Pacific sector of the Arctic Ocean in summer 2016 is shown by the black line in Figure 1. The cruise was restricted by the ice-breaking ability of the vessel, local ice conditions, and the locations of ice and oceanographic stations. Furthermore, the vessel's captain tended to navigate to regions with wide leads. Sea ice was first encountered at around 71.7°N on 25 July. After entering into the sea ice zone, the vessel sailed northward to 82.7°N on 7 August for a long-term ice camp (nine days), and then drifted back to 82.2°N. The southward navigation began on 15 August and lasted until 23 August, when the vessel sailed out of the sea ice zone at around 74.7°N. The cruise passed through Chukchi Sea, Beaufort Sea, and the Central Arctic, all of which are important seas along the Arctic Passage. Shipborne P-OBS was initiated once sea ice was first encountered and suspended temporarily during the long-term ice camp. When the southward navigation began, P-OBS was resumed and finally terminated as the vessel sailed out of the sea ice zone. A threshold of 60% SIC was used to define the boundary between the marginal ice zone (MIZ) and the pack ice zone (PIZ) [43]. The region north of 82°N on the northward leg was taken as the PIZ (see Section 3.1). In the PIZ, surface wave actions were weak because they were largely constrained by sea ice. In the MIZ, waves were enhanced with large open water areas present. While it was found that the wavelengths were less than the size of most floes, so that sea ice seldom fluctuated with waves [44]. The weather conditions were recorded every minute by the shipborne weather instruments. During ice navigation, air temperature varied between −5.2 °C and 7.3 °C, 82% of the time in the range of from −3 to 0 °C. Wind velocity was normally distributed approximately with the average and standard deviation of 8.0 m/s and 4.0 m/s, respectively. Relative humidity ranged between 77% and 100%, and exceeded 98% in half of the sailing duration.

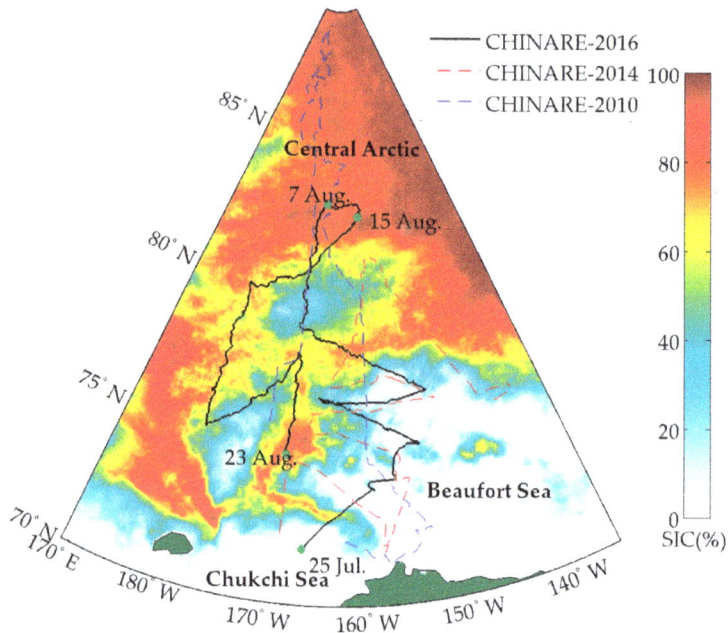

Figure 1. The cruise path of R/V *Xuelong* (black solid line) during Chinese National Arctic Research Expedition (CHINARE)-2016 with mean sea ice concentration (SIC) derived using the AMSR2-ASI (Advanced Microwave Scanning Radiometer 2-Arctic radiation and turbulence interaction study sea ice) algorithm between 25 July 2016 and 23 August 2016 shown in the background. Also included are the dates of the main turning points (green dots), and the cruise tracks of CHINARE-2014 (red dashed line) [43] and CHINARE-2010 (blue dashed line) [30] with ice observations in the same sector.

2.2. Shipborne Photographic Observations

The shipborne P-OBS was carried out using oblique-oriented cameras to automatically capture the sea surface. Two digital cameras were mounted on the port and starboard sides of the R/V *Xuelong* to observe the surface morphology during the cruise. Each camera was installed in a plastic box inlaid with a glass window at the front to protect it from snow and wind (the portside example is shown in Figure 2). The boxes were attached to the ship's hull firmly, so that cameras could only shake with the vessel motion. To gain a wide view, the cameras were placed on the bridge deck (27.3 m above the sea surface) and oriented obliquely at an angle of 25° to the horizon to ensure that the view encompassed the range from shipside to the skyline. The focal length of the cameras was fixed throughout the cruise. The cameras worked automatically with a time interval of 1 min and their time format was set to the Coordinated Universal Time, which was adopted in the shipborne Global Position System, so that the locations of the photographs could be determined from the moment of capture.

The first step to derive P-OBS SIC from oblique-oriented pictures is image partitioning. We used the band-thresholding method, which has been adopted in several investigations in Arctic [27,35,36,45,46]. The accuracy of this method has been assessed by Li et al. [46] using the confusion matrix, and the results showed that the overall accuracy and kappa coefficient were 87%–91% and 0.80–0.86, respectively. We used the same assessment method as Li et al. [45] in this study and the overall accuracy and kappa coefficient turned out to be 94% and 0.91, respectively, which indicates high reliability. As shown in Figure 3a, melt ponds can be distinguished from the surrounding ice because they look bluish, and because of its lower reflectance, open water can be identified by its darker appearance [47]. According to these criteria, sea surface morphology can be partitioned into three categories: sea ice, melt ponds, and open water, by selecting red, green, and blue thresholds based on the color distribution histograms within each picture, as shown in Figure 3b. Because the color of these sea surface categories varies depending on the weather conditions (sunlight or cloud cover) and viewing angle, it was impossible to select common thresholds for the whole image set. Therefore, the threshold level for each surface category in each image was independently determined by hand. Especially, melt ponds were usually misidentified during image partitioning. Manual intervention was a necessary procedure to check the boundary of melt ponds based on the threshold calculated by computer and justified by naked eyes. Then the targets mistaken as melt ponds could be excluded and corrected.

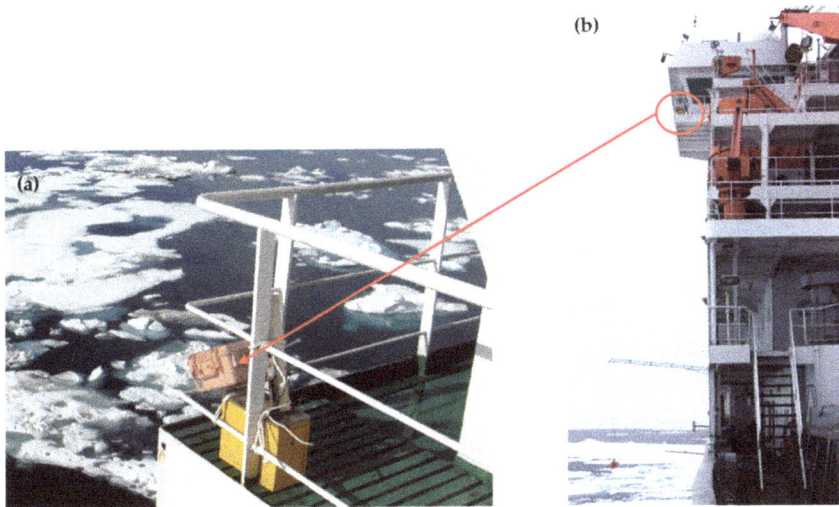

Figure 2. The camera used to observe the sea surface (**a**) on the portside of the R/V *Xuelong* (**b**).

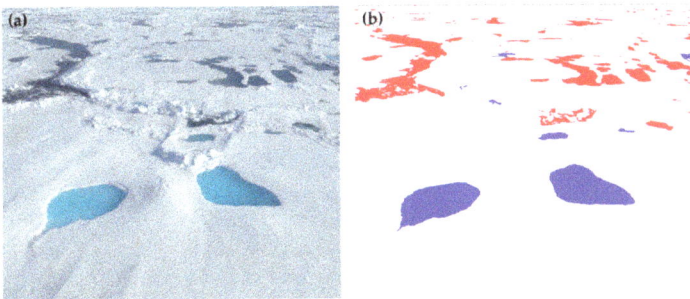

Figure 3. An original oblique-oriented picture (**a**) and its corresponding image after partitioning (**b**) in which the white, red, and blue parts are sea ice, open water, and melt ponds, respectively.

In oblique-oriented cameras, unless the image is rectified, geometric distortion can cause errors in the areal fractions of the sea surface categories. Therefore, the second step in deriving P-OBS SIC is geometric orthorectification. The simplified method proposed by Lu and Li [40] was adopted in this study. In this method, the actual image pixel size is calculated using camera height H, camera tilt α, and focal length f. H and α were obtained during camera installation, and f could be obtained from image information. The ship motion, especially the roll of vessel, influences the geometric distortion. While the error can be effectively controlled by setting an upper limit of the viewing angle. To maintain the balance between the avoidance of large errors and the full utilization of the image, the upper limit of the viewing angle was set to 85°. The relative error induced per degree of the ship's roll was, therefore, less than 30%, and approximately 78% of the image could be used.

After the above two steps, the areal fractions of open water, melt ponds, and sea ice were determined by dividing the area of the pixels of each sea surface category by the total area of the pixels. The results are denoted as A_w, A_p, and A_i, respectively, for open water, melt ponds and sea ice. SIC is the sum of A_p and A_i. Of the total 53,200 pictures captured, 19.7% were not processed due to poor visibility caused by heavy fog or precipitation. Photographs from both port and starboard sides were selected at 10-min intervals to reduce our workload, and a total of 4792 pictures were used for the analysis in this study. The average of the areal fractions obtained from both sides at the same location was used as the local value. In all, we obtained the sea ice conditions at 2396 locations along the cruise path.

2.3. Passive Microwave Data

The PM data used in this work are from SSMIS launched onboard the Defense Meteorological Satellite Program satellites and AMSR2 launched onboard the Global Change Observation Mission 1st-Water satellite. These satellites are on sun-synchronous, near-circular polar orbits and cover the Polar Regions every day. The SSMIS sensor is a 24-channel PM radiometer with frequencies ranging from 19.4 to 183.3 GHz, and the AMSR2 sensor is a 7-channel PM radiometer with frequencies ranging from 6.9 to 89.0 GHz. We selected several commonly used algorithms at their highest spatial resolution, including Bootstrap and NT algorithms based on SSMIS data and ASI, Bootstrap, SICCI, and NT2 algorithms based on AMSR2 data (Table 1).

Table 1. Information on the passive microwave (PM) SIC used in the study. NT: NASA Team; NT2: NASA Team 2; SICCI: Sea Ice Climate Change Initiative.

PM Sensor	Algorithm	Channels Used for Retrieval (GHz)	Resolution (km)
SSMIS	Bootstrap	19.4 V [1], 37.0 V	25 × 25
	NT	19.4 V, 19.4 H, 37.0 V	25 × 25
AMSR2	ASI	89.0 V, 89.0 H	3.125 × 3.125
	Bootstrap	18.7 V, 36.5 V	6.25 × 6.25
	NT2	18.7 V, 18.7 H, 36.5 V, 36.5 H, 89.0 V, 89.0 H	12.5 × 12.5
	SICCI	18.7 V, 18.7 H, 36.5 V, 36.5 H	25 × 25

[1] V and H denote vertical and horizontal polarization, respectively.

2.4. Auxiliary Data

Alongside CHINARE-2016 data, shipborne V-OBS SIC data from CHINARE-2010 and shipborne P-OBS data from CHINARE-2014 were used to characterize the interannual variability of SIC in the Central Arctic Passage. During CHINARE-2010, V-OBS was conducted from the bridge deck of R/V *Xuelong* at time intervals of half an hour [30]. The main information related to SIC, floe size, ice type, melt pond coverage, and ice and snow thickness were recorded according to the ASSIST standard; we used SIC for this work. During CHINARE-2014, P-OBS was carried out using the same method as in this study to measure SIC [43], and these SIC data were therefore also used for comparison.

3. Results

3.1. Sea Ice Concentration Derived from Shipborne Photographic Observations

The SIC distribution along the cruise path is presented in Figure 4a. P-OBS was suspended temporarily during the long-term ice camp (nine days) and also for a short time when the instruments were repaired, as shown by the discontinued sections of the curve with colored points on the northward and southward legs. P-OBS shows that there were two ice belts at the beginning of the northward leg between 71.7° and 72.6°N and between 73.0° and 73.7°N, with a mean SIC of 70.3% and 95.7%, respectively. The sea surface in these ice belts was mainly covered by small floes with melt ponds and melt holes spreading onto the ice surface (Figure 4b). There was not much open water present, but leads were found between small floes. In the 73.7°–78.1°N region, SIC decreased rapidly with an average of 10.4%. Although the sea ice still appeared mainly as small floes, a lot of open water area was observed (Figure 4c). Sailing northward, the open water area diminished, and floe size increased with a mean SIC of 53.5% between 78.1° and 82.7°N (Figure 4d,e). Within the high-latitude part of the southward leg, 80.0°–82.2°N, the latitude 81°N appeared to be a demarcation point. The SIC between 81.0° and 82.2°N was less than that in the same latitude on the northward leg; while the SIC between 80.0° and 81.0°N was similar to that in the same latitude on the northward leg. SIC decreased abruptly with a mean value of 7.6% moving southward and large open water areas reappeared with small floes and brash ice (Figure 4f).

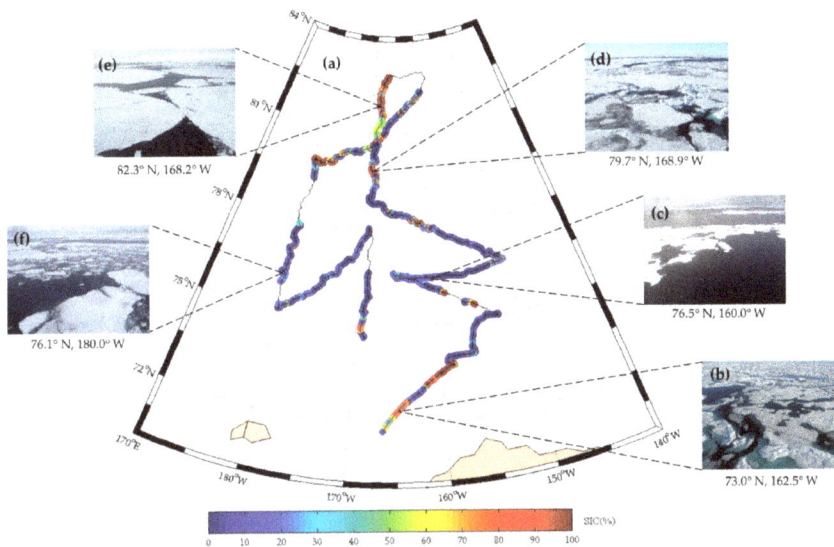

Figure 4. The distribution of sea ice concentration (SIC) (colored points) derived from shipborne photographic observation along the cruise track (black line) in 2016 (**a**) with photographs of typical ice conditions and corresponding locations (**b–f**).

The frequency distribution of SIC along the cruise path is shown in Figure 5, with a 10% bin size. SIC was not normally distributed, but bimodal, and tended to show an approximate U-shape. Because the cruise path was biased towards wide leads, most data were in the 0%–10% bin with a frequency of 0.55. The frequency in the 90%–100%-bin was the second highest, with a value of 0.19. In the other bins, the frequencies in the low SIC class interval (10%–30%-bin) were marginally more than those in the high interval (30%–90%-bin), with frequencies of approximately 0.02.

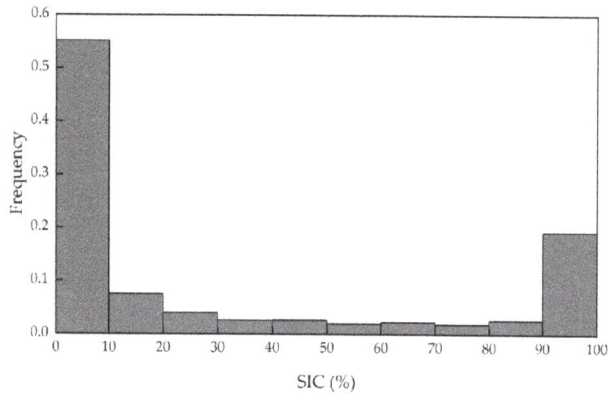

Figure 5. Histogram of the frequency distribution of sea ice concentration (SIC) along the cruise path with a 10% bin size.

3.2. Sea Surface Categories Distribution along the Cruise Path

Figure 6 presents the distribution of each sea surface category (open water, melt ponds, and sea ice) along the cruise path. There were three regions with a small amount of water along the cruise track (Figure 6a): 72.0°–73.7°N (mean A_w = 39.8%) and 80.6°–82.7°N (mean A_w = 15.8%) on the northward leg and 79.9°–80.1°N (mean A_w = 11.9%) on the southward leg. In most of the other regions, open water accounted for the largest fraction. Observations also showed that along the whole cruise path, A_p fluctuated within a relatively narrow range from 0% to 39.8% with an average of 1.2%. The distribution of A_i was the opposite of A_w; the three regions with small A_w mentioned above all had large A_i, with mean A_i values of 60.0%, 82.9%, and 85.6%, respectively.

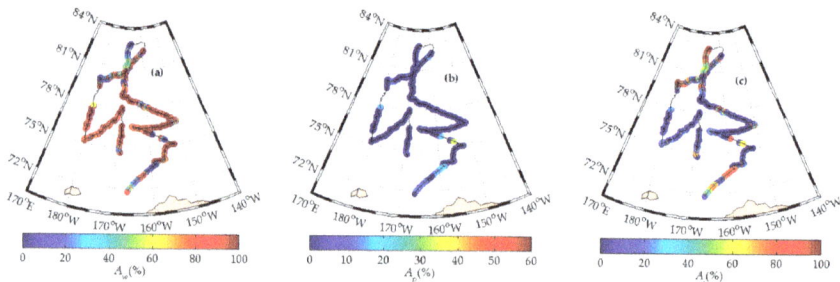

Figure 6. The distribution of areal fractions of open water (A_w) (**a**), melt ponds (A_p) (**b**), and sea ice (A_i) (**c**) (colored points) along the cruise path (black line) based on the photographic observation data.

To better depict the spatial variability in the sea ice conditions, the mean fractional coverage of the sea surface categories and SIC on the northward and southward legs at each latitude are shown in Figure 7. On the northward leg (Figure 7a), the mean A_w was approximately 60% at the southern edge of the sea ice zone (< 74°N), and increased to more than 90% in the 74°–77°N latitude bands. The mean A_w decreased with a slope of 14.8% per degree of latitude in the 78°–82°N latitude bands, although there was a peak at 80°N. The spatial variability of A_i and SIC on the northward leg showed similar trends that were opposite to the A_w. Both the mean A_i and SIC decreased from approximately 35% and 40%, respectively, to 7% as the latitude increased from 72°–73°N to 74°–77°N, and then increased by 14.8% per degree of latitude between 78° and 82°N. Overall, the mean A_p remained fairly small, measuring only 5% for 72°–73°N, and then decreased to less than 2.5% for 74°–82°N. On the southward leg (Figure 7b), the spatial variability of the sea ice conditions was similar to that on the northward cruise track, but at 81°N the mean A_w was higher and the mean A_i and SIC were lower. The mean A_p was never more than 1.8% on the southward leg.

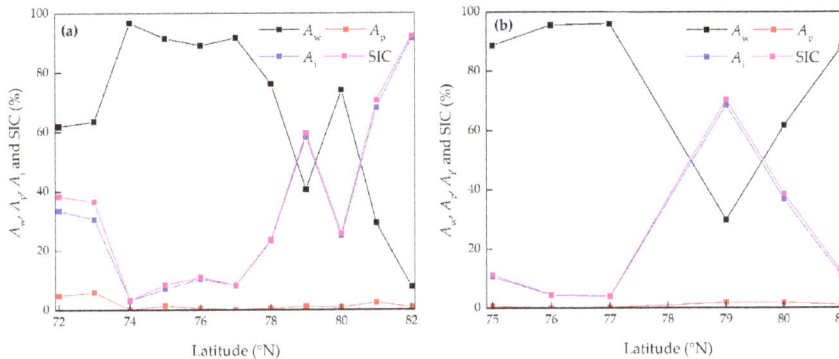

Figure 7. Average areal fractions of open water, melt ponds, and sea ice (A_w, A_p, and A_i, respectively) and sea ice concentration (SIC) on the northward (**a**) and southward (**b**) legs along the cruise path.

3.3. Comparison of Sea Ice Concentration Derived from Two Sources

Because of the fixed PM data grid, there is a large difference in the spatial resolution between PM SIC and P-OBS SIC. According to the equipment installation parameters (H and α) and the upper limit angle set in the image process algorithm, the actual P-OBS area is a 624-m-wide strip along the cruise track. Furthermore, the average sailing speed of R/V *Xuelong* in the ice zone was 4.47 knots (8.28 km/h), and therefore, the ship advanced a distance of 1.38 km in each 10-min interval. It is therefore questionable whether a direct comparison between PM SIC and P-OBS SIC is viable given the different resolutions [27]. However, a moving average (MA) method was adopted to provide a better comparison. The PM SIC of the same day as the P-OBS was interpolated to each image location using the value of the closest grid point. The displacement between the locations of two adjoining P-OBS images was calculated to determine the average window, which was close to but no more than the corresponding PM resolution (Table 1). The P-OBS SIC within the average window was then averaged. Finally, the SIC derived from the two sources with different resolutions were compared. The standard deviation of the MA P-OBS SIC was calculated. It was found that in areas with MA P-OBS SIC < 50%, the standard deviation increased from 0 to approximate 40% with SIC increasing from 0 to 50%; while in areas with MA P-OBS SIC $\geq$ 50%, the standard deviation decreased from approximate 40% to 0 with SIC increasing from 50% to 100%.

However, a number of uncertainties regarding spatial and temporal differences also warrant discussion before moving onto the comparison. First, because the P-OBS area is only a 624-m-wide strip along the cruise path, and although it is averaged, the result does not exactly match the PM pixels of 3.125–25 km in both dimensions. With the objective to evaluate the performances of PM algorithms in estimating SIC along ship routes, it was assumed that the ice conditions were isotropic and homogeneous on the scale of the PM grid cells [40]. Second, P-OBS SIC is instantaneous while PM SIC is a daily average. An error is introduced due to the temporal difference, which can be divided into two main components: the varied sea ice surface condition due to dynamic and thermodynamic processes and the varied sea ice position due to drift. The former component is difficult to estimate because of the varying dynamic-thermodynamic sea ice processes in Arctic summer, while the latter can be estimated qualitatively. Considering that the drift speed of Arctic sea ice is generally no more than 0.1 m/s [48–50], the influence caused by sea ice drift is expected to have minor effects on the comparison of PM SIC with sparser resolutions such as 12.5 km and 25 km. Above all, the error in the comparison is objective because of the distinct resolutions and sampling time of the data. In fact, this issue occurs when comparing the PM SIC with any field observations, not only P-OBS as in our study, but also SIC from aerial observations [27,46] and visual surveys [19,42]. To the authors' best knowledge, a good solution to this issue is absent so far. Maybe improving the field observation technique is a possible way to access the problem in the future.

The comparison between PM SIC and the corresponding MA P-OBS SIC for different resolutions along the cruise path are shown in Figure 8. All the PM SIC series exhibited similar characteristics. PM SIC fluctuated at the start of the northward leg (distance < 1000 km), followed by a peak at a distance of approximately 1000 km (region A). Between 1000 and 2000 km the PM SIC series again fluctuated and reached a peak at approximately 2000 km (region B). Next, all the PM SIC series decreased abruptly, except for AMSR2-NT2 SIC (red circle), and then increased up to approximately 100% when the ship arrived at the long-term ice camp. At the start of the southward leg, all the PM SIC series fluctuated between 3000 and 4000 km, followed by a gradual decline, and again increased abruptly in region C and at the end of the cruise.

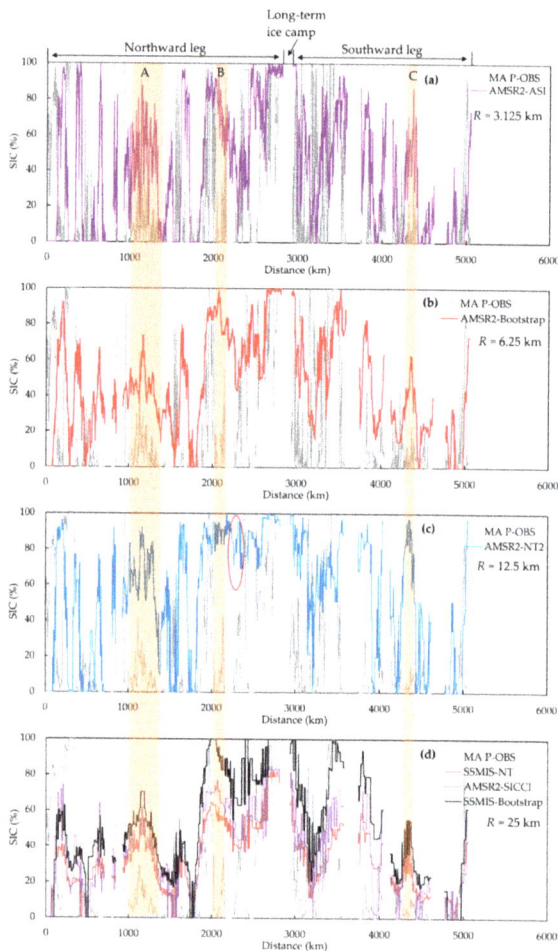

Figure 8. Comparison between passive microwave (PM) sea ice concentration (SIC) and moving average (MA) photographic observation (P-OBS) SIC along the cruise path for each algorithm (**a–d**). The orange columns represent the regions with a large difference between PM SIC and MA P-OBS SIC, and *R* denotes the spatial resolution. The sections of the northward leg, long-term ice camp, and southward leg are also indicated.

Compared with the MA P-OBS SIC, the PM SIC series showed similar variability. There were three regions (A, B, and C shown in Figure 8) where the PM SIC data were much higher than the corresponding MA P-OBS SIC. AMSR2-NT2 SIC had the largest bias with respect to MA P-OBS SIC in these three regions, reaching 60.6%, 79.8%, and 72.5% for regions A, B, and C, respectively. Examining the image set, the three regions showed similar ice conditions, where open water covered almost the whole sea surface with only a few floes in the far range of the image.

The results of the frequency distributions between PM SIC and MA P-OBS SIC are shown in Figure 9. In general, the frequencies of low SIC derived from PM data were much less than those observed by P-OBS, while the opposite occurred for high SIC. This demonstrates that it is difficult for PM to detect low levels of SIC. AMSR2-ASI, AMSR2-Bootstrap, SSMIS-NT, AMSR2-SICCI, and SSMIS-Bootstrap all underestimated the SIC frequency in the < 20%-bin, and AMSR2-NT2 underestimated the SIC frequency in the < 40%-bin. In the middle section (40%–80%-bin for AMSR2-NT2 and 20%–80%-bin for the others), the frequencies were higher for PM SIC than for MA P-OBS SIC. In the > 80%-bin, individual differences occurred. SSMIS-NT and AMSR2-SICCI did not detect SIC beyond 90%, and only SSMIS-NT SIC had a lower frequency than the MA P-OBS SIC in the 80%–90%-bin. Apart from SSMIS-NT, the other PM SIC data showed higher frequencies than MA P-OBS SIC in the > 80%-bin.

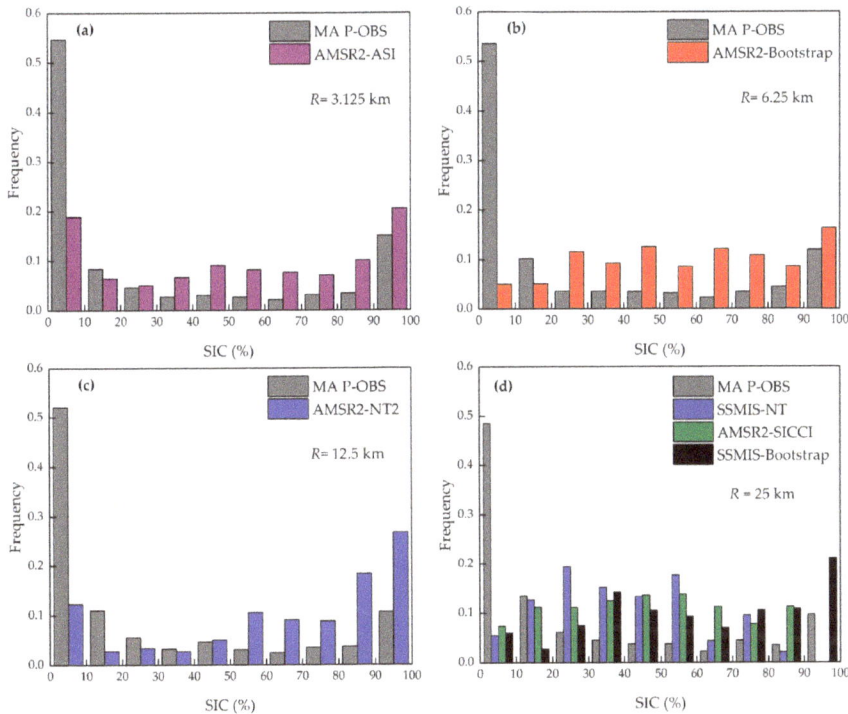

Figure 9. Comparison of the frequency distributions between passive microwave (PM) sea ice concentration (SIC) and moving average (MA) photographic observation (P-OBS) SIC for each algorithm (**a–d**), where *R* denotes the spatial resolution.

As MA P-OBS SIC reflects the true situation in the field, the difference between PM SIC and MA P-OBS is due to an error in the PM algorithms. To further evaluate the agreement of the SIC derived from PM algorithms with the SIC along a ship cruise path, we calculated the mean error and the root-mean-square-error (RMSE) between MA P-OBS SIC and PM SIC. Figure 10 shows the comparison of the mean error and RMSE. The PM algorithms with lower resolutions do not always have poorer accuracy, e.g., PM SIC with a resolution of 25 km did not perform the worst in the comparisons. SSMIS-NT (mean = 10.3%, RMSE = 32.0%) and AMSR2-SICCI (mean = 15.3%, RMSE = 35.9%) were the best two and even better than AMSR2-ASI (mean = 22.1%, RMSE = 47.9%), which had the highest resolution. AMSR2-Bootstrap had a slightly larger mean error than AMSR2-ASI, whereas it had a lower RMSE. SSMIS-Bootstrap and AMSR2-NT2 had the poorest performance, and AMSR2-NT2 (mean = 34.7%, RMSE = 54.9%) was worse than SSMIS-Bootstrap (mean = 30.2%, RMSE = 45.4%). AMSR2-ASI, AMSR2-Bootstrap, AMSR2-NT2, and SSMIS-Bootstrap obtained SIC values between 0 and 100%. However, for SSMIS-NT the upper limits were 64% in the regime with MA P-OBS SIC < 64%, and 82% in the regime with MA P-OBS SIC > 64%, and for AMSR2-SICCI the upper limit was 83.8% in the whole SIC regime; and the lower limit of PM SIC was 0 for both algorithms.

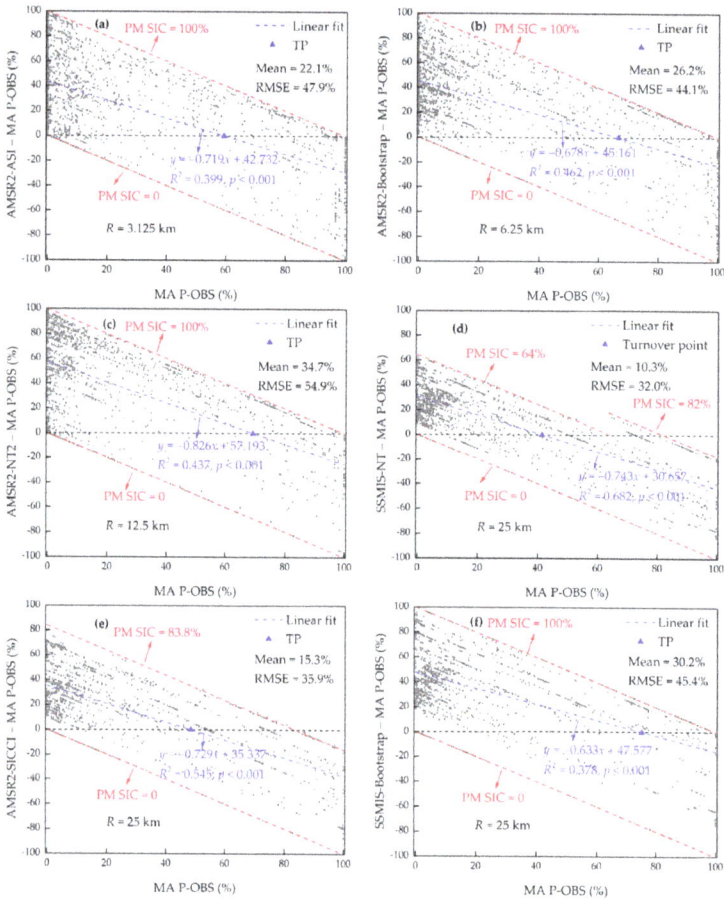

Figure 10. The difference in passive microwave (PM) sea ice concentration (SIC) with respect to moving average (MA) photographic observation (P-OBS) SIC for each algorithm (**a–f**). The gray dashed lines represent a difference equal to 0. The blue dashed lines represent the linear fit line of the difference. The red dashed lines represent the upper and lower limits of the PM SIC. The blue triangles represent the turnover points (TPs) of SIC. Also shown are the mean error, root-mean-square-error (RMSE), equations of linear fit, coefficient of determination R^2, and the significance level p.

The linear fit as shown in Figure 10 can be used to further describe the general variation of the mean error; the over- and underestimation of the PM SIC compared with MA P-OBS SIC can be seen clearly.

In general, with increasing SIC, PM SIC shifts from an overestimate to an underestimate compared with P-OBS SIC. For example, for AMSR2-ASI (Figure 10a), the linear fit turned from positive to negative at SIC = 59.4%, i.e., an overestimate at SIC < 59.4% and an underestimate at SIC ≥ 59.4%. The turnover point (TP) of the SIC for each PM algorithm is listed in Table 2. SSMIS-NT and SSMIS-Bootstrap had the lowest and highest TP values, respectively. Each PM algorithm performed differently in the SIC regimes below and above the TP. Table 3 summarizes the mean error and RMSE of each PM algorithm in these two SIC regimes. Almost all algorithms had a higher mean error below the TP than the absolute value of the mean error above the TP. This result was also seen in the RMSE. The exception was SSMIS-NT, where the mean error below the TP was slightly less than the absolute value of the mean error above the TP. For SSMIS-NT and AMSR2-SICCI, the RMSE values below the TP were 8.8% and 0.8% less than above the TP, respectively. It is noteworthy that the TP values can be used to statistically determine the over- and underestimation of the PM algorithms, while it does not mean that the SIC derived from the PM measurement must be greater or less than that derived from P-OBS in the SIC regime below or above the TP.

Table 2. The turnover point (TP) of the sea ice concentration (SIC) from an overestimate to an underestimate for the passive microwave (PM) algorithms compared with moving average photographic observation SIC.

PM Algorithm	AMSR2-ASI	AMSR2-Bootstrap	AMSR2-NT2	SSMIS-NT	AMSR2-SICCI	SSMIS-Bootstrap
TP (%)	59.4	66.6	69.2	41.3	48.5	75.2

Table 3. The mean error and root-mean-square-error (RMSE) of the passive microwave (PM) sea ice concentration (SIC) with respect to moving average photographic observation SIC below and above the turnover point (TP) of the SIC.

PM Algorithm	<TP		≥TP	
	Mean Error (%)	RMSE (%)	Mean Error (%)	RMSE (%)
AMSR2-ASI	36.5	49.1	−23.6	42.8
AMSR2-Bootstrap	37.3	46.3	−15.9	34.3
AMSR2-NT2	46.9	58.0	−19.5	38.0
SSMIS-NT	23.9	29.4	−26.6	38.2
AMSR2-SICCI	27.8	35.7	−24.5	36.5
SSMIS-Bootstrap	38.1	47.2	−14.6	33.2

4. Discussion

4.1. Factors Influencing the Difference of Sea Ice Concentration Derived from Two Sources

The resolution of the PM data and the physical properties of summer sea ice and snow are the two contributing factors responsible for the difference between the shipborne observed and PM SIC [27]. On the one hand, because vessel cruise paths are typically biased towards areas with leads and isotropic ice conditions are assumed, P-OBS always underestimates the mean regional SIC. It is difficult to detect leads from PM measurements because the spatial resolution is much coarser than the lead scale, which has resulted in overestimates by the PM SIC algorithms in regions with leads compared with P-OBS data. On the other hand, sea ice and snow cover become saturated and flooded with liquid water during the melting season, and hence their physical properties, especially emissivity, change and the surface appears as a mixture of ice and open water to the PM sensors. This causes an underestimate of PM SIC [19].

As shown in Figure 10, PM SIC shifts from an overestimate to an underestimate compared with P-OBS SIC. Therefore, different factors play major roles in the SIC regimes below and above the TP. The coarse resolution dominates mainly below the TP, because more open water leads exist in areas with low SIC. With increasing SIC, less leads occur, and overestimation of the PM SIC decreases. However, the physical properties of summer sea ice and snow dominate above the TP. In areas with high SIC, the sea surface is mostly covered by sea ice and fewer leads are present. In these areas the bias is controlled by the physical properties of wet ice and snow, which results in an underestimate of PM SIC. Similar observations were reported by Knuth and Ackley [27], who compared the SIC derived from the SSM/I-NT algorithm (25 km) with V-OBS SIC in the summer and fall in the Antarctic, respectively. In the summer, there is a similar trend whereby PM SIC shifts from an overestimate to an underestimate compared with V-OBS SIC with a TP of approximately 57%. In the fall, however, PM SIC underestimates V-OBS data at any SIC.

In addition to PM resolution and physical properties of snow and ice, there are several other factors influencing the difference between MA P-OBS SIC and PM SIC, such as the inherent characteristics (e.g., channels and their sensitivities to atmospheric conditions and tie-points) of PM algorithms. Of the investigated algorithms, NT2 and ASI algorithms based on the AMSR2 sensor are very sensitive to atmospheric effects because of using the 89 GHz channel. The cloud liquid water and water vapor can reduce the polarization difference over open water and near ice edge [21]. NT algorithm shows some sensitivity to wind-roughened ocean surface [21]. Algorithms based solely on the 19 and 37 GHz vertically polarized channels display the smallest sensitivity to the atmospheric conditions [22]. Each algorithm uses a set of tie-points to retrieve SIC, while the brightness temperature may have a range of variability for the same ice type or open water due to varying emissivity, atmospheric conditions, and the temperature of the emitting layer, which in turn affects the retrieval accuracy. To compare the superiority of the PM algorithms without the influence of resolution, we selected the algorithms with the same resolution of 25 km, i.e., SSMIS-Bootstrap, SSMIS-NT, and AMSR2-SICCI. Comparison in Figure 10 shows that although the SSMIS-Bootstrap had the smallest sensitivity to the atmospheric conditions, its performance was still lower than SSMIS-NT which accuracy is affected by wind, indicating that the selection of tie-points has heavier effects on Bootstrap algorithm than NT algorithm. SICCI algorithm was suggested as a good choice for SIC retrieval [21], and it is found from the comparison that the performance of AMSR2-SICCI was better than most of other algorithms, but was slightly poorer than SSMIS-NT. The characteristics of SSMIS and AMSR2 sensors may be the causes of the differences. While a thorough study on the characteristics of PM algorithms and sensors is beyond the scope of the current work, as this research has paid more attention to the application of PM SIC algorithms in guiding Arctic navigation. Systemic evaluation can be seen in Ivanova et al. [21] and Andersen et al. [22].

4.2. Melt Pond Effects on the Mean Error of Passive Microwave Sea Ice Concentration

Melt ponds are known to be an important source of error for PM algorithms in detecting Arctic summer SIC, and thus, have attracted much attention [21,51,52]. The previous works provide a qualitative understanding of the melt pond effects; however, to the authors' best knowledge, only very limited studies have quantified the melt pond effects. Only Kern et al. [53] presented that AMSR-E algorithms underestimated MODIS SIC by 20–30% in areas with a high melt pond fraction of 50%. To explore the melt pond effects in a quantitative way, the data shown in Figure 10 were analyzed further. The MA A_P of each data point was first determined, and then the MA A_P range was divided at 1% intervals. The mean error at each interval was defined using the average difference of the PM SIC with respect to MA P-OBS SIC. The variations of the mean error versus MA A_P are shown in Figure 11.

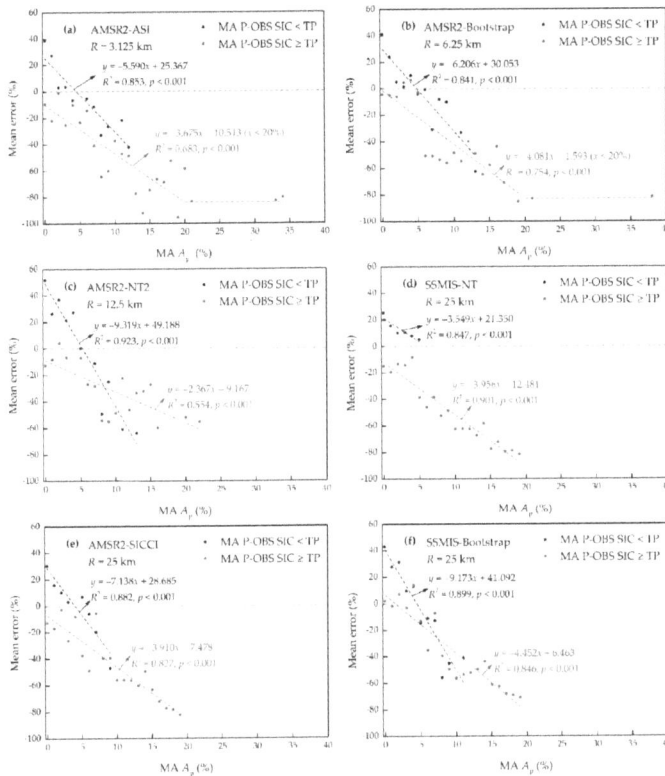

Figure 11. The mean error of the passive microwave (PM) sea ice concentration (SIC) with respect to moving average (MA) photographic observation (P-OBS) SIC with the MA areal fraction of melt ponds (A_p) for each algorithm (**a–f**), where R denotes the spatial resolution. The gray dashed lines represent the mean error equal to 0. The black and blue dashed lines represent the linear fit line of the mean error below and above the turnover point (TP) for each PM SIC algorithm. Also shown are the equations of the linear fit, the coefficient of determination R^2, and the significance level p.

Below the TP, all the PM SIC methods overestimated the MA P-OBS SIC when the MA A_p was approximately less than 5%, mainly because the PM measurement has difficulty in detecting leads due to the coarser spatial resolution. As the MA A_p exceeded 5%, the PM SIC shifted from an overestimate to an underestimate. Because of the similar microwave radiometric nature between open water and melt ponds (the presence of liquid water) as well as the limited penetration depth of microwave radiation into liquid water, melt ponds are easily interpreted as open water by PM measurements [21]. The MA A_p regime below the TP for SSMIS-NT was only 0%–5%, while according to the trend shown in Figure 11d, SSMIS-NT SIC could also underestimate the MA P-OBS SIC if the MA A_p exceeded 5%. Above the TP, almost all PM SIC underestimated the MA P-OBS SIC when the MA A_p varied between approximately 0 and 20%, and the underestimation increased with increasing MA A_p. The exception was SSMIS-Bootstrap, where the mean error was positive with a value of 2.0% when MA A_p was 0. For AMSR2-ASI and AMSR2-Bootstrap above the TP, when the MA A_p exceeded 20%, the mean error was independent of A_p and reached the lower limit, as the horizontal curves show in Figure 11a,b. We also assumed that there are critical values of A_p for the other PM algorithms in the SIC regime above the TP and all the PM algorithms in the SIC regime below the TP, above which the mean error of PM SIC reaches the lower limit, at least –100%, and is independent of A_p.

The trends of the mean error with the MA A_p in the observed A_p regime below and above the TP of SIC were quantitatively described using linear regression (Figure 11). The results show a good linear relationship between the mean error and MA A_p. The slope of the linear fit indicates the influence level of A_p on the mean error. The influence level of A_p below the TP was greater than that above the TP for most PM SIC, with the ratio ranging between 1.5 and 3.9. Only SSMIS-NT had a relatively smaller A_p influence level below the TP than that above the TP. Among these PM algorithms, AMSR2-NT2 had both the highest and lowest influence level of A_p below and above the TP, respectively. SSMIS-NT had the lowest A_p influence level below the TP and SSMIS-Bootstrap had the highest A_p influence level above the TP.

4.3. An Inter-Comparison of Sea Ice Concentration from 2010 to 2016

Shipborne observations of SIC were also conducted during CHINARE-2010 [30] and CHINARE-2014 [43]. The expedition areas covered 178.8°E–152.5°W, 75.6°–88.5°N in 2010 and 139.0°–169.0°W, 72.3°–82.7°N in 2014 (Figure 1); both were similar to the area covered in 2016. The observations were carried out from 21 July–28 August 2010 and from 2 August–1 September 2014, similar to the duration of the observation in 2016. Therefore, the comparison of SIC between these three expeditions can be helpful to depict the variation of summer sea ice in the Central Arctic Passage.

The results show that there were clear differences in SIC between 2010, 2014, and 2016. The boundary between PIZ and MIZ was clear in 2010 and 2014: the area north of 75°N on the northward leg and north of 80°N on the southward leg can be defined as the PIZ in 2010. For 2014, PIZ covered the area north of 76°N on both of the northward and southward legs. It was difficult to identify the boundary between MIZ and PIZ in 2016 due to the dramatically varying SIC, and only the region north of 82°N on the northward leg could be taken as the PIZ. Therefore, the boundary was farther north in 2016 than in 2010 and 2014. In 2010, the mean SIC in the PIZ was 66% for the northward leg and 71% for the southward leg, and the mean SIC in the MIZ was 30% (Table 4). The numbers were slightly higher in 2014: the mean SIC was 76% in the PIZ and 48% in the MIZ. The mean SIC in the MIZ decreased to 20% in 2016. However, in the PIZ, the SIC remained at a similar level to that in 2010 and 2014, with an average of 70%.

Table 4. The mean sea ice concentration along the cruise path in marginal ice zone (MIZ), pack ice zone (PIZ), Chukchi Sea, Beaufort Sea, and Central Arctic in 2010, 2014, and 2016, respectively.

Year	MIZ	PIZ	Chukchi Sea	Beaufort Sea	Central Arctic
2010	30%	66% (northward leg) 71% (southward leg)	/	/	/
2014	48%	76%	56%	59%	98%
2016	20%	70%	21%	8%	56%

Based on the Arctic Sea area division provided by the National Snow and Ice Data Center, the SIC in the R/V *Xuelong* cruise region covered parts of the Chukchi Sea, Beaufort Sea, and the Central Arctic. Therefore, the SIC in these seas can be compared, while due to a lack of detailed information, the SIC in 2010 was not included in the following comparison. Our results show that the SIC in Chukchi Sea and Beaufort Sea were similar in 2014, with averages of 56% and 59%, respectively, while there were larger areas of open water in these two seas in 2016. The mean SIC was only 21% in the Chukchi Sea and even less in the Beaufort Sea (8%). In the Central Arctic, the sea surface in 2014 was almost fully covered with sea ice (or ponded sea ice), with a mean SIC of 98%, while in 2016 the mean SIC was only 56%.

5. Conclusions

With the objective to perform a detailed evaluation on the performances of PM algorithms in estimating SIC along the vessel routes from the perspective of guiding Arctic navigation, a shipborne photographic observation (P-OBS) program was conducted as a part of the Chinese National Arctic Research Expedition in the summer of 2016. When sailing in the sea ice zone, the vessel's cruise path was biased towards wide ice leads. P-OBS was performed using two cameras mounted obliquely to map the sea surface along a strip beside the ship's track. The sea ice concentration (SIC) and areal fractions of open water, melt ponds, and sea ice (A_w, A_p, and A_i, respectively) along the strip were calculated based on the collected images, which were processed using image partitioning and geometric orthorectification. Using these P-OBS data, we compared several commonly used passive microwave (PM) SIC algorithms, including Bootstrap (25 km resolution) and NASA Team (NT) (25 km) algorithms based on the Special Sensor Microwave Imager/Sounder (SSMIS) data and Arctic radiation and turbulence interaction study (ARTIST) sea ice (ASI) (3.125 km), Bootstrap (6.25 km), Sea Ice Climate Change Initiative (SICCI) (25 km), and NASA Team 2 (NT2) (12.5 km) algorithms based on the Advanced Microwave Scanning Radiometer 2 (AMSR2) data.

The observations showed that the distribution of SIC along the cruise path was U-shaped, where the SIC in the SIC classes 0%–10% and 90%–100% accounted for a large proportion of the probability mass. PM SIC mainly underestimated the proportion of SIC in the 0%–10%-bin and overestimated the proportion of SIC in the 90%–100%-bin, indicating that PM measurements have difficulties in detecting very low SIC values. Observations also showed that there were only three areas with small A_w values along the whole cruise track, while in the other parts of the cruise, open water accounted for the largest proportion of the sea surface. Throughout the whole cruise track, A_p fluctuated within a relatively narrow range from 0 to 39.8% with an average of 1.2%.

To provide an optimal option for guiding Arctic navigation, the PM SIC algorithms investigated in this paper were ranked based on the agreement compared with the MA P-OBS SIC (mean error and root-mean-square-error, RMSE). SSMIS-NT produces the smallest mean error and RMSE, and thus performs best out of the chosen algorithms. AMSR2-SICCI produces a slightly larger mean error and RMSE than SSMIS-NT, and is the second best. AMSR2-ASI and AMSR2-Bootstrap have similar performances, while AMSR2-ASI exhibits a lower mean error and ranks as the third choice. The relatively high mean error makes the performance of AMSR2-Bootstrap poorer than that of AMSR2-ASI. Both SSMIS-Bootstrap and AMSR2-NT2 perform worse; AMSR2-NT2 has a larger mean error and RMSE, and is therefore ranked as the last.

The moving average (MA) P-OBS SIC is considered to reflect the true situation, and the difference between PM SIC and MA P-OBS is due to an error in the PM algorithms. A turnover point (TP) in the SIC was determined using the variations in the PM SIC error. Below the TP, PM SIC overestimated P-OBS SIC due to the much coarser PM resolution, whereas it underestimated P-OBS SIC above the TP because the emissivity of the saturated and flooded sea ice and snow cover made the surface signatures appear as a mixture of ice and open water to the PM sensors. The effects of melt ponds on the mean error of PM SIC were analyzed quantitatively. Below the TP, the mean error shifted from positive to negative with increasing A_p. Above the TP, the underestimation of PM SIC increased as A_p increased from 0% to 20%. A linear trend was found for the mean error varying with MA A_p, which was used to quantify the effects of melt ponds on PM SCI.

Arctic sea ice is changing rapidly, which results in increased ship traffic on the Arctic ship routes. Therefore, there will be a greater need for information on the sea ice conditions along ship cruise paths for Arctic ice management. The present research focused on to evaluate the performances of PM algorithms in estimating SIC along the ship routes. Thus, the results will facilitate navigation in the Polar Regions in which only PM satellite data are available. In addition, more factors influencing the accuracy of PM SIC retrieval, such as weather conditions, will be evaluated further in future. Field observation technique also needs improvement to diminish the error caused by temporal and spatial differences in comparisons.

Author Contributions: Conceptualization, Q.W.; Methodology, P.L. and Z.L.; Formal analysis, Q.W. and Y.Z.; Investigation, Q.W.; Writing—original draft preparation, Q.W.; Writing—review and editing, P.L., M.L. and G.Z.

Funding: This research was funded by the National Major Research High Resolution Sea Ice Model Development Program of China (grant number 2018YFA0605903), the National Key Research and Development Program of China (grant number 2017YFE0111400), the National Natural Science Foundation of China (grant numbers 41876213, 51639003 and 51579028), the High Technology of Ship Research Project of the Ministry of Industry and Information Technology (grant number 350631009), the National Postdoctoral Program for Innovative Talent (grant number BX20190051), and the Academy of Finland (grant number 325363).

Acknowledgments: In this study, both AMSR2-ASI and AMSR2-Bootstrap SIC were downloaded from the archive provided by the University of Bremen (https://seaice.uni-bremen.de/start/data-archive/). The SSMIS-Bootstrap, SSMIS-NT, and AMSR2-NT2 SIC were downloaded from the archive provided by the National Snow and Ice Data Center (https://nsidc.org). The AMSR2-SICCI SIC was downloaded from the archive provided by the University of Hamburg (http://icdc.cen.uni-hamburg.de/1/projekte/esa-cci-sea-ice-ecv0.html). The Arctic sea area division was provided by the National Snow and Ice Data Center (https://nsidc.org/data/masie). We also wish to acknowledge the crews of the R/V *Xuelong* for their assistance during the shipborne observations.

Conflicts of Interest: The authors declare no conflict of interest.

References

1. Renner, A.H.H.; Gerland, S.; Haas, C.; Spreen, G.; Beckers, J.F.; Hansen, E.; Nicolaus, M.; Goodwin, H. Evidence of Arctic sea ice thinning from direct observations. *Geophys. Res. Lett.* **2014**, *41*, 5029–5036. [CrossRef]
2. Lindsay, R.; Schweiger, A. Arctic sea ice thickness loss determined using subsurface, aircraft, and satellite observations. *Cryosphere* **2015**, *9*, 269–283. [CrossRef]
3. Comiso, J.C.; Parkinson, C.L.; Gersten, R.; Stock, L. Accelerated decline in the Arctic sea ice cover. *Geophys. Res. Lett.* **2008**, *35*, L01703. [CrossRef]
4. Stroeve, J.C.; Kattsov, V.; Barrett, A.; Serreze, M.; Pavlova, T.; Holland, M.; Meier, W.N. Trends in Arctic sea ice extent from CMIP5, CMIP3 and observations. *Geophys. Res. Lett.* **2012**, *39*, L16502. [CrossRef]
5. Laxon, S.W.; Giles, K.A.; Ridout, A.L.; Wingham, D.J.; Willatt, R.; Cullen, R.; Kwok, R.; Schweiger, A.; Zhang, J.; Haas, C.; et al. CryoSat-2 estimates of Arctic sea ice thickness and volume. *Geophys. Res. Lett.* **2013**, *40*, 732–737. [CrossRef]
6. Kwok, R.; Cunningham, G.F. Contribution of melt in the Beaufort Sea to the decline in Arctic multiyear sea ice coverage: 1993–2009. *Geophys. Res. Lett.* **2010**, *37*, 79–93. [CrossRef]
7. Comiso, J.C. Large decadal decline of the Arctic multiyear ice cover. *J. Clim.* **2012**, *25*, 1176–1193. [CrossRef]
8. Deser, C.; Teng, H. Evolution of Arctic sea ice concentration trends and the role of atmospheric circulation forcing, 1979–2007. *Geophys. Res. Lett.* **2008**, *35*, L02504. [CrossRef]
9. Lei, R.; Xie, H.; Wang, J.; Leppäranta, M.; Jónsdóttir, I.; Zhang, Z. Changes in sea ice conditions along the Arctic Northeast Passage from 1979 to 2012. *Cold Reg. Sci. Technol.* **2015**, *119*, 132–144. [CrossRef]
10. Zhou, C.; Zhang, T.; Zheng, L. The characteristics of surface albedo change trends over the Antarctic sea ice region during recent decades. *Remote Sens.* **2019**, *11*, 821. [CrossRef]
11. Serreze, M.C.; Barry, R.G. Processes and impacts of Arctic amplification: A research synthesis. *Glob. Planet. Chang.* **2011**, *77*, 85–96. [CrossRef]
12. Takimoto, T.; Kanada, S.; Shimoda, H.; Wako, D.; Uto, S.; Izumiyama, K. Field measurements of local ice load on a ship hull in pack ice off the southern Sea of Okhotsk. In Proceedings of the OCEANS 2008-MTS/IEEE Kobe Techno-Ocean, Kobe, Japan, 8–11 April 2008.
13. Langlois, A.; Barber, D.G.; Hwang, B.J. Development of a winter snow water equivalent algorithm using in situ passive microwave radiometry over snow-covered first-year sea ice. *Remote Sens. Environ.* **2007**, *106*, 75–88. [CrossRef]
14. Langlois, A.; Barber, D.G. Advances in seasonal snow water equivalent (SWE) retrieval using in situ passive microwave measurements over first-year sea ice. *Int. J. Remote Sens.* **2008**, *29*, 4781–4802. [CrossRef]
15. Kim, J.; Kim, K.; Cho, J.; Kang, Y.; Yoon, H.; Lee, Y. Satellite-based prediction of Arctic sea ice concentration using a deep neural network with multi-model ensemble. *Remote Sens.* **2019**, *11*, 19. [CrossRef]
16. Strong, C.; Golden, K. Filling the polar data gap in sea ice concentration fields using partial differential equations. *Remote Sens.* **2016**, *8*, 442. [CrossRef]
17. Comiso, J.C.; Meier, W.N.; Gersten, R. Variability and trends in the Arctic sea ice cover: Results from different techniques. *J. Geophys. Res. Oceans* **2017**, *122*, 6883–6900. [CrossRef]
18. Spreen, G.; Kaleschke, L.; Heygster, G. Sea ice remote sensing using AMSR-E 89-GHz channels. *J. Geophys. Res.* **2008**, *113*, C02S03. [CrossRef]

19. Beitsch, A.; Kern, S.; Kaleschke, L. Comparison of SSM/I and AMSR-E sea ice concentrations with ASPeCt ship observations around Antarctica. *IEEE Trans. Geosci. Remote Sens.* **2015**, *53*, 1985–1996. [CrossRef]

20. Rees, G. *Remote Sensing of Snow and Ice*; Taylor & Francis: London, UK, 2006.

21. Ivanova, N.; Pedersen, L.T.; Tonboe, R.T.; Kern, S.; Heygster, G.; Lavergne, T.; Sørensen, A.; Saldo, R.; Dybkjær, G.; Brucker, L.; et al. Inter-comparison and evaluation of sea ice algorithms: Towards further identification of challenges and optimal approach using passive microwave observations. *Cryosphere* **2015**, *9*, 1797–1817. [CrossRef]

22. Andersen, S.; Tonboe, R.; Kern, S.; Schyberg, H. Improved retrieval of sea ice total concentration from spaceborne passive microwave observations using numerical weather prediction model fields: An intercomparison of nine algorithms. *Remote Sens. Environ.* **2006**, *104*, 374–392. [CrossRef]

23. Comiso, J.C.; Cavalieri, D.J.; Parkinson, C.L.; Per, G. Passive microwave algorithms for sea ice concentration: A comparison of two techniques. *Remote Sens. Environ.* **1997**, *60*, 357–384. [CrossRef]

24. Heygster, G.; Wiebe, H.; Spreen, G.; Kaleschke, L. AMSR-E geolocation and validation of sea ice concentrations based on 89 GHz data. *J. Remote Sens. Soc. Jpn.* **2009**, *29*, 226–235.

25. Ivanova, N.; Johannessen, O.M.; Pedersen, L.T.; Tonboe, R.T. Retrieval of Arctic sea ice parameters by satellite passive microwave sensors: A comparison of eleven sea ice concentration algorithms. *IEEE Trans. Geosci. Remote Sens.* **2014**, *52*, 7233–7246. [CrossRef]

26. Worby, A.P.; Allison, I. A technique for making ship-based observations of Antarctic sea ice thickness and characteristics, Part I: Observational techniques and results. In *Antarctic CRC Research Report*; Antarctic CRC: Hobart, Australia, 1999; Volume 14, pp. 1–23.

27. Knuth, M.A.; Ackley, S.F. Summer and early-fall sea-ice concentration in the Ross Sea: Comparison of in situ ASPeCt observations and satellite passive microwave estimates. *Ann. Glaciol.* **2006**, *44*, 303–309. [CrossRef]

28. Lei, R.; Li, Z.; Li, N.; Lu, P.; Cheng, B. Crucial physical characteristics of sea ice in the Arctic section of 143°−180°W during August and early September 2008. *Acta Oceanol. Sin.* **2012**, *31*, 65–75. [CrossRef]

29. Lei, R.; Tian-Kunze, X.; Li, B.; Heil, P.; Wang, J.; Zeng, J.; Tian, Z. Characterization of summer Arctic sea ice morphology in the 135°−175°W sector using multi-scale methods. *Cold Reg. Sci. Technol.* **2017**, *133*, 108–120. [CrossRef]

30. Xie, H.; Lei, R.; Ke, C.; Wang, H.; Li, Z.; Zhao, J.; Ackley, S.F. Summer sea ice characteristics and morphology in the Pacific Arctic sector as observed during the CHINARE 2010 cruise. *Cryosphere* **2013**, *7*, 1057–1072. [CrossRef]

31. Ozsoy-Cicek, B.; Ackley, S.F.; Worby, A.; Xie, H.; Lieser, J. Antarctic sea-ice extents and concentrations: Comparison of satellite and ship measurements from International Polar Year cruises. *Ann. Glaciol.* **2011**, *52*, 318–326. [CrossRef]

32. Pang, X.; Pu, J.; Zhao, X.; Ji, Q.; Qu, M.; Cheng, Z. Comparison between AMSR2 sea ice concentration products and pseudo-ship observations of the Arctic and Antarctic sea ice edge on cloud-free days. *Remote Sens.* **2018**, *10*, 317. [CrossRef]

33. Worby, A.P.; Comiso, J.C. Studies of the Antarctic sea ice edge and ice extent from satellite and ship observations. *Remote Sens. Environ.* **2004**, *92*, 98–111. [CrossRef]

34. Hall, R.J.; Hughes, N.; Wadhams, P. A systematic method of obtaining ice concentration measurements from ship-based observations. *Cold Reg. Sci. Technol.* **2002**, *34*, 97–102. [CrossRef]

35. Perovich, D.K.; Tucker, W.B., III; Ligett, K.A. Aerial observations of the evolution of ice surface conditions during summer. *J. Geophys. Res.* **2002**, *107*. [CrossRef]

36. Inoue, J.; Curry, J.A.; Maslanik, J.A. Application of Aerosondes to melt-pond observations over Arctic sea ice. *J. Atmos. Ocean. Technol.* **2008**, *25*, 327–334. [CrossRef]

37. Lu, P.; Li, Z.; Cheng, B.; Lei, R.; Zhang, R. Sea ice surface features in Arctic summer 2008: Aerial observations. *Remote Sens. Environ.* **2010**, *114*, 693–699. [CrossRef]

38. Weissling, B.; Ackley, S.; Wagner, P.; Xie, H. EISCAM—Digital image acquisition and processing for sea ice parameters from ships. *Cold Reg. Sci. Technol.* **2009**, *57*, 49–60. [CrossRef]

39. Lu, W.; Zhang, Q.; Lubbad, R.; Loset, S.; Skjetne, R. A shipborne measurement system to acquire sea ice thickness and concentration at engineering scale. In Proceedings of the Arctic Technology Conference, St. John's, NL, Canada, 24–26 October 2016.

40. Lu, P.; Li, Z. A method of obtaining ice concentration and floe size from shipboard oblique sea ice images. *IEEE Trans. Geosci. Remote Sens.* **2010**, *48*, 2771–2780. [CrossRef]

41. Worby, A.P.; Geiger, C.A.; Paget, M.J.; Van Woert, M.L.; Ackley, S.F.; DeLiberty, T.L. Thickness distribution of Antarctic sea ice. *J. Geophys. Res.* **2008**, *113*. [CrossRef]

42. Alekseeva, T.A.; Frolov, S.V. Comparative analysis of satellite and shipborne data on ice cover in the Russian Arctic seas. *Izv. Atmos. Ocean. Phys.* **2013**, *49*, 879–885. [CrossRef]

43. Wang, Q.; Li, Z.; Lu, P.; Lei, R.; Cheng, B. 2014 summer Arctic sea ice thickness and concentration from shipborne observations. *Int. J. Digit. Earth* **2018**, 1–17. [CrossRef]

44. Mcgovern, D.J.; Bai, W. Experimental study on kinematics of sea ice floes in regular waves. *Cold Reg. Sci. Technol.* **2014**, *103*, 15–30. [CrossRef]

45. Huang, W.; Lu, P.; Lei, R.; Xie, H.; Li, Z. Melt pond distribution and geometry in high Arctic sea ice derived from aerial investigations. *Ann. Glaciol.* **2016**, *57*, 105–118. [CrossRef]

46. Li, L.; Ke, C.; Xie, H.; Lei, R.; Tao, A. Aerial observations of sea ice and melt ponds near the North Pole during CHINARE2010. *Acta Oceanol. Sin.* **2017**, *36*, 64–72. [CrossRef]

47. Lu, P.; Leppäranta, M.; Cheng, B.; Li, Z.; Istomina, L.; Heygster, G. The color of melt ponds on Arctic sea ice. *Cryosphere* **2018**, *12*, 1331–1345. [CrossRef]

48. Martin, T.; Augstein, E. Large-scale drift of Arctic Sea ice retrieved from passive microwave satellite data. *J. Geophys. Res. Oceans* **2000**, *105*, 8775–8788. [CrossRef]

49. Sumata, H.; Kwok, R.; Gerdes, R.; Kauker, F.; Karcher, M. Uncertainty of Arctic summer ice drift assessed by high-resolution SAR data. *J. Geophys. Res. Oceans* **2015**, *120*, 5285–5301. [CrossRef]

50. Leppäranta, M. *The Drift of Sea Ice*, 2nd ed.; Springer-Praxis: Heidelberg, Germany, 2011.

51. Cavalieri, D.J.; Onstott, R.G.; Burns, B.A. Investigation of the effects of summer melt on the calculation of sea ice concentration using active and passive microwave data. *J. Geophys. Res.* **1990**, *95*, 5359–5369. [CrossRef]

52. Comiso, J.C.; Kwok, R. Surface and radiative characteristics of the summer Arctic sea ice cover from multisensor satellite observations. *J. Geophys. Res.* **1996**, *101*, 28397–28416. [CrossRef]

53. Kern, S.; Rösel, A.; Pedersen, L.T.; Ivanova, N.; Saldo, R.; Tonboe, R.T. The impact of melt ponds on summertime microwave brightness temperatures and sea-ice concentrations. *Cryosphere* **2016**, *10*, 2217–2239. [CrossRef]

remote sensing

MDPI

Technical Note

Radar Scatter Decomposition to Differentiate between Running Ice Accumulations and Intact Ice Covers along Rivers

Karl–Erich Lindenschmidt * and Zhaoqin Li

Global Institute for Water Security, University of Saskatchewan, 11 Innovation Blvd., Saskatoon, SK S7N 3H5, Canada; zhaoqinli@gmail.com
* Correspondence: karl-erich.lindenschmidt@usask.ca; Tel.: +1-(306)-966-6174

Received: 3 January 2019; Accepted: 31 January 2019; Published: 3 February 2019

Abstract: For ice-jam flood forecasting it is important to differentiate between intact ice covers and ice runs. Ice runs consist of long accumulations of rubble ice that stem from broken up ice covers or ice-jams that have released. A water wave generally travels ahead of the ice run at a faster celerity, arriving at the potentially high flood–risk area much sooner than the ice accumulation. Hence, a rapid detection of the ice run is necessary to lengthen response times for flood mitigation. Intact ice covers are stationary and hence are not an immediate threat to a downstream flood situation, allowing more time for flood preparedness. However, once ice accumulations are moving and potentially pose imminent impacts to flooding, flood response may have to switch from a mitigation to an evacuation mode of the flood management plan. Ice runs are generally observed, often by chance, through ground observations or airborne surveys. In this technical note, we introduce a novel method of differentiating ice runs from intact ice covers using imagery acquired from space-borne radar backscatter signals. The signals are decomposed into different scatter components—surface scattering, volume scattering and double-bounce—the ratios of one to another allow differentiation between intact and running ice. The method is demonstrated for the breakup season of spring 2018 along the Athabasca River, when an ice run shoved into an intact ice cover which led to some flooding in Fort McMurray, Alberta, Canada.

Keywords: Athabasca River; decomposition; Fort McMurray; ice run; MODIS; RADARSAT-2

---

1. Introduction

Ice-jam releases can be quite detrimental to flood-prone areas for a number of reasons. The water wave and running ice accumulation created by the release of impounded water and ice can travel at high velocities of up to almost 11 m/s [1,2] leading to high rates of water level rise with very little notice in high flood risk areas. This is particularly difficult for emergency measure coordination when evacuations in such high flood hazard areas are required. The extra water and ice can also exasperate an existing ice-jam situation by adding ice to the ice-jam volume and flow to increase backwater staging. The momentum of the additional water and ice flow can shove into the existing jam to thicken its cover, which in turn increases the jam's underside roughness to augment backwater staging. Hence, from an ice-jam flood forecasting perspective, it is important to distinguish between ice upstream of a flood-prone area, that is; (i) intact or (ii) running and accompanying a water wave. Intact ice provides some delay in the time when the ice can potentially become a hazard to the downstream-lying areas of flood risk. The running ice generally stems from broken up ice covers or ice-jam releases. The arrival of the ice accumulation in the flood-prone area is more imminent, reducing the time for flood preparedness and potentially having to switch from flood mitigation activities to an evacuation mode of the flood management plan.

Much work has already been carried out to study the behaviour of ice-jam releases and their effects on existing downstream ice covers, ice-jams and open water stretches. Earlier work established empirical criteria to determine the onset of breakup. Guo [3] successfully used the shear stress of the water flow on the ice cover as a criterion for ice cover breakup. Ettema [4] emphasised the important role confluences play as conducive locations for the jamming and release of ice accumulations. Jasek [5] studied ice-jam releases and their effects along the Porcupine and Yukon rivers in the Yukon Territory of Canada. Shen and Liu [6] studied a rare occurrence of an ice-jam on the Shokotsu River in Japan showing that, not only the rapid flow increase from snowmelt but also geomorphological characteristics at the jam location (rapid reduction in slope and width in the flow direction) were responsible for the mechanical breakup that led to jamming of the resulting ice accumulations. Beltaos and Burrell [7,8] studied field data of ice-jam release waves (javes) on the Little Southwest Miramichi, Restigouche and Saint John rivers in New Brunswick to characterise parameters describing the waveforms. Once the parameters have been calibrated, particularly celerities of the leading edge and crest of the waves, flow velocities, discharges and shear stress induced by the javes can be quantified. Kowalczyk Hutchinson and Hicks [2] and She et al. [9] investigated ice-jam releases along the Athabasca River and provided a protocol for the collection of data required for numerical modelling of ice-jam releases. Nafziger et al. [10] studied the effects of ice runs in sequence along the Hay River and paid particular attention to the changes in shape and celerity of water wave and ice run hydrographs. Beltaos [1] investigated ice-jam release waves in the lower Mackenzie River and the upper channels of the river's delta. Shen et al. [11] applied the model DynaRICE, a two-dimensional flow and ice model, to study the behaviour of ice-jam releases and highlighted the importance of upstream flood waves in instigating ice-jam releases. Kolerski and Shen [12] applied the model to calculate bed shear stress during the release of an ice-jam along the St. Clair River, flowing from the Laurentian great lake Lake Huron to Lake St. Clair. They established that there needs to be a substantial increase in the stress for backwater from ice-jamming and subsequent ice-jam release to occur. Knack et al. [13] applied the same model to the Saint-John River in New Brunswick, Canada, to simulate ice cover breakup and found that much of the breakup occurrences along that river are due to accumulations from upstream ice-jam release surges.

In general, several insights can be gleaned and summarized from these studies in regard to an ice-jam flood forecasting context. Firstly, both water waves and ice runs disperse as they travel along the river. The celerity of the water waves is generally faster than those of the ice runs. This adds complexity to flood forecasting efforts since the rising flood waters precede the arrival of the running ice accumulation. The ice run can be readily detected by observations from the ground or aerial surveys, which do not provide enough warning for the arrival of the water wave. Secondly, fluvial geomorphology has a marked influence on the hydrograph shapes and travel as they advance along the river. For instance, ice run surface concentrations and celerities can increase when the ice accumulation passes through a width constriction along the river and river segments that are deep. As the slope of the river decreases, as in delta channels, breakup of the ice cover is mostly caused by upstream-originating ice-jam release waves. Thirdly, there are differences in the surge behaviours depending if they are impeded (the water and ice flow into an existing ice cover or jam) or unimpeded (the water and ice flow along an open stretch of water). Stalled jams, when the flow of ice temporarily stops or slows down allowing backwater staging to increase before releasing again, can accentuate wave peaks potentially resulting in more havoc in the flood-prone area.

All of these studies relied on ground observations, either through inspection, recording instrumentation, photography (trail cameras), aerial surveys (helicopter or fixed-wing airplane), numerical modelling and space-borne remote sensing. The use of satellite imagery to distinguish between intact and running ice has thus far not been reported in the scientific literature.

In this technical note, we introduce a novel methodology of differentiating between the two ice covers, intact and running ice, using microwave backscatter signals from the RADARSAT-2 sensor. The sensor allows the transmission and receiving of electromagnetic waves polarised in both horizontal (H) and vertical (V) planes, from which four different transmission-receiving combinations (quad-polarisation) can be acquired; HH, HV, VH, and VV. The four backscatter signals (HH, HV, VH, and VV) can be decomposed into surface backscattering, volume scattering, and double bounce scattering components using the Freeman–Durden decomposition technique [14]. The Freeman–Durden decomposition is a physical-based model, first developed to estimate the contribution of volume scattering from randomly oriented dipoles, surface scattering of first-order Bragg surface, and double bounce scattering, to the total backscatter [14]. The backscattering components of C-band SAR (Synthetic Aperture Radar) images decomposed using the Freeman–Durden decomposition technique have been used to distinguish between solid and macro-porous ice covers and monitoring river ice cover development along the Slave River [15,16]. According to the authors' knowledge, no other applications of this method to river ice has been reported in the scientific literature, since it has only recently been applied in the river ice context.

For snow-covered freshwater ice covers, double bounce scattering of SAR–freshwater ice interaction is typically in a small and negligible magnitude. For a SAR sensor with specific SAR parameters (i.e., wavelength and look angle), the surface scattering component is controlled by the roughness and the permittivity (ε) contrast between air ($\varepsilon \sim 1.0$), snow (dry snow: ε: 1.0–2.0), ice ($\varepsilon \sim 3.7$), and water ($\varepsilon \sim 80$), and volume scattering is controlled by the physical properties of snow and ice [17,18]. For an interface with similar magnitude of surface roughness, the larger the permittivity contrast, the larger the reflection coefficient at the interface thus resulting in the larger surface backscattering. The discussion in this paper is based on C-band SAR. Dry snow on the ice surface and the air/snow interface have limited contribution to the backscattering decomposition components of C-band SAR image [19,20]. The surface scattering component of SAR–freshwater ice interactions is mainly determined by the effective roughness of the ice–water interface due to the large dielectric contrast of water and ice, followed by the roughness of the snow–ice–water interface [15,18,20,21]. However, the ice–water interface in a river typically is smoothed due to thermal erosion and thus has limited contribution to the surface scattering. In addition, the C-band radar pulse usually cannot penetrate the ice layers of consolidated ice covers and reach the bottom ice because of the absorption and reflection of different mediums with variable permittivity [21]. Hence, the main contributor of surface scattering of river ice is the roughness of the snow–ice interface. When the surface of ice covers becomes wet during the ice breakup season, the increased dielectric contrast between the snow/air–ice interface would result in the increase in surface backscattering. The volume scattering is mainly determined by physical properties of ice covers, which mainly refer to the inclusions, including air bubbles, dead vegetation, and sediments in ice covers, as well as the porosity of ice covers.

Ice runs typically have a wetter surface and a greater porosity than intact ice covers, and therefore ice runs are expected to have higher surface scattering due to the increased dielectric contrast of the snow/air–ice interface and a larger volume scattering because of increased porosity of ice covers. Hence, in this paper, we apply the surface backscattering and volume backscattering of the Freeman–Durden decomposition, and the ratio of volume and surface scattering to differentiate between intact and running ice to provide valuable information for ice-jam predictions in a flood forecasting context.

2. Study Site

The study site is the Athabasca River near Fort McMurray, Alberta, Canada extending approximately 200 km from the mouth of the House River to approximately 15 km downstream of the Clearwater River mouth (see Figure 1). The reach upstream of Fort McMurray is relatively steep (slope ≈ 0.001), narrow (width of 150–250 m) and sinuous, interspersed with many rapids making this stretch conducive to the generation of a thick consolidated ice cover during river freeze-up and to sequences of ice-jamming and release during ice cover breakup. The fluvial geomorphology changes abruptly at Fort McMurray when the river flows into a reach with a bed that is much flatter (slope ≈ 0.0003) and wider (width = 300–700 m). Although less sinuous, this reach is interlaced with islands providing many sites for the arrest of ice flow and the formation of ice-jams. The inflow from the Clearwater River tributary in this area also provides an additional source of ice and water. However, the tributary also buffers flood staging when Athabasca River water backs up into the Clearwater River when flood waves and ice runs flow along the Athabasca River through and past the Fort McMurray area. This buffering can slow down the flood wave and ice run enough to allow ice to jam downstream of the Clearwater River mouth, exasperating backwater staging. The rising water levels also extend upstream along the Clearwater River to pose a flood hazard to the downtown area of Fort McMurray.

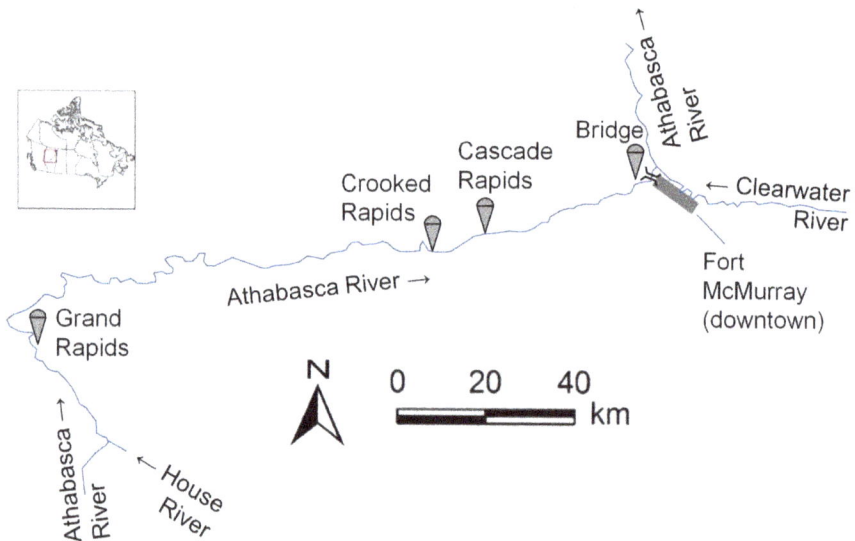

Figure 1. Athabasca River near Fort McMurray, Alberta.

Ice cover breakup monitoring is carried out every spring by scientists and engineers from Alberta Environment and Park's (AEP) River Forecasting Centre. Important Athabasca River gauges are used to track water level elevations, shown in Figure 1, including the one just upstream of Grand Rapids, those just downstream of Crooked and Cascade rapids and the one at the Athabasca River Bridge. Other gauges along the Athabasca River are usually available, but failed during the spring of 2018 breakup period, hence are not shown in Figure 1.

Trail cameras, aerial surveys and satellite images are also used to monitor the progression of spring breakup. Products from RADARSAT-2 and SENTINEL-1 satellite imagery provide classifications of the ice cover into sheet or rubble ice, both intact ice covers, and open water stretches [22]. However, a classification between intact ice covers and moving/floating ice that constitute ice runs is not yet integrated in the classification protocol. Oftentimes, the satellite imagery is not available because acquisitions conflict with other end users or due to long revisit intervals, hence some optical satellite sensors such as the Terra and Aqua sensors of the MODIS (Moderate Resolution Imaging Spectroradiometer) mission have been drawn upon to fill in time gaps. Optical imagery is limited, though, to cloud-free and daytime observations. It is hoped that with the launch of the Radarsat Constellation Mission (RCM) in February 2019, a more reliable and frequent imagery source will be made available which is desperately needed in an ice-jam flood forecasting context. The reader is referred to Lindenschmidt et al. [23] for an example of the requirements of an ice-jam flood forecasting system.

3. Methods

To differentiate the ice runs from intact ice, the Freeman–Durden decomposition [14] was applied to the SQ5W RADARSAT-2 images acquired at 06:57 on 26 April 2018. The incidence angle of a SQ5W RADARSAT-2 image ranges from 22.5° to 26.0° and the nominal range resolution is between 20.6 and 23.6 m. We used the Freeman–Durden decomposition because it has performed well in monitoring ice cover development [16]. Mermoz et al. [21] also applied a similar but mathematics-based decomposition approach, the Cloude–Pottier decomposition method [24], to determine ice thicknesses. The scattering mechanism of surface scattering and volume scattering on differentiating the ice covers of ice runs and intact ice covers dominated by consolidated ice are proposed in Figure 2.

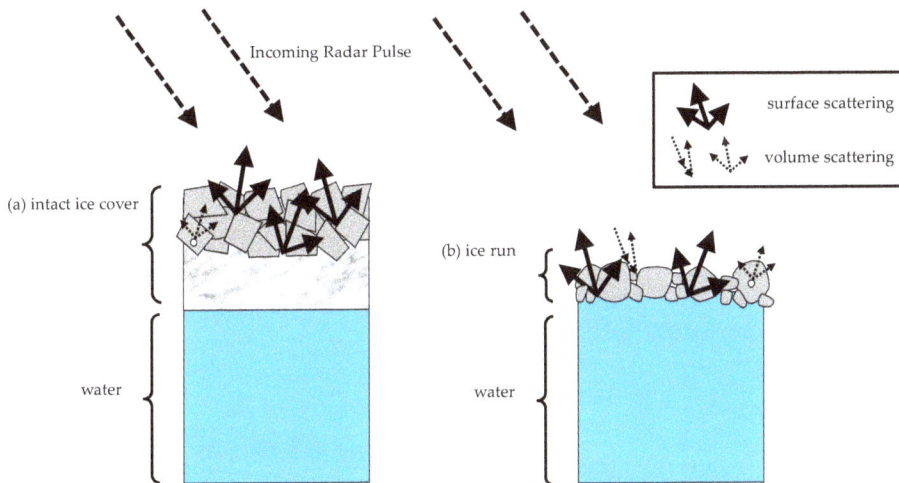

Figure 2. The scheme of surface and volume scattering of fully polarimetric SAR (Synthetic Aperture Radar) images for differentiating (**a**) intact ice covers and (**b**) ice runs (adapted from Gherboudj et al. [19]).

A mask was applied to isolate the river. Prior to decomposition, the SQ5W image was orthorectified in PCI Geomatica 2017 using the Radar-Specific Model and the ASTER Global Digital Elevation Model Version 2 (GDEM V2). The pixel size of the image was resampled to 25 m using the nearest neighbour method and Sigma Nought radiometric correction was also applied. The orthorectified and radiometrically corrected image was then despeckled by applying a boxcar filter within a 5 × 5 pixel moving window using the PoLSAR Boxcar Filter in the Focus Module of PCI Geomatica 2017. Finally, the asymmetric covariance matrix of the despeckled SQ5W image was converted to a symmetric covariance matrix and the Freeman–Durden decomposition was applied to the symmetric covariance matrix to decompose the total power to the contribution of surface scattering, volume scattering and double bounce in the Focus Module of PCI Geomatica 2017.

The decomposed volume, surface backscattering, and their ratio were used to differentiate ice covers of ice runs from intact ice covers. MODIS images (used for validation of ice cover conditions when cloud free), aerial photographs, and gauge hydrographs at the Grand Rapids, Crooked Rapids, Cascade Rapids, and Fort Murray Bridge, taken in the ice breakup time of the winter of 2017–2018 were used to support the analysis. A flowchart has been provided in Figure 3.

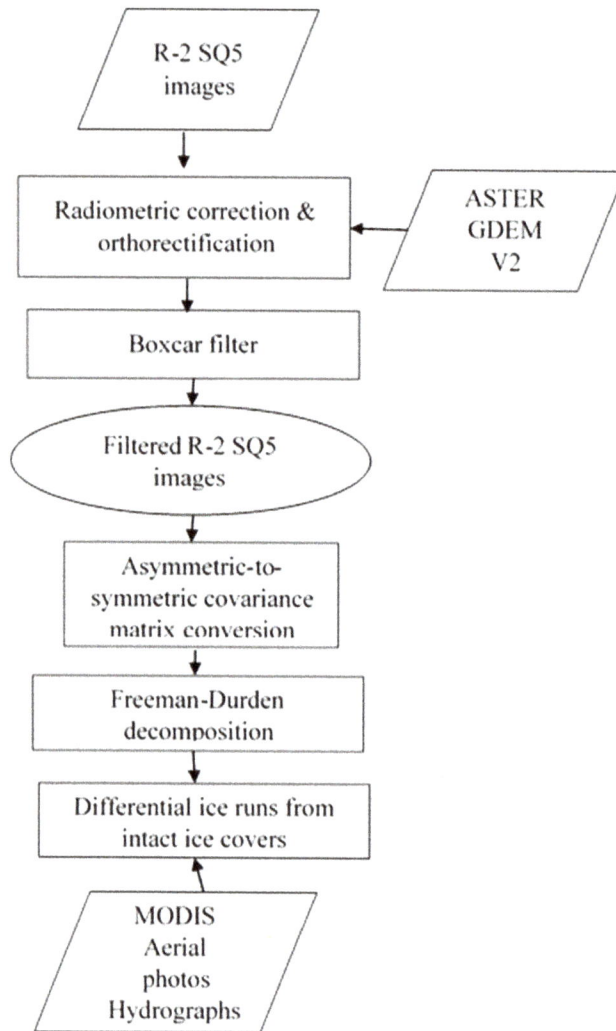

Figure 3. The procedures of image processing and differentiation of ice covers of ice runs and intact ice covers.

4. Ice Cover Breakup of Spring 2018

On 24 April 2018, the ice cover along the river stretch of interest was intact, as evidenced in the Terra MODIS imagery acquired on that day (figure not shown). The following day, deterioration of the ice cover was quite advanced with some areas exhibiting the opening up of water leads in the ice cover, particularly near rapids.

As reported by Alberta Parks and Environment [25,26] and referring to the ice map in Figure 4, some sections along the intact ice cover (orange) between Grand Rapids and Crooked Rapids had moved and broken up resulting in accumulations of rubble ice at the downstream ends of open water stretches (dark blue in Figure 4) approximately 1 to 3 km in length. Some stretches where the ice had broken but not moved were flooded with water (light blue in Figure 4) or had open leads (Figure 5b). Some ice accumulations had resulted in the formation of small ice-jams approximately 2 km in length (red in Figure 4, Figure 5a). The ice cover between Crooked Rapids and Fort McMurray remained intact (Figure 5c). The river downstream of the Clearwater River mouth was open for approximately 15 km (Figure 5d).

Some movement of the ice cover became first evident in the water level hydrographs recorded at Grand Rapids at 10:15 on 25 April 2018 (Figure 6a). Breakup of the ice cover was first recorded at Grand Rapids at midnight, during the night from 25 to 26 April 2018 (Figure 6b). The wave of this breaking front propagated downstream to arrive at the Crooked Rapids gauge 7 hours later (Figure 6c), at the Cascade Rapids gauge (Figure 6d) and at the bridge gauge (Figure 6e), all during the same morning. The breaking front wave tripped the ice movement indicators at 07:00 and 07:45, respectively at Crooked and Cascade rapids (Figure 6f,g). The RADARSAT-2 image was acquired three minutes before ice movement was recorded at Crooked Rapids, dedicating Crooked Rapids to a threshold location, upstream of which the ice was running and downstream of which the ice cover was still intact. The breakup reached the Fort McMurray bridge at approximately 10:00 (Figure 6e) with a large amount of ice and water being forced into the Fort McMurray area. Celerities of this impeded breaking front (propagating downstream into an intact ice cover) decreased in the downstream direction from an average of 3.87 m/s between Grand and Crooked rapids, 3.67 m/s between Crooked and Cascade rapids, and 3.12 m/s between Cascade Rapids and the bridge.

An ice run that began far upstream of our study site, indicated as ice run #2 in Figures 7 and 8, arrived at the Crooked Rapids gauge at 18:30 on 26 April 2018 (Figure 6h) and progressed further downstream, arriving at the Cascade Rapids gauge approximately a half hour later (Figure 6i). The celerity of this unimpeded ice run front (propagating downstream into open water or water with free-floating ice) between these two gauges averaged 4.85 m/s, which is faster than the previously propagating impeded breaking front. This ice run added water and ice (Figure 6j) to the already high water and ice flows at the bridge.

Both ice movement indicators at Crooked and Cascade rapids recorded movement of the ice run for approximately 26.5 h, until 09.35 and 10:00 on 27 April 2018 at the respective gauges (Figure 6k,l).

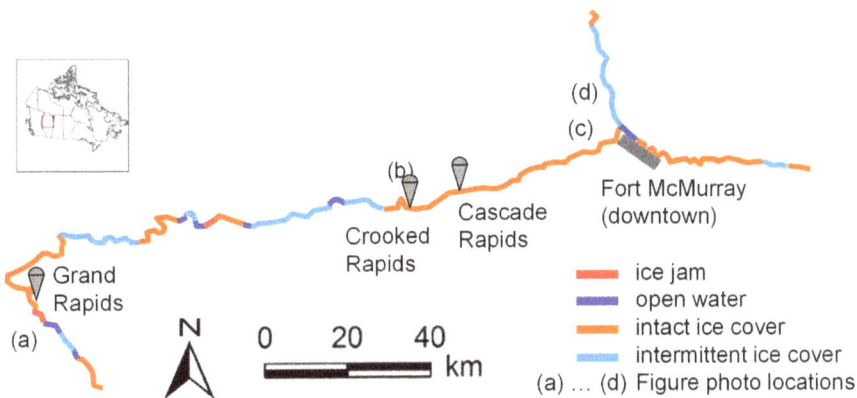

Figure 4. Ice map of the Athabasca and Clearwater rivers near Fort McMurray from 25 April 2018 (data source: [26]).

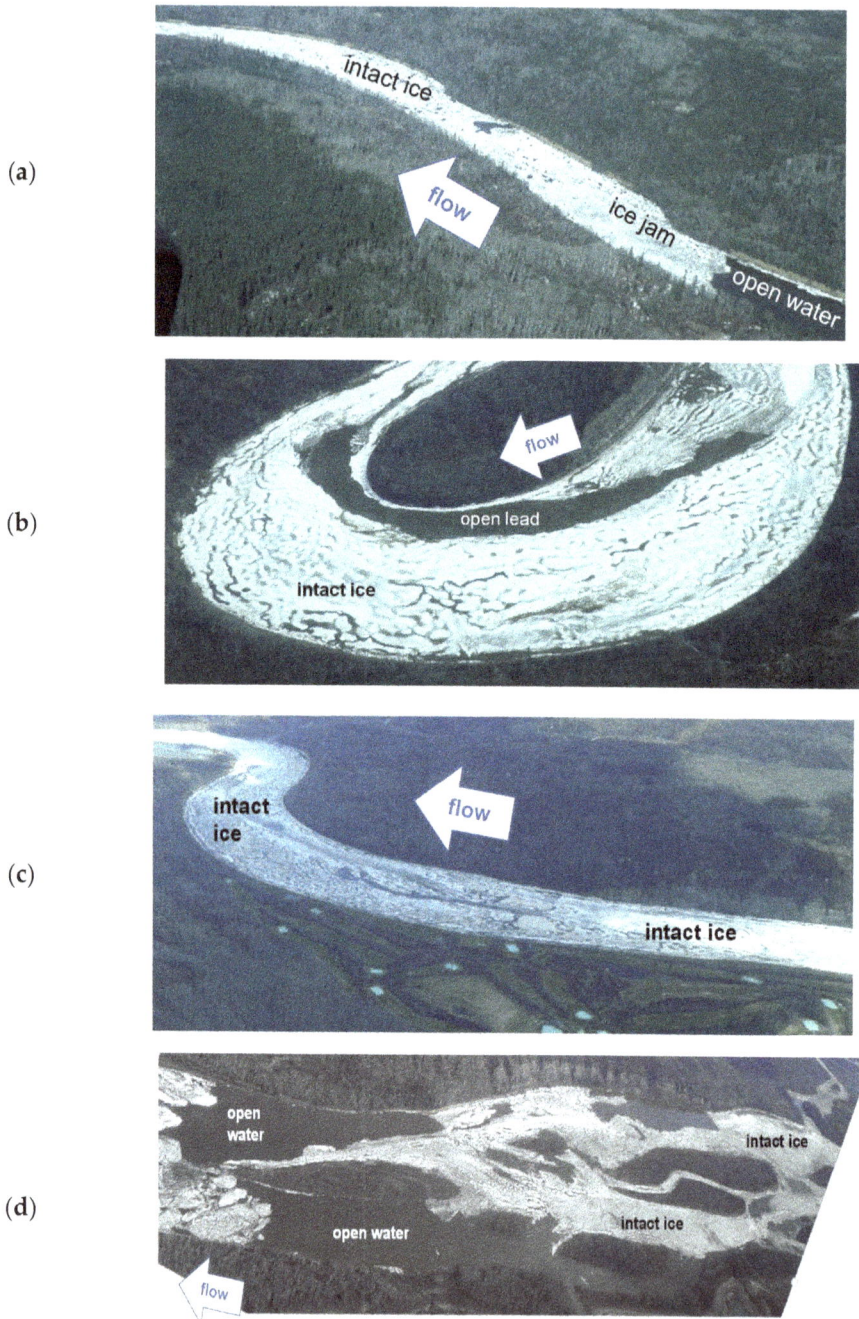

Figure 5. (**a**) Small ice-jam downstream of Grand Rapids; (**b**) Open lead in intact ice cover at Crooked Rapids; (**c**) Intact ice extending 6 km upstream of bridges; (**d**) Intermittent ice cover downstream of Clearwater River mouth. (source: [26]).

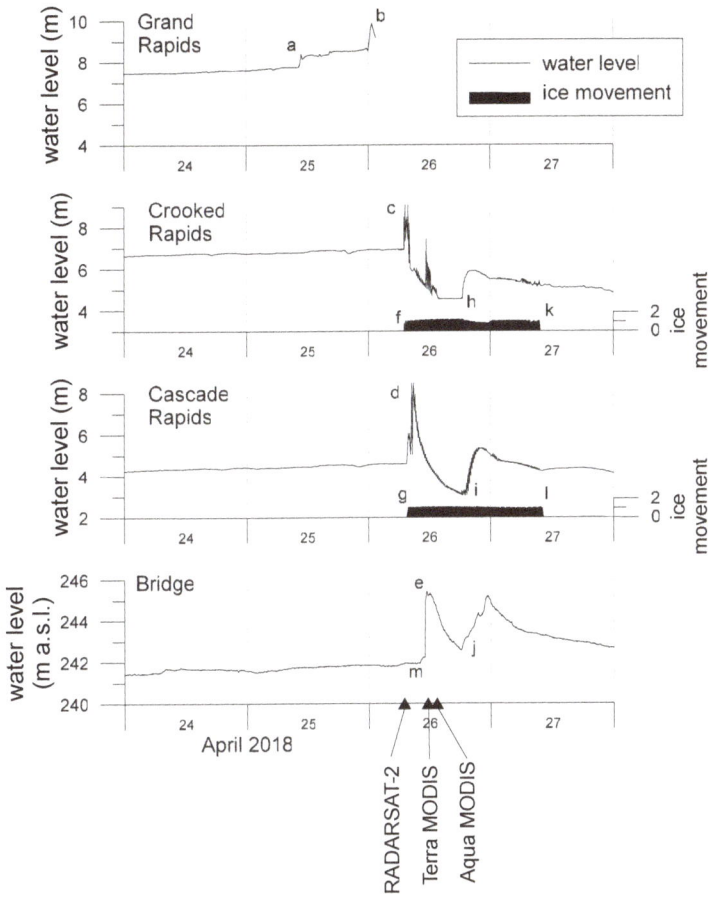

Figure 6. Gauges from hydrographs during the April 2018 breakup event.

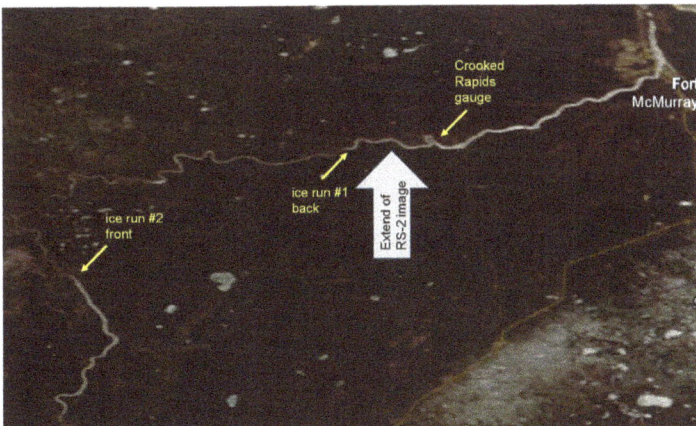

Figure 7. MODIS Terra image acquired 26 April 2018 11:35 MST.

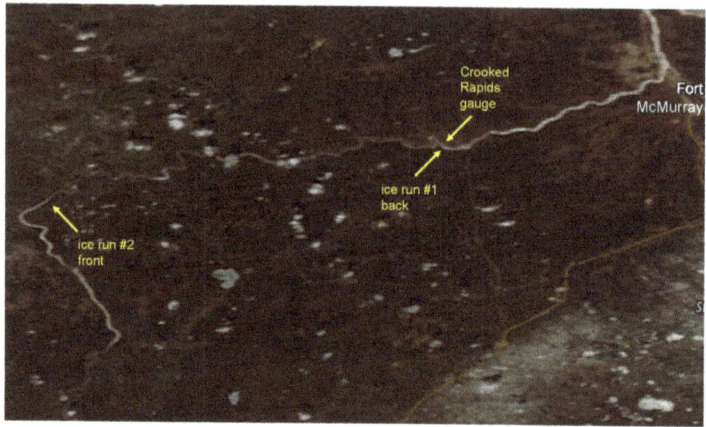

Figure 8. MODIS Aqua image acquired 26 April 2018 13:23 MST.

5. Results and Discussion

The top panel of Figure 9 shows longitudinal profiles of the surface scattering, volume scattering, and double bounce averaged along the course of the river. The double bounce was very small in relation to the other two scatter components. Surface scattering overpowered volume scattering for most of the stretch, in particular the stretch with intact ice. However, in the stretch with running ice, upstream of the Crooked Rapids gauge (35 to 48 river km), the volume scatter did appear to approach the same power contribution as surface scattering. This was more differentiated in the ratio volume scattering/surface scattering as shown in the bottom panel of Figure 9. Intact ice has a very low volume to surface scattering ratio, while an ice run has a much larger volume/surface ratio.

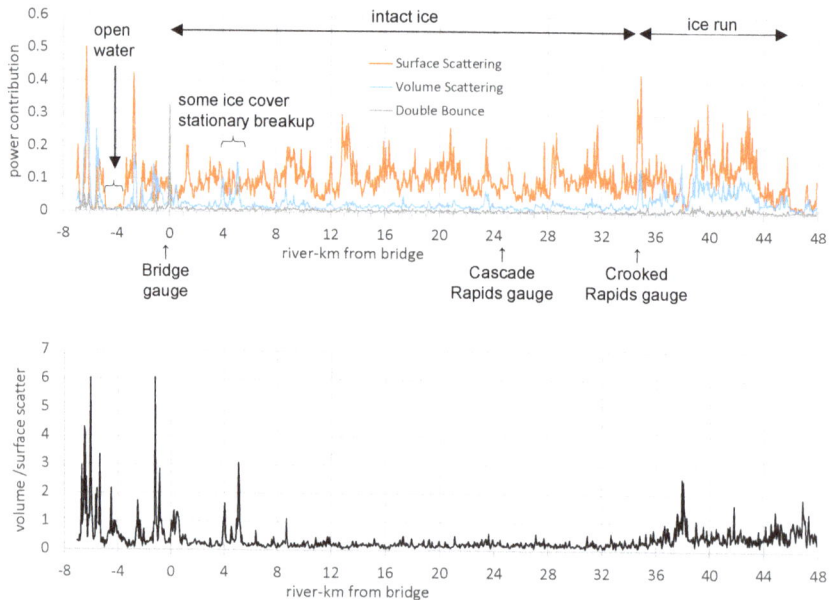

Figure 9. Longitudinal profile of averaged backscatter components (**top panel**) and ratio of volume scattering/surface scattering (**bottom panel**).

The power contribution of the volume scatter did become substantial downstream of the bridge where the local ice cover may have broken up and formed consolidated ice covers. Unfortunately, photographs were not taken during this time to provide evidence, however, a stretch of open water between −3.5 and −5 river-km did point to an area where the ice cover had broken up from which the rubble ice would had to have travelled downstream to shove into the downstream ice cover to form a jam between −5 and −8 river-km. Water levels recorded at the bridge (Figure 6m) did indicate some staging at the time the RADARSAT-2 image was acquired. For open water, almost all of the transmitted microwaves were reflected and scattered forward from the water surface, hence little scatter components remained.

Some higher volume scatter relative to the surface scatter was shown a few kilometres upstream of the bridge, between 4 and 6 river-km (top panel of Figure 9), with a relative strong volume scatter/surface scatter ratio (bottom panel of Figure 9). This may be due to some breakup of the local ice cover, however, this was difficult to verify since no photographic evidence was available for this stretch at that time.

Figure 10 shows a spatial representation of the decomposed scatter components at Crooked Rapids. This area represents the transition from running ice and intact ice. Again, the double bounce signal was minute and need not to be shown in the figure. However, both the surface and volume scatter components, respectively the top and middles panels of the figure, had higher values in the ice run stretch (upstream of the gauge) than the intact ice stretch (downstream of the gauge), which coincides with the results shown in Figure 10. More distinction in the transition from running to intact ice can be seen in the map of the volume scatter/surface scatter ratio, shown in the bottom panel of Figure 10. The running ice will be more porous and wetter, hence exhibit more surface scattering, whereas the intact ice will retain more volume scattering due to its continuous, solid medium.

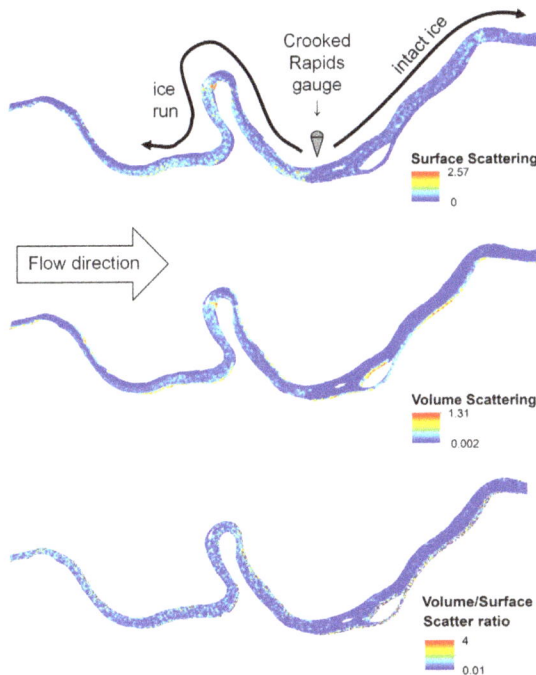

Figure 10. Surface scatter (top panel), volume scatter (middle panel) and volume scatter/surface scatter ratio (bottom panel) (RADARSAT-2 Data and Products © MacDonald, Dettwiler and Associates Ltd—All Rights Reserved. RADARSAT is an official trademark of the Canadian Space Agency).

The river reach with ice runs showed higher volume and surface scattering and a larger volume to surface scattering ratio than the reach covered by intact ice. The dramatically increased volume scattering of ice runs is probably attributed to the interactions of radar pulse between adjacent broken ice in ice runs (increased porosity of ice runs). However, numerical modelling and more images at ice breakup are needed to confirm this finding in the future. For intact ice, volume scattering mainly stems from the inclusions (e.g., air bubbles, sediments, and air bubbles) in ice covers [19,21]. The amount of such inclusions determines the contribution of volume scattering to the total backscattering. The large surface scattering was attributed to the rough air–ice interface [17], while the contribution of the ice–water interface in a high flow river channel was negligible because the bottom of ice is typically smooth in a river with high water flow due to the thermal erosion. In this regard, the larger surface scattering of ice runs may result from the increased surface roughness, compared to intact ice. In addition, broken ice covers of ice runs tend to have a wetter surface, compared to the intact ice covers, which would also account for the higher surface scattering as a result of an increased dielectric contrast of the air–ice interface.

Applying Freeman–Durden decomposition to C-band quad-pol RADARSAT-2 images has demonstrated great potential to differentiate ice runs from intact ice in this study. Nevertheless, further research or observation data is needed to investigate the sources of the high volume scattering and large ratio of volume to surface scattering near and downstream of the bridge.

6. Conclusions

Using the Freeman–Durden decomposition method, intact ice and running ice were successfully differentiated from quad-pol RADARSAT-2 imagery. This novel method of using space-borne imagery to provide wider coverage of ice characteristics along rivers can be very useful in an ice-jam flood forecasting context. Characterising the difference between running and stationary ice covers helps determine the potential timing of the arrival of water and ice run waves which can lead to ice-jam backwater staging and flooding in high flood risk areas. This methodology will also refine an ice-jam flood forecasting approach using a stochastic modelling approach first applied to the town of Fort McMurray in the spring of 2018. A second attempt of stochastic ice-jam flood forecasting is planned for the spring breakup of 2019 and it is hoped that differentiating between running ice and intact ice can help better quantify the volumes of ice available for ice-jamming in the town.

Author Contributions: Z.L. carried out the decomposition analyses. K.–E.L. drafted the technical note with contributions from Z.L.

Funding: This activity was undertaken with financial support of the Canadian Space Agency. Additional funding was provided through the University of Saskatchewan's Global Water Futures research program.

Acknowledgments: The authors are grateful to Jennifer Nafziger and Nadia Kovachis Watson from Alberta Environment and Parks for providing photographs and data of the Athabasca River ice breakup event of 2018. They are also grateful to the Canadian Space Agency's Science and Operational Applications Research—Education (SOAR-E) initiative through which the RADARSAT-2 imagery was made available. They also acknowledge the use of imagery from the NASA Worldview application which is part of the NASA Earth Observing System Data and Information System (EOSDIS) (https://worldview.earthdata.nasa.gov).

Conflicts of Interest: The authors declare no conflict of interest.

References

1. Beltaos, S. Hydrodynamic properties of ice-jam release waves in the Mackenzie Delta, Canada. *Cold Reg. Sci. Technol.* **2014**, *103*, 91–106. [CrossRef]
2. Kowalczyk Hutchinson, T.; Hicks, F. Observations of ice jam release waves on the Athabasca River near Fort McMurray, Alberta. *Can. J. Civ. Eng.* **2007**, *34*, 473–484.
3. Guo, Q. Applicability of criterion for onset of river ice breakup. *J. Hydraul. Eng.* **2002**, *128*, 1023–1026. [CrossRef]

4. Ettema, R.; Muste, M.; Kruger, A. Ice jams in river confluences. In *Cold Regions Research and Engineering Laboratory Report 99-6*; U.S. Army Corps of Engineers: Hanover, NH, USA, 1999.

5. Jasek, M. Ice jam release surges, ice runs, and breaking fronts: Field measurements, physical descriptions, and research needs. *Can. J. Civ. Eng.* **2003**, *30*, 113–127. [CrossRef]

6. Shen, H.T.; Liu, L. Shokotsu River ice jam formation. *Cold Reg. Sci. Technol.* **2003**, *37*, 35–49. [CrossRef]

7. Beltaos, S.; Burrell, B. Determining ice-jam-surge characteristics from measured wave forms. *Can. J. Civ. Eng.* **2005**, *32*, 687–698. [CrossRef]

8. Beltaos, S.; Burrell, B. Field measurements of ice-jam-release surges. *Can. J. Civ. Eng.* **2005**, *32*, 699–711. [CrossRef]

9. She, Y.; Andrishak, R.; Hicks, F.; Morse, B.; Stander, E.; Krath, C.; Keller, D.; Abarca, N.; Nolin, S.; Nzokou Tanekou, F.; et al. Athabasca River ice jam formation and release events in 2006 and 2007. *Cold Reg. Sci. Technol.* **2009**, *55*, 249–261. [CrossRef]

10. Nafziger, J.; She, Y.; Hicks, F. Celerities of waves and ice runs from ice jam releases. *Cold Reg. Sci. Technol.* **2016**, *123*, 71–80. [CrossRef]

11. Shen, H.T.; Gao, L.; Kolerski, T.; Liu, L. Dynamics of Ice Jam Formation and Release. *J. Coast. Res.* **2008**, *52*, 25–31. [CrossRef]

12. Kolerski, T.; Shen, H.T. Possible effects of the 1984 St. Clair River ice jam on bed changes. *Can. J. Civ. Eng.* **2015**, *42*, 696–703. [CrossRef]

13. Knack, I.M.; Shen, H.T. A numerical model study on Saint John River ice breakup. *Can. J. Civ. Eng.* **2018**, *45*, 817–826. [CrossRef]

14. Freeman, A.; Durden, S.L. A three-component scattering model for polarimetric SAR data. *IEEE Trans. Geosci. Remote Sens.* **1998**, *36*, 963–973. [CrossRef]

15. Lindenschmidt, K.-E.; Das, A.; Chu, T. Air pockets and water lenses in the ice cover of the Slave River. *Cold Reg. Sci. Technol.* **2017**, *136*, 72–80. [CrossRef]

16. Lindenschmidt, K.-E.; Li, Z. Monitoring river ice cover development using the Freeman–Durden decomposition of quad-pol RADARSAT-2 images. *J. Appl. Remote Sens.* **2018**, *12*, 026014. [CrossRef]

17. Ulaby, F.T.; Moore, R.K.; Fung, A.K. *Microwave Remote Sensing: Active and Passive. Volume 1—Microwave Remote Sensing Fundamentals and Radiometry, Addison-Wesley, Reading, Massachusetts*; Artech House Publishers: Norwood, MA, USA, 1981.

18. Gunn, G.E. Re-Evaluating Scattering Mechanisms in Snow-Covered Freshwater Lake Ice Containing Bubbles Using Polarimetric Ground-Based and Spaceborne Radar Data. Ph.D. Thesis, University of Waterloo, Waterloo, ON, Canada, 2015.

19. Gherboudj, I.; Bernier, M.; Leconte, R. A backscatter modeling for river ice: Analysis and numerical results. *IEEE Trans. Geosci. Remote Sens.* **2010**, *48*, 1788–1798. [CrossRef]

20. Atwood, D.K.; Gunn, G.E.; Roussi, C.; Wu, J.; Duguay, C.; Sarabandi, K. Microwave backscatter from Arctic lake ice and polarimetric implications. *IEEE Trans. Geosci. Remote Sens.* **2015**, *53*, 5972–5982. [CrossRef]

21. Mermoz, S.; Allain-Bailhache, S.; Bernier, M.; Pottier, E.; Van Der Sanden, J.J.; Chokmani, K. Retrieval of river ice thickness from C-band PolSAR data. *IEEE Trans. Geosci. Remote Sens.* **2014**, *52*, 3052–3062. [CrossRef]

22. Puestow, T.; Cuff, A.; Richard, M.; Tolszczuk-Leclerc, S.; Proulx-Bourque, J.-S.; Deschamps, A.; van der Sanden, J.; Warren, S. The River Ice Automated Classifier Tool (RIACT). CGU HS Committee on River Ice Processes and the Environment. In Proceedings of the 19th Workshop on the Hydraulics of Ice Covered Rivers, Whitehorse, YT, Canada, 9–12 July 2017.

23. Lindenschmidt, K.-E.; Carstensen, D.; Fröhlich, W.; Hentschel, B.; Iwicki, S.; Kögel, M.; Kubicki, M.; Kundzewicz, Z.W.; Lauschke, C.; Łazarów, A.; et al. Development of an ice-jam flood forecasting system for the lower Oder River—Requirements for real-time predictions of water, ice and sediment transport. *Water* **2019**, *11*, 95. [CrossRef]

24. Cloude, S.; Pottier, E. An entropy based classification scheme for land applications of polarimetric SAR. *IEEE Trans. Geosci. Remote Sens.* **1997**, *35*, 68–78. [CrossRef]

25. Alberta Environment and Parks. 2018 Athabasca River Report No. 5 – River Ice Observation Report. River Forecasting Centre, Alberta Environment and Parks, 25 April 2018. Available online: http://environment. alberta.ca/forecasting/RiverIce/pubs/rfs_ice_observation_report_20180425_173000.pdf (accessed on 2 February 2019).

26. Alberta Environment and Parks. 2018 Athabasca River Report No. 5 – River Ice Observation Map. River Forecasting Centre, Alberta Environment and Parks, 25 April 2018. Available online: https://environment. alberta.ca/forecasting/RiverIce/pubs/rfs_ice_observation_map_20180425_174500.pdf (accessed on 2 February 2019).

remote sensing

MDPI

Article

Development of a Snow Depth Estimation Algorithm over China for the FY-3D/MWRI

Jianwei Yang [1], Lingmei Jiang [1,*], Shengli Wu [2], Gongxue Wang [1], Jian Wang [1] and Xiaojing Liu [1]

[1] State Key Laboratory of Remote Sensing Science, Jointly Sponsored by Beijing Normal University and Institute of Remote Sensing and Digital Earth of Chinese Academy of Sciences, Beijing Engineering Research Center for Global Land Remote Sensing Products, Faculty of Geographical Science, Beijing Normal University, Beijing 100875, China; yangjianwei@mail.bnu.edu.cn (J.Y.); wanggongxue@mail.bnu.edu.cn (G.W.); wjian@mail.bnu.edu.cn (J.W.); lxjing@mail.bnu.edu.cn (X.L.)

[2] National Satellite Meteorological Center, China Meteorological Administration, Beijing 100081, China; wusl@cma.gov.cn

* Correspondence: jiang@bnu.edu.cn; Tel.: +86-10-5880-5042

Received: 5 March 2019; Accepted: 20 April 2019; Published: 24 April 2019

Abstract: Launched on 15 November 2017, China's FengYun-3D (FY-3D) has taken over prime operational weather service from the aging FengYun-3B (FY-3B). Rather than directly implementing an FY-3B operational snow depth retrieval algorithm on FY-3D, we investigated this and four other well-known snow depth algorithms with respect to regional uncertainties in China. Applicable to various passive microwave sensors, these four snow depth algorithms are the Environmental and Ecological Science Data Centre of Western China (WESTDC) algorithm, the Advanced Microwave Scanning Radiometer for Earth Observing System (AMSR-E) algorithm, the Chang algorithm, and the Foster algorithm. Among these algorithms, validation results indicate that FY-3B and WESTDC perform better than the others. However, these two algorithms often result in considerable underestimation for deep snowpack (greater than 20 cm), while the other three persistently overestimate snow depth, probably because of their poor representation of snowpack characteristics in China. To overcome the retrieval errors that occur under deep snowpack conditions without sacrificing performance under relatively thin snowpack conditions, we developed an empirical snow depth retrieval algorithm suite for the FY-3D satellite. Independent evaluation using weather station observations in 2014 and 2015 demonstrates that the FY-3D snow depth algorithm's root mean square error (RMSE) and bias are 6.6 cm and 0.2 cm, respectively, and it has advantages over other similar algorithms.

Keywords: snow depth; FY-3D/MWRI; regional algorithms; China

1. Introduction

Seasonal snow cover is an important component of the Earth's hydrologic cycle, energy balance, and climate system [1–4]. Snow cover parameters—including the snow water equivalent (SWE), snow cover extent (SCE), and snow albedo (SA)—are vital to initialize numerical weather prediction models, hydrologic models, and land surface process models [5–7]. SWE, which is determined by integrating snow density over snow depth, describes how much water would be released if snowpack melted completely at once [8,9]. Manual snow surveys are time-consuming and expensive, and observations from widely spaced weather stations cannot represent the detailed spatial distribution of snow depth. Fortunately, passive microwave (PMW) sensors offer the advantages of all-weather capability and all-year coverage at good temporal (daily) and moderate spatial resolutions (~25 km). Another advantage of microwave over optical sensors is the ability to obtain dry snow's volume information, not just the surface [10–12]. These advantages make snow depth estimation with satellite PMW sensors an attractive option. The extraction of snow depth from satellite observations requires algorithms relating the snow's physical properties to the microwave signal. Most of the widely used inversion algorithms are based on empirical relationships between snow depth and multifrequency spaceborne satellite brightness temperature gradients [13]. Various linear coefficients were derived empirically for specific areas and based on assumed fixed snow properties, such as density and grain size, to derive snow depth from spaceborne measurements [13–17]. The accuracy, however, was affected by uncertainties in the assumptions. One such assumption is that snow grain size (radius) and snow density are assumed to be static in all layers of snowpack. In nature, snowpack varies in density and grain size with depth. The microstructure—the size, shape, and bonding—of snow grains, mainly defines how microwave radiation is scattered in snowpack [18]. Therefore, simplifying snowpack as one homogeneous layer may result in significant errors in snow depth retrieval. Another source of uncertainty is that these algorithms did not account for the effects of forest canopy and atmosphere, which attenuate the signals emitted from the surface and emit their own energy toward the satellite. These impacts on snow depth retrieval are reported to lead to underestimation [19–24]. Later, more advanced algorithms were developed for global [25–29] and regional [30–38] applications. The algorithm designed for the Advanced Microwave Scanning Radiometer for Earth Observing System (AMSR-E) accounts for the influence of forest cover and snow grain growth and also takes advantage of the expanded range of channels available on the AMSR-E/2 instruments [25,26]. This algorithm retrieves the snow depth from moderate snow accumulations using the 37 GHz channel and from deep snow using the 19 and 10 GHz channels. Additionally, there are approaches that use theoretical or semi-empirical radiative transfer models, coupled with atmospheric and vegetation models, to simulate microwave emissions and inversely calculate snow parameters from satellite measurements, such as the European Space Agency (ESA) Global Snow Monitoring for Climate Research (GlobSnow) SWE product, which combines synoptic weather station data with satellite passive microwave radiometer data though the forward model (Helsinki University of Technology snow emission model, HUT) [27–29]. Note that the GlobSnow SWE highly relies on weather station data. To avoid spurious or erroneous deep snow observations, a mask is used in mountainous areas [36]. Meanwhile, the algorithm may not be as feasible as those empirical algorithms to operate in real time because of its sophisticated procedure and diverse inputs (auxiliary data). In addition to the traditional algorithms, machine learning approaches (e.g., artificial neural networks, support vector regression, random forest) to estimate snow depth have emerged in recent years [39–41]. Though machine learning techniques can present good results without requiring users to have much knowledge, it is difficult to retrieve real-time snow depth with PMW measurements.

In order to avoid the deficiencies of empirical algorithms, we first validate five well-known snow depth algorithms with in situ snow depths and PMW measurements over China. Concerning the need for a feasible and reliable retrieval algorithm specifically for FY-3D, regional snow depth retrieval algorithms that perform well over China are proposed in this paper. The remote sensing and auxiliary data as well as snow depth methods are described in Section 2. Section 3.1 presents their performance in detail. The purpose is to determine which one is best and identify their problems and advantages. Then, the regional algorithms are validated and analyzed in Section 3.2. A discussion is presented in Section 4, and in Section 5 we give the conclusions of this study.

2. Materials and Methods

2.1. Data

2.1.1. Satellite Passive Microwave Measurements

The FY-3D satellite was launched on 15 November 2017 with the goal of observing global atmospheric and geophysical features around the clock. It is in a sun-synchronous orbit with local ascending overpasses at about 2:00 p.m. The microwave radiation imager (MWRI) loaded in the FY-3D satellite is a 10-channel, 5-frequency, 2-polarization radiometer system that measures brightness temperatures ranging from 10.65 GHz to 89 GHz at horizontal and vertical polarizations. The FY-3D and FY-3C satellites make up a series of Chinese polar-orbit meteorological satellites and form a constellation network. Because of the limited amount of FY-3D/MWRI data until now, AMSR-E and FY-3C/MWRI brightness temperature data (L1 product, 25 km) were used in development and validation. Table 1 shows the main parameters of passive microwave remote sensing sensors. The MWRI and AMSR-E differ in four important ways: (a) the MWRI has no C-band channel; (b) the MWRI has a coarser footprint size and slightly narrower orbital swath for all frequencies relative to the AMSR-E sensor; (c) there is a satellite overpass time difference between MWRI and AMSR-E of approximately 30 min (FY-3D) or dozens of hours (FY-3C); and (d) the MWRI has an approximate 53° earth incident angle instead of 55° for the AMSR-E sensor. Fortunately, intercalibration results indicated that snow depth bias caused by instrumental differences was very low [42]. To eliminate brightness temperature uncertainties caused by snow humidity in the daytime, only data collected at night (FY-3C, 22:00; AMSR-E, 01:30) were used.

Table 1. Summary of main passive microwave remote sensing sensors.

Sensor	AMSR-E	MWRI	
Satellite	EOS Aqua	FY-3C	FY-3D
Incident angle	55	53	53
Equator crossing time (Local time zone)	A: 01:30 D: 13:30	A: 22:00 D: 10:00	A: 14:00 D: 02:00
Frequency: footprint (GHz: km × km)	6.925: 43 × 75 10.65: 29 × 51 18.7: 16 × 27 23.8: 18 × 32 36.5: 8 × 14 89: 4 × 6	10.65: 51 × 85 18.7: 30 × 50 23.8: 27 × 45 36.5: 18 × 30 89: 9 × 15	

AMSR-E, Advanced Microwave Scanning Radiometer for Earth Observing System; MWRI, microwave radiation imager; FY-3C/3D, FengYun-3C/3D; A, ascending; D, descending.

2.1.2. In Situ Measurements

Weather station data were acquired from the National Meteorological Information Centre, China Meteorology Administration (CMA). The dataset of snow depth measurements from 753 stations throughout China spans from 2002 to 2015 in temporal coverage (Figure 1, left). Recorded variables include the site name, observation time, geolocation (latitude and longitude), elevation (m), near-surface soil temperature (measured at 5 cm depth; °C), and snow depth (cm). Quality control steps were conducted prior to comparison with the satellite product. The first step was to select records only when the near-surface soil temperature was lower than 0 °C. The second step was to remove any sites where the areal fraction of open water exceeded 30% in corresponding pixels. This is because a water body acts an emitter rather than a scatterer and confuses the relationship between brightness temperature difference (TBD) and snow depth. Finally, only ground-measured snow depths greater than 3 cm were used in the validation, because microwave response to thinner snow cover at 37 GHz is basically negligible. In addition, Chinese snow surveys were conducted from December 2017 to March 2018. Figure 1 shows the four snow course routes in Xinjiang (routes 1 and 2, 143 samples) and Northeast China (routes 3 and 4, 154 samples). The parameters include snow depth, air temperature, and snow density measured every 10–20 km. Table 2 shows the statistics of air temperature, snow depth, and snow density, including maximum, minimum, and mean.

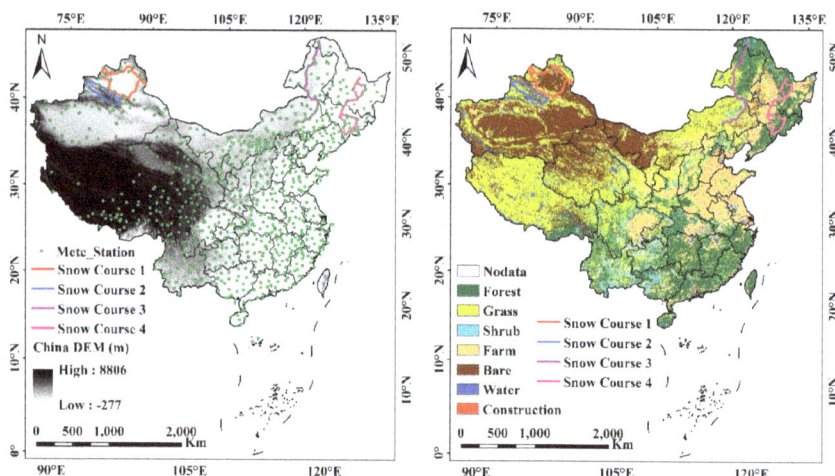

Figure 1. Spatial distribution of weather stations (left) and land cover (right) in China. Green points are sites and colored lines are snow course routes spanning from December 2017 to March 2018. The base map on the left shows the elevation (m) in China.

Table 2. Summary of snow course data (location, air temperature, snow depth, density, and number of samples).

Snow Course Route	Location (lat, lon)	Air Temperature (°C)			Snow Depth (cm)			Snow Density (g/cm³)			Samples
		Max	Min	Mean	Max	Min	Mean	Max	Min	Mean	
1	43.90°N–48.06°N 82.97°E–89.88°E	−1.7	−34.0	−18.8	50.0	3.0	13.2	0.30	0.10	0.18	70
2	42.97°N–44.50°N 80.83°E–88.97°E	−0.6	−29.5	−12.9	63.0	3.0	19.8	0.12	0.41	0.21	73
3	45.10°N–53.46°N 118.30°E126.96°E	−1.5	−33.8	−15.8	51.5	3.2	16.4	0.31	0.06	0.16	100
4	41.88°N–48.17°N 125.73°E–130.31°E	−3.1	−30.6	−12.4	45.2	4.1	16.6	0.24	0.15	0.18	54

2.1.3. Land Cover Fraction

A 1 km land use/land cover (LULC) map (Figure 1, right) derived from 30 m Thematic Mapper (TM) imagery classification was provided by the Data Center for Resources and Environmental Sciences, Chinese Academy of Sciences (http://www.resdc.cn/). Because the 1 km LULC map was derived from 30 m TM imagery, it can be recalculated as percentage of each land cover type in 25 km grid cells. Then it was used to produce a 25 km land cover fraction dataset of main land cover types: grassland, barren, farmland, forest, water body, and construction. The dataset is not reviewed here; see Jiang et al. (2014) for more details [17].

2.2. *Methodology*

In order to better develop the FY-3D algorithm, we introduced and validated five well-known operational snow depth algorithms. Then regional FY-3D algorithms were built with weather station snow depths and PMW measurements over China. Finally, they were quantitatively evaluated using weather station observations and satellite brightness temperature data obtained from the FY-3C/MWRI in 2014 and 2015 (winter season: January, February, March, November, and December). The in situ snow depth is from weather stations, measured every morning at 08:00 a.m. If there was more than one site in a pixel, those sites were averaged. The estimated snow depth was retrieved with different algorithms. To remove the scattering signals of frozen ground, cold desert, and rainfall, this study applied Li's snow cover identification method [43] based on Liu et al.'s (2018) assessment of snow cover mapping methods [44]. It also should be noted that in this study the validation was conducted with brightness temperatures at 10.65, 18.7, 36.5, and 89 GHz from FY-3C/MWRI. However, some algorithms were developed based on 18 (18.7) GHz and 37 (36.5) GHz channels. The difference was ignored in this paper [42].

2.2.1. Well-Known Operational Algorithms

Although numerous snow depth estimation algorithms have been proposed, we only chose five well-known operational algorithms to validate their performance in China. The first method is the Chang equation (Chang algorithm)

$$SD = 1.59 \times (TB_{18h} - TB_{37h}), \tag{1}$$

where SD is snow depth in cm, and TB_{18h} and TB_{37h} are brightness temperature in K at horizontally polarized ~18 and ~37 GHz channels, respectively. The coefficient value 1.59 was determined by assuming a grain radius of 0.30 mm and a snowpack density of 0.30 g/cm^3 [13].

The second algorithm was initially developed for the AMSR-E sensor [26]. It includes a measure of forest cover fraction and density and uses the ~10 GHz channel and both vertically and horizontally polarized ~19 and ~37 GHz channels to retrieve data from shallow, moderate, and thick snow (AMSR-E algorithm)

$$SD = ff \times SD_f + (1 - ff) \times SD_o, \tag{2}$$

where *ff* is forest fraction (unitless) ranging from 0 to 1, $(1 - ff)$ is the nonforested component, SD_f is snow depth (cm) in forested areas, and SD_o is snow depth (cm) in nonforested areas. SD_f and SD_o are calculated using the equations

$$SD_f = 1/\log_{10}(pol_{37}) \times (TB_{19v} - TB_{37v})/(1 - 0.6 \times fd) \tag{3}$$

$$SD_o = 1/\log_{10}(pol_{37}) \times (TB_{10v} - TB_{37v}) + [1/\log_{10}(pol_{19}) \times (TB_{10v} - TB_{19v})], \tag{4}$$

where *fd* is the forest density, pol_{37} is the polarization difference at 37 GHz (i.e., $TB_{37v} - TB_{37h}$), and pol_{19} is the polarization difference at 19 GHz. TB_{10v}, TB_{19v}, and TB_{37v} are brightness temperatures in K at vertically polarized ~10, ~19, and ~37 GHz channels, respectively.

The third algorithm was developed based on Chinese weather station observations and PMW brightness temperatures [16]. This is the improved Chang algorithm in terms of specific snowpack conditions and satellite data. It has been used to generate a long-term snow dataset for the algorithm of the Environmental and Ecological Science Data Centre of Western China (WESTDC algorithm). Its equation is

$$SD = 0.66 \times (TB_{19h} - TB_{37h}), \tag{5}$$

where TB_{19h} and TB_{37h} are brightness temperatures at horizontally polarized ~19 and ~37 GHz channels, respectively. The coefficient value was changed from 1.59 to 0.66 based on a relationship between Chinese in situ snow depths and brightness temperatures.

The fourth algorithm was established by Foster et al. [14] in 1997. The linear fitting coefficient is 0.78, and combining the forest cover parameter yields the Foster algorithm

$$SD = 0.78 \times (TB_{18h} - TB_{37h})/(1 - ff), \tag{6}$$

where *ff* is the fractional forest cover and TB_{19h} and TB_{37h} are brightness temperature at horizontally polarized ~18 and ~37 GHz channels, respectively.

The last algorithm is a mixed-pixel method for the FY-3B meteorological satellite in China [17]. Frequencies of 10.7, 18.7, 36.5, and 89 GHz with both polarizations were used to develop the regressions of the empirically derived algorithm. The estimates were the sums of values from four individual land-cover algorithms, weighted by the percentage of each type (FY-3B algorithm):

$$SD = ff_{grass} \times SD_{grass} + ff_{barren} \times SD_{barren} + ff_{forest} \times SD_{forest} + ff_{farmland} \times SD_{farmland}, \tag{7}$$

where ff is the fractional land cover. The subscripts denote grass, barren, forest, and farmland. SD_{xx} is snow depth in pure pixels where the land cover fraction is greater than 85%. The pure-pixel functions are

$$SD_{farmland} = -4.235 + 0.432 \times (TB_{18h} - TB_{36h}) + 1.074 \times (TB_{89v} - TB_{89h}) \tag{8}$$

$$SD_{grass} = 4.320 + 0.506 \times (TB_{18h} - TB_{36h}) - 0.131 \times (TB_{18v} - TB_{18h}) + 0.183 \times (TB_{10v} - TB_{89h}) - 0.123 \times (TB_{18v} - TB_{89h}) \tag{9}$$

$$SD_{barren} = 3.143 + 0.532 \times (TB_{36h} - TB_{89h}) - 1.424 \times (TB_{10v} - TB_{89v}) + 1.345 \times (TB_{18v} - TB_{89v}) - 0.238 \times (TB_{36v} - TB_{89v}) \tag{10}$$

$$SD_{forest} = 11.128 - 0.474 \times (TB_{18h} - TB_{36v}) - 1.441 \times (TB_{18v} - TB_{18h}) + 0.678 \times (TB_{10v} - TB_{89h}) - 0.649 \times (TB_{36v} - TB_{89h}) \tag{11}$$

2.2.2. Development of FY-3D Algorithm

Numerous studies have demonstrated that no single standard algorithm can describe snow cover characteristics well everywhere [19,29,32,45]. Thus, regional algorithms that have been calibrated at a local scale might be capable of providing a reasonable snow depth estimation. Similar studies have been carried out over the years by several scholars [46–48]. They divided Chinese snow cover into different regions based on topography, land cover, and snow cover duration, e.g., Xingjiang, Qinghai–Tibetan Plateau, Northeast, and others. Based on these previous studies, China's snow cover is divided into three regions (Figure 2):

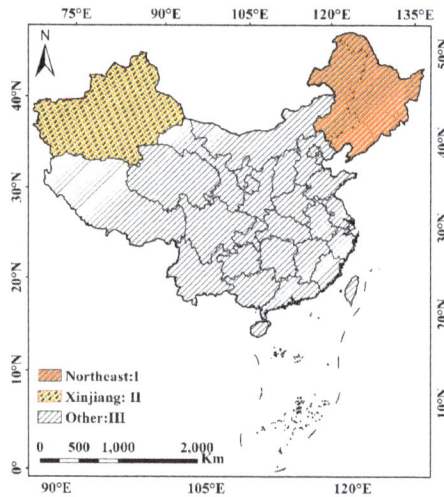

Figure 2. Three regions for regional algorithms: Region I: Xinjiang; Region II: Xinjiang; Region III: Others.

(1) Region I: Northeast China

Northeast China consists of Liaoning, Jilin, Heilongjiang, and eastern Inner Mongolia. Various land cover types are unevenly distributed. Cultivated and forest land predominate in Northeast China, and large uncertainties in snow depth retrieval are associated with forest cover. Foster et al. [14] developed an algorithm that accounts for the influence of forest cover on brightness temperature. Unfortunately, it tends to overestimate snow depth in China, especially in densely forested areas. Therefore, the Foster algorithm can be improved by the use of $1/(1 - ff)$ to alleviate overestimation.

Referring to the study by Kelly et al. [26] performed in 2009, incorporating a weight factor that limits $1/(1 - ff)$ within reasonable intervals can reduce overestimation. Thus, we modified the Foster algorithm as follows:

$$SD = 0.38 \times (TB_{19h} - TB_{37h})/(1 - 0.7 \times ff), \tag{12}$$

where the constant 0.38 is a regression fitting coefficient between AMSR-E brightness temperatures and weather station observations, and 0.7 is the weight factor that keeps the term $1/(1 - ff)$ within a range of 1 to 5 (Figure 3). For a weight factor value of 0.5 or 0.9, the algorithm still overestimates snow depth in dense forest areas.

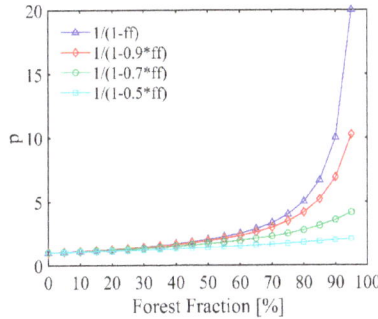

Figure 3. Relationship between forest fraction ff and p. The p is a correction factor calculated with $p = 1/(1 - a \times ff)$. The coefficient a is the weight factor. Blue line: a = 1; red line: a = 0.9; green line: a = 0.7; cyan line: a = 0.5.

(2) Region II: Xinjiang

The Xinjiang Uyghur Autonomous Region is located in northwest China. The southern and northern parts of Xinjiang are dominated by grass, and the land cover in the central part is mainly bare ground dominated by desert (Figure 1). Snow cover is relatively thick in northern Xinjiang, where underestimation usually occurs (e.g., by the FY-3B and WESTDC algorithms). Based on scatter diagrams of station snow depth versus satellite BTD from the AMSR-E sensor in the 2002–2009 period (not shown here), the combination of cross-polarization at 19 GHz and 37 GHz was selected. The shorter 37 GHz wavelength emissions from the ground will be scattered more by the snowpack than the longer 19 GHz wavelength emissions. The brightness temperature at vertical polarization is less affected by incidence angle [34]. Moreover, the brightness temperature of cross-polarization is more sensitive to snow depth than that of co-polarization, owing to the effects of depth hoar [15,17,32,37,38,49]. The regression equation is

$$SD=0.48 \times (TB_{19v} - TB_{37h}), \tag{13}$$

where the constant 0.48 is a regression fitting coefficient between AMSR-E brightness temperatures and weather station observations. At the bottom of snowpack, the snow has evolved with time and undergone compaction due to the overburden and freeze/melt cycle, which makes grain size larger. Snow grain size increases with layer depth (0.3–3 mm). Especially for the depth hoar, the radius can be larger than 3 mm because of snow metamorphism [19]. Moreover, the temperature brightness gradient for cross-polarization is higher than that for co-polarization because of a better penetration capacity of vertical polarization. These factors explain why the fitting coefficient is 0.48 rather than 1.59 in Chang's algorithm for which the assumptions fail in Xinjiang.

(3) Region III: Others

In areas other than Xinjiang and northeastern China, a mixed-pixel method is suitable because of complex land cover and thin snow cover, and the original FY-3B method performs well at retrieving snow depth from shallow snowpack [17]. Therefore, the FY-3B algorithm was used to estimate snow depth in Region III.

3. Results

The main framework of the paper is to first show the deficiencies and advantages of current empirical algorithms and then develop the FY-3D algorithm, which complements their strengths. Comparisons and validations of five well-known algorithms are shown in Section 3.1. Section 3.2 displays the validation and analysis of the FY-3D algorithm.

3.1. Comparisons and Validations of Five Well-Known Algorithms

The validation results of five algorithms are shown in Figure 4. The WESTDC and FY-3B algorithms performed better than the other three methods (Figure 4c,e), primarily because they were developed based on Chinese weather station measurements. However, there are still many problems and doubts. For example, the FY-3B version tends to underestimate snow depth for thick snow (greater than 20 cm). The error probably originates from the nonuniform training samples. FY-3B estimates were the sum of values from four individual pure-pixel algorithms, weighted by the land cover fraction. Figure 5 shows the spatial distribution of pure-pixel samples (with a certain land cover fraction greater than 85%), including forest, grass, farm, and barren. The base map is snow types based on snow cover days (instantaneous snow cover: 0–10 days; unstable snow cover: 10–60 days; stable snow cover: 60–365 days). The stable snow cover areas usually are covered with deep snow, such as Xinjiang and Northeast China. As shown in Figure 5, many pure pixels are mainly distributed in thin snow dominated areas, while there are few samples in Northeast China and Xinjiang. Therefore, underestimation occurs for empirical relationships presented in Equation (7).

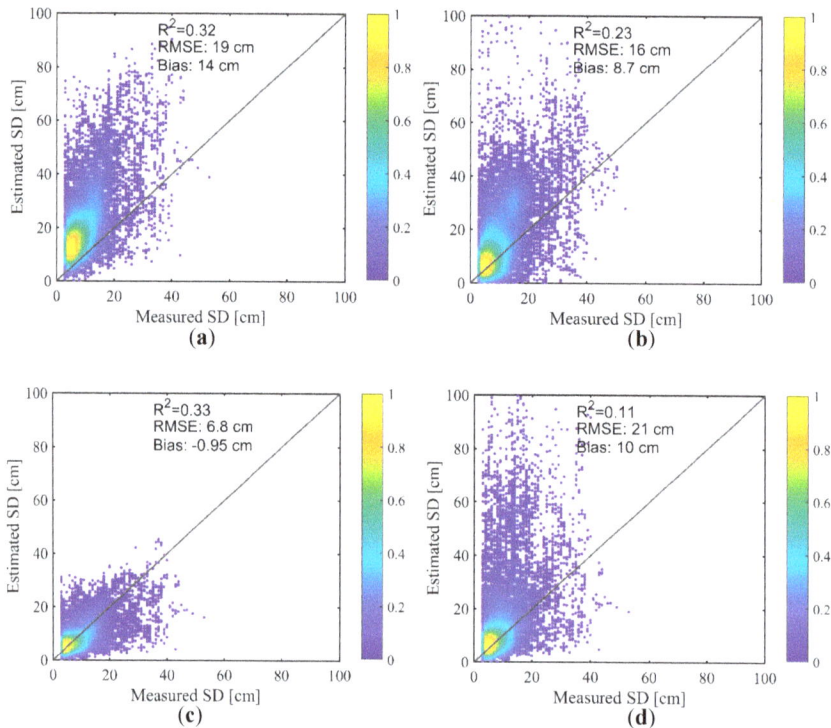

Figure 4. *Cont.*

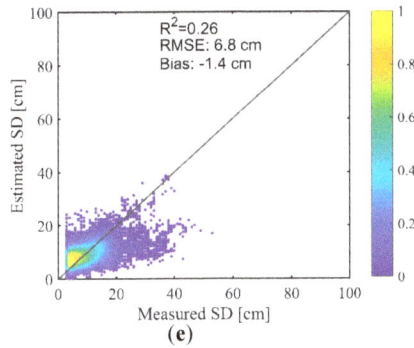

Figure 4. Color-density scatterplots of estimated and measured snow depth for five algorithms: (**a**) Chang; (**b**) AMSR-E; (**c**) WESTDC; (**d**) Foster; (**e**) FY-3B. Color scale represents data density of scattered points, ranging from 0 to 1. Number of samples is 8495. RMSE, root mean square error.

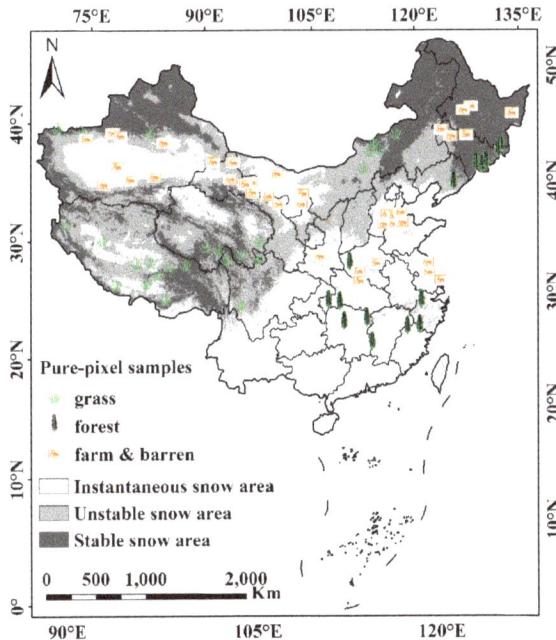

Figure 5. Spatial distribution of pure-pixel samples (fractional land cover greater than 85%). Grass samples are from 25 meteorological stations; forest samples are from 15 meteorological stations, mostly in South China; farm and barren samples are from 36 meteorological stations.

Figure 4b shows that the AMSR-E algorithm generally tends to overestimate snow depth in China compared with ground meteorological station observations. The main cause is that the dynamic coefficient, polarization factor $pol36 = T_{b36V} - T_{b36H}$ or $pol18 = T_{b18V} - T_{b18H}$, does not clearly indicate the variation of snow grain size and may need adjustment with further testing [26]. Figure 4a shows that the Chang algorithm produces larger errors of overestimation in China. The fitting coefficient value is 1.59, based on the assumption that the snow density is 0.30 g/cm^3 and snow grain size is 0.30 mm. In situ measurements, however, show that these assumptions fail in China. The grain radius of fresh snow in the uppermost layer is approximately 0.30 mm, while the size within the middle or bottom layers is up to 4 mm [37,38]. Figure 6 shows a simulation of the single-layer HUT model with different inputs. The input variables are snow grain radius and snow density, and fixed parameters are snow temperature, atmospheric temperature, and forest fraction. The result shows that the fitting coefficients vary with different inputs. The fitting coefficients are 1.5979 and 1.6800, respectively, for horizontal and vertical polarization under the assumptions of the Chang algorithm (snow density, 0.30 g/cm^3; snow grain radius, 0.30 mm). However, the coefficients are 0.7014 and 0.6793, respectively, for a snow density of 0.18 g/cm^3 and snow grain radius of 0.80 mm, based on Chinese field work measurements in 2018 (Figures 1 and 7). Figure 4c also shows that the WESTDC algorithm has better performance than Chang's when the fitting coefficient is 0.66 rather than 1.59.

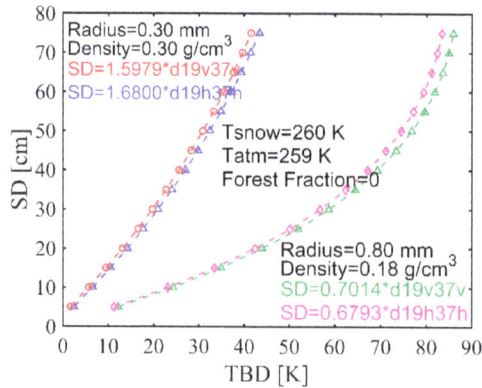

Figure 6. Relationship between brightness temperature difference (TBD) (19–37 GHz) and snow depth based on single-layer HUT model with varying inputs. d19v37v and d19h37h are the TBD in K between vertically and horizontally polarized ~19 and ~37 GHz channels, respectively. The fixed parameters are snow temperature (Tsnow), atmosphere temperature (Tatm), and forest fraction (~0). Red and blue lines represent modeling results with input variables (snow grain radius, 0.30 mm; snow density, 0.30 g/cm^3). Green and magenta lines are modeling results with input variables (snow grain radius, 0.80 mm; snow density, 0.18 g/cm^3).

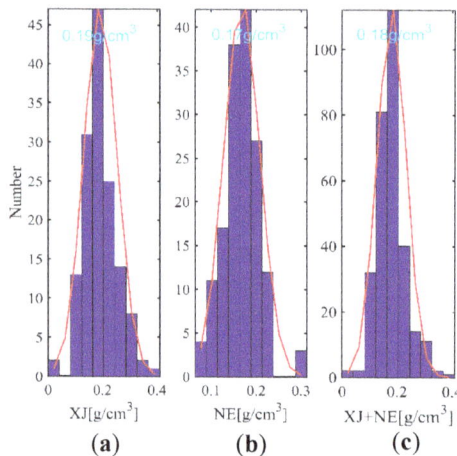

Figure 7. In situ snow densities based on snow surveys (winter season) in (**a**) Xinjiang (XJ) and (**b**) Northeast (NE), and (**c**) total (XJ + NE). Snow surveys were conducted from December 2017 to March 2018. Figure 1 shows the four snow course routes in Xinjiang (143 samples) and Northeast China (154 samples).

Figure 4d shows the performance of the Foster algorithm. The coefficient was changed from 1.59 to 0.78. The algorithm also accounts for the tendency of forest cover to reduce the sensitivity of brightness temperature to snow depth. However, it is noteworthy that false high retrievals occur. Owing to the equation form $p = 1/(1 - ff)$ and high fitting coefficient (0.78) in China, the snow depth shows explosive growth when the forest fraction (ff) is greater than 60% (Figure 3, blue line). Thus, the Foster algorithm does not perform well on areas covered by dense forest in China. Figure 3 also shows that a weight factor (a) that limits $1/(1 - ff)$ within reasonable intervals can reduce overestimation. This study offers a new idea for improving the Foster algorithm.

To illustrate the performance of various algorithms in three areas (Xinjiang, Northeast China, and others; Figure 2), the regional validation is shown in Figure 8. The pattern is similar to that of the whole validation (Figure 4). Regardless of location, the WESTDC and FY-3B algorithms perform best based on their root mean square errors (RMSEs). However, they tend to underestimate the snow depth in Xinjiang and Northeast China and overestimate it in other areas. As mentioned earlier, the Foster algorithm performs well in open or sparsely vegetated areas, such as in Xinjiang, where the land cover is mainly barren or grassland. Conversely, it yields the poorest estimates in forested areas, such as Northeastern China. Thus, the five well-known operational snow depth retrieval algorithms cannot fully capture the temporal and spatial distribution of snow cover in China. It is essential to develop a suite of algorithms based on China's snow cover characteristics, rather than directly implement FY-3B/MWRI's operational empirical retrieval algorithm.

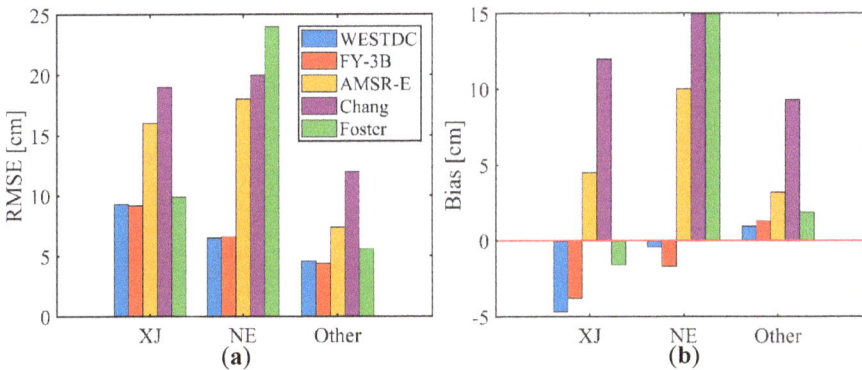

Figure 8. Reginal validation results: (**a**) RMSE; (**b**) bias. Histograms present performance (RMSE and bias) of five algorithms. XJ, Xinjiang; NE, Northeast China. Three regions are based on Figure 2.

3.2. Validation and Analysis of FY-3D Algorithm

Section 2.2 presents the FY-3D algorithm. In this paper, the snow depth product retrieved using the FY-3D algorithm is validated rather than the algorithms because the product's performance without any auxiliary data is what we are interested in. To mitigate any distinct borders in retrievals between adjacent pixels from different regions, a moving-average filter was used to perform smoothing. To demonstrate the performance of the FY-3D algorithm, three products retrieved from the FY-3B, WESTDC, and FY-3D algorithms were validated and compared.

As shown in Figure 9, the FY-3D algorithm's RMSE and bias are 6.6 cm and 0.2 cm, respectively. The correlation coefficient is 0.71, which is greater than those of the other algorithms. The RMSEs of the products retrieved with the WESTDC and original FY-3B algorithms are 8.9 cm and 9.0 cm, respectively. The WESTDC and original FY-3B algorithms encounter underestimation at snowpack deeper than 13 cm. The bias of the WESTDC and original FY-3B algorithms is −2.6 cm, while it is just 0.2 cm for the FY-3D algorithm, which also shows that the FY-3D algorithm performs better than the others. Owing to limited FY-3D/MWRI data (1 January to 31 March 2018), there are only about

3800 samples to validate the new algorithm. Figure 10 shows that snow depths estimated with FY-3D brightness temperature are closer to the ground truth.

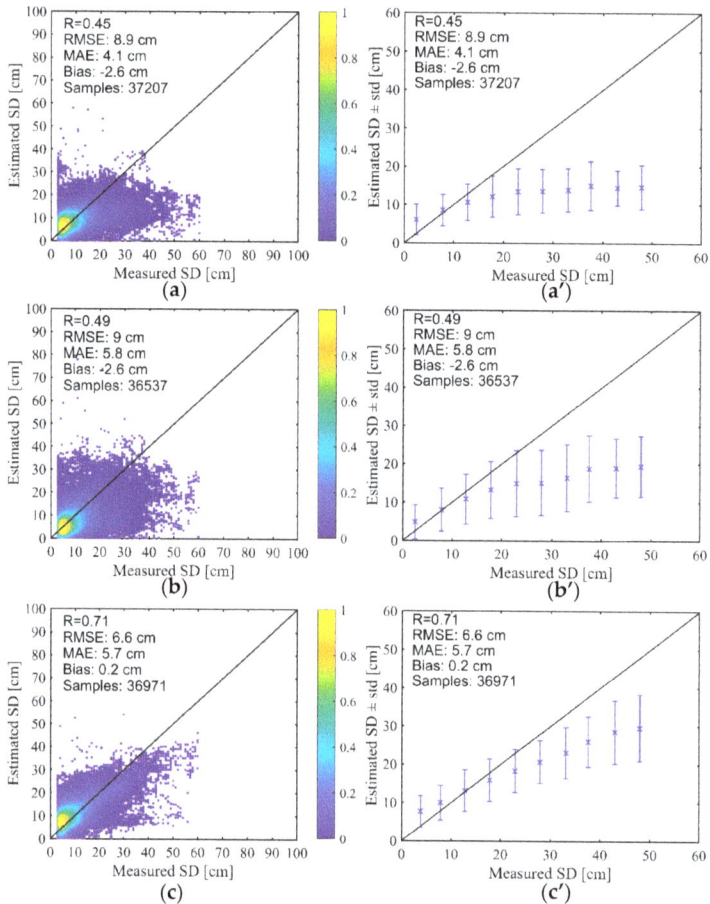

Figure 9. Scatter diagrams of estimated vs. measured snow depth using (**a**) FY-3B algorithm, (**b**) WESTDC algorithm, and (**c**) FY-3D algorithm, and error bars of (**a′**) FY-3B algorithm, (**b′**) WESTDC algorithm, and (**c′**) FY-3D algorithm. The 'x' marks the mean snow depth computed at each corresponding bin, while upper and lower blue bars indicate one standard deviation from the mean.

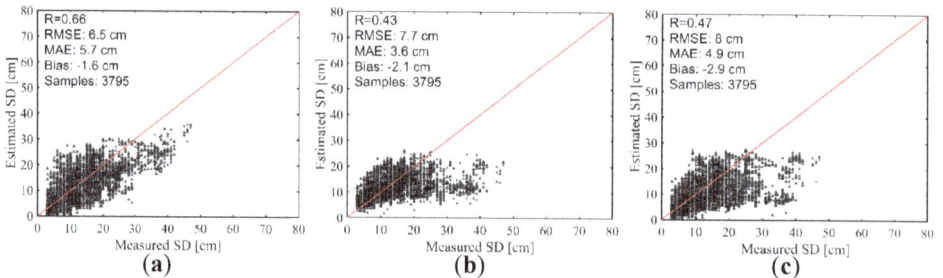

Figure 10. Validation of three algorithms with FY-3D/MWRI measurements: (**a**) FY-3D algorithm; (**b**) FY-3B algorithm; (**c**) WESTDC algorithm.

Figure 11 shows the spatial distribution of RMSE and bias of pixels where the sites are located. The RMSE distribution indicates that the three algorithms perform similarly in Region III. However, there are large differences in Xinjiang and Northeast China. The FY-3D algorithm has an advantage over the others in northern Xinjiang and Heilongjiang Provinces because of low RMSEs. The results also confirm that the FY-3D algorithm is more sensitive to deep snow. In terms of bias (difference between retrieved and measured snow depth), Figure 11 shows that underestimation occurs mostly in the North China Plain and South China, where the snow is often thin and wet. Wet snow usually corresponds to low BTD, resulting in underestimation [32,45,50]. In northern Xinjiang and Northeast China, the original FY-3B algorithm produces the lowest bias, as low as −10 cm. The WESTDC algorithm performs better than the original FY-3B algorithm in deep snow cover, as shown in Figure 11. Interestingly, there is overestimation in the Qinghai–Tibetan Plateau. There, the snow cover differs from that in other seasonally snow-covered regions; it is often shallow, patchy, and of short duration [50–52]. A distinct meteorological characteristic is the large diurnal temperature range, which causes snow to undergo frequent freeze–thaw cycles. Note that these cycles lead to rapid grain growth and consequently to a low brightness temperature [45,50]. Frozen soil is also a factor that reduces the accuracy of estimates in the Qinghai–Tibetan Plateau. Both snow and frozen ground are volume-scattering materials, and they have similar microwave radiation characteristics, making them difficult to distinguish.

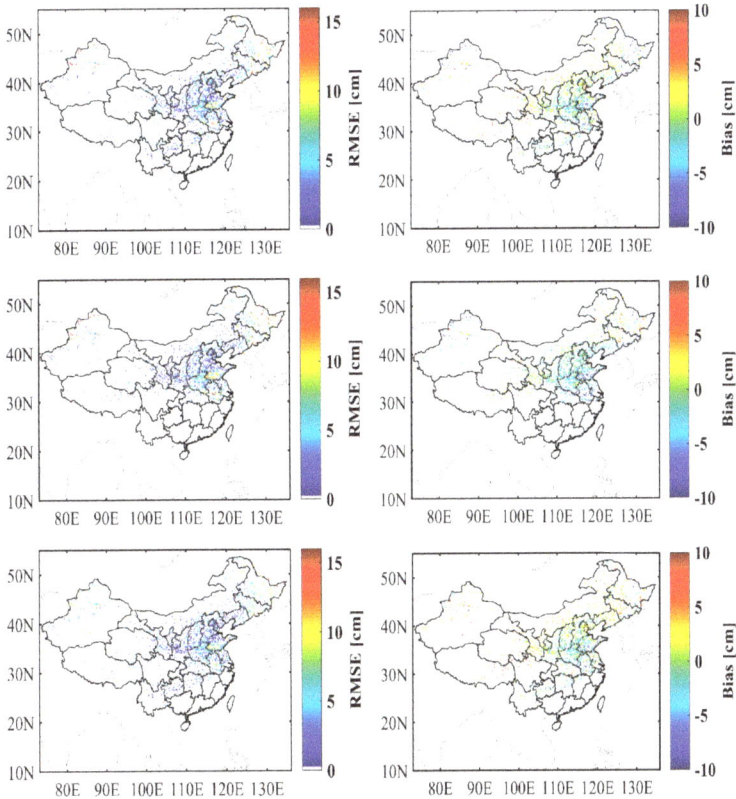

Figure 11. Spatial patterns of RMSE and bias produced by the FY-3B algorithm (top), WESTDC algorithm (middle), and FY-3D algorithm (bottom). Left and right columns represent RMSE and bias, respectively. Each point represents one pixel (spatial resolution: 25 × 25 km).

In view of the heterogeneity of snow depth in the three regions, RMSE and bias cannot fully explain where an algorithm consistently performs well or poorly. Thus, the spatial distribution of relative error (RMSE divided by mean snow depth) is shown in Figure 12. First, the error in areas of shallow snow cover is higher than that in thick snow areas. This pattern is caused by the different mean snow depth. Similarly, a high RMSE does not mean poor performance. What is certain, however, is that the FY-3D algorithm performs best in Xinjiang and the northeast regardless of RMSE, bias, or error.

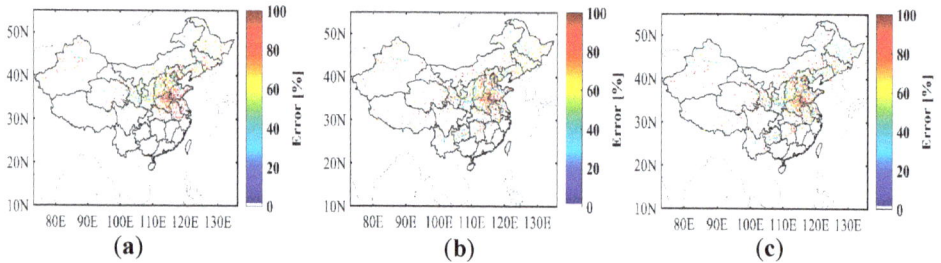

Figure 12. Spatial distributions of relative error corresponding to the (**a**) WESTDC algorithm; (**b**) FY-3B algorithm; (**c**) FY-3D algorithm.

To evaluate the monthly performance of the algorithms, RMSE and bias were calculated independently. The results are shown in Figure 13. The maximal RMSE occurs in March. This poor performance is associated with the confounding effect of snow grain size and stratigraphy. Another reason is thick snow cover. The minimum RMSE occurs in November. On the one hand, the snow parameters are stable and have no evolution. Therefore, the relationship between snow depth and brightness temperature is relatively strong. On the other hand, the snow cover is shallow in November. Bias ranges from −1 to 2 for the FY-3D algorithm, and it performs better in the snowy season, except in March.

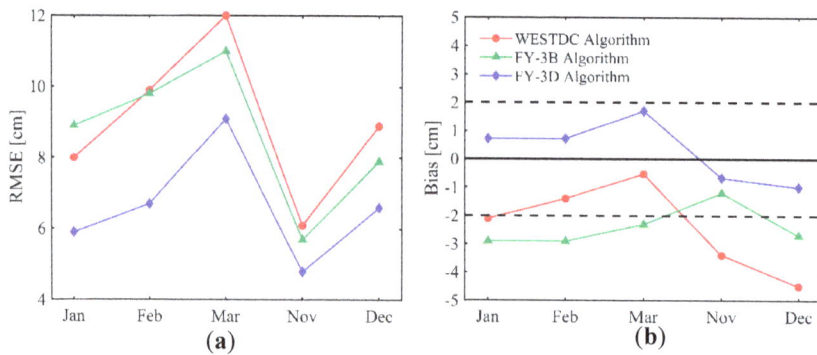

Figure 13. Monthly statistics of the three algorithms: (**a**) RMSE and (**b**) bias. Blue, green, and red lines represent FY-3D, FY-3B, and WESTDC algorithms, respectively. Dashed black lines denote bias at ±2 cm.

4. Discussion

4.1. Influence of Snow Microphysical Properties

PMW retrievals are plagued by a number of challenges, which often disrupt the relationship between snow depth and brightness temperature, resulting in poor estimation [34,38,45,53,54]. Neglecting to account for snow's microphysical properties (mostly grain size and density) tends to cause retrieval errors. Figure 14 shows time series of snow depth (station observations, 2002–2009) and TBD (19 GHz and 37 GHz, AMSR-E, 2002–2009) at Aletai station (deep snow, maximum snow depth 57 cm, mean snow depth 26 cm) and Mingshui station (thin snow, maximum snow depth 23 cm, mean snow depth 10 cm). When the snow depth is invariable, however, the TBD still increases at Aletai station, as shown within the dashed black frames in Figure 14a. Figure 14b shows that the TBD increases with decreasing snow depth at Mingshui station. Simulation with the single-layer microwave emission model of layered snowpack (MEMLS) model shows that the TBD increases with increasing snow grain correlation length (Figure 15, top) and snow density (Figure 15, middle) and decreases with increasing humidity except for cross-polarization (Figure 15, bottom). TBD for cross-polarization ($TB_{19v} - TB_{37h}$) increases with increased snow liquid water content. This is mainly because the cross-polarization difference at 37 GHz ($TB_{37v} - TB_{37h}$) is large due to water dielectric properties, and ($TB_{19v} - TB_{37h}$) can be expressed as ($TB_{19v} - TB_{37v}$) + ($TB_{37v} - TB_{37h}$). It is clear that ($TB_{19v} - TB_{37v}$) is small (Figure 15, bottom left), so the major contributions are from ($TB_{37v} - TB_{37h}$). Therefore, the factor that can be used to explain the anomalous pattern in Figure 14 is the evolution in snow grain size and snow density. Although empirical and physical models have been developed to predict the growth of snow crystals and increase in density [25,26], they are not suited to regional- or global-scale conditions.

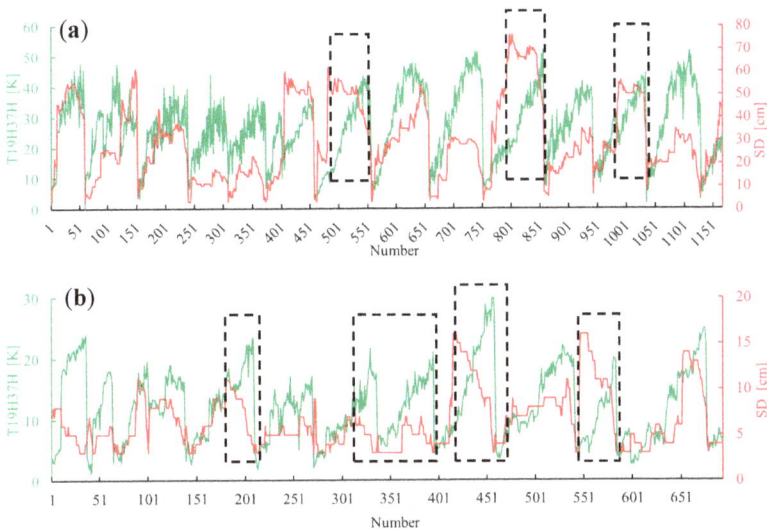

Figure 14. Time series of snow depth (solid red line) and T19H37H = $TB_{19h} - TB_{37h}$ (solid green line) at (**a**) Aletai station (grassland) and (**b**) Mingshui station (farmland). Boxes marked with dashed black lines show the anomalous relationship between snow depth and brightness temperature difference.

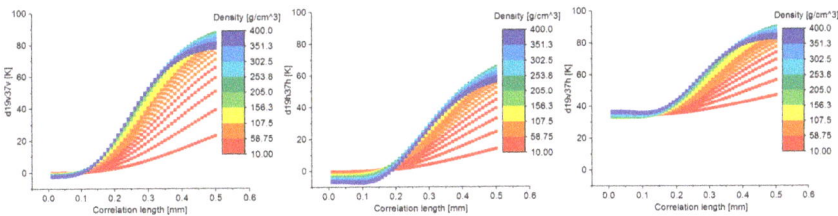

Figure 15. *Cont.*

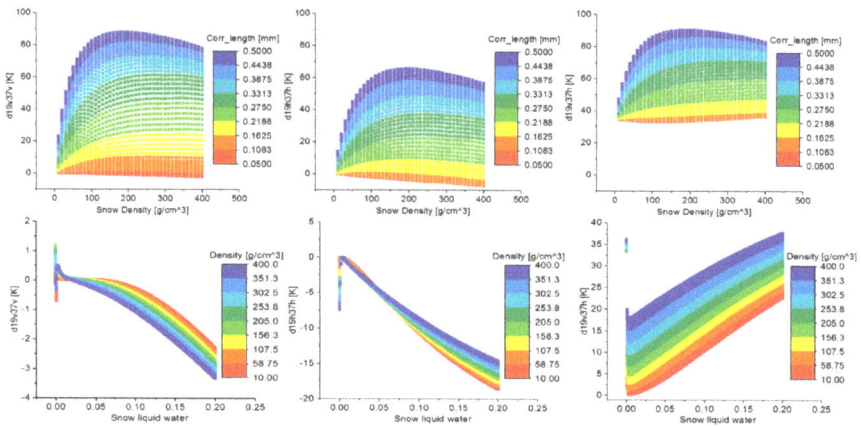

Figure 15. Sensitivity analysis of brightness temperature difference to snow parameters, including snow grain correlation length (top), snow density (middle), and snow liquid water (bottom). Left: d19v37v = $TB_{19v} - TB_{37v}$; middle: d19h37h = $TB_{19h} - TB_{37h}$; right: d19v37h = $TB_{19v} - TB_{37h}$.

4.2. Influence of Snow Density on SWE Mapping

Snow density is a key parameter that not only confounds the relationship between TB and snow depth, but also influences SWE [17,32,33]. Much SWE retrieval processing is based on snow depth information

$$SWE = SD \times \rho_s/\rho_w \times 10, \tag{14}$$

where SD is the snow depth (cm), ρ_s is the snow density (g/cm^3), ρ_w is the density of water (1 g/cm^3), and SWE is measured in millimeters of water. The first implementation of the FY-3B SWE retrieval scheme utilized reference snow density of Sturm's climatological snow classes [55,56]. Based on in situ snow course data in Xinjiang and Northeast China, however, the snow density is generally around 0.18 g/cm^3 (Table 2, Figure 7). The average snow densities in the northeast and Xinjiang are similar, 0.19 g/cm^3 and 0.17 g/cm^3, respectively. Weather station observations from 1992 to 2009 include snow pressure, which can be converted to SWE. The relationship between snow depth and SWE is shown in Figure 16. The slope of the fitting is 0.16, meaning that snow density is about 0.16 g/cm^3.

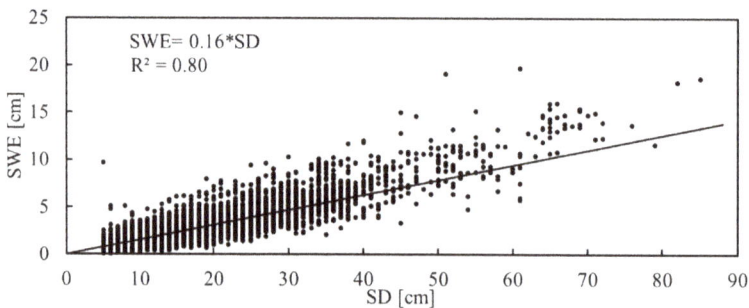

Figure 16. Snow density based on the relationship between snow depth and SWE (weather station observations from 1992 to 2009, 13,462 samples). The solid black line is the fitted linear relationship obtained from Equation (14) with regression coefficient 0.16.

Figure 17a is a Kriging interpolation map of snow densities obtained from weather stations during the winters of 2002–2009. It is masked with instantaneous snow cover in Figure 5. Figure 17b shows the spatial distribution of snow density based on Sturm's climatological snow classes [55,56]. It is clear that snow density based on Sturm's climatological snow classes (minimum ~ 0.21) tends to be bigger than that in China (maximum ~ 0.20), which results in systematic overestimation of SWE. Therefore, the snow density used in the FY-3D SWE version is 0.18 g/cm^3 rather than that in the Kriging interpolation because of uncertainties resulting from unevenly distributed weather stations (Figure 1). In future work, the temporospatial distribution of snow density in China will be mapped based on field measurements from 2018 to 2021 and weather station observations.

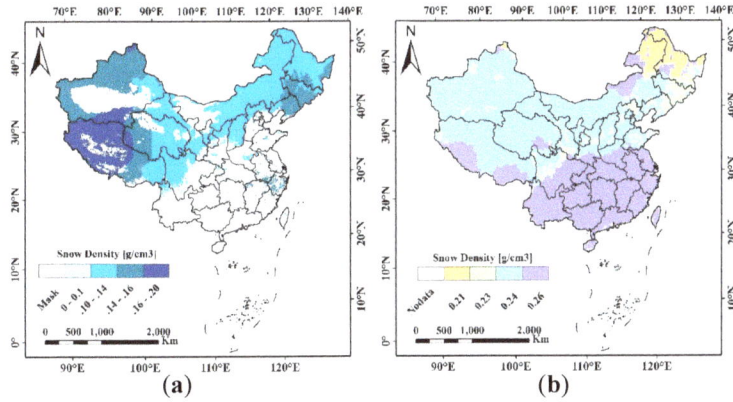

Figure 17. Spatial distribution of snow density in China: (**a**) Kriging interpolation and (**b**) based on Sturm's climatological snow classes.

4.3. Influence of Forest Cover Fraction

Forest cover represents a significant source of error in satellite PMW snow depth and SWE retrieval algorithms. In this work, we improved the Foster algorithm by employing a weight factor to avoid overestimation in densely forested areas. In fact, this method results in overestimation in sparsely forested areas. Figure 18 shows that RMSE increases with increasing forest cover fraction, and bias overestimation is serious as forest fractions range from 20% to 60% (Figure 18, magenta line), mainly caused by mixed pixels [38]. When the forest fraction is greater than 80%, serious underestimation occurs because of minimal penetration depth of the microwave signal in the forest canopy. Because snow depth varies in different areas, relative error is more reasonable to show the algorithm's performance. Although RMSEs in densely forested areas are large, relative errors are still smooth, not larger than those in sparsely forested areas. In future work, we will study the influence of forest on brightness temperature with a physically based radiative transfer model and the influence of snow microphysical properties on brightness temperature with a snow forward model, then calibrate microwave signals to improve snow depth estimation [53,54].

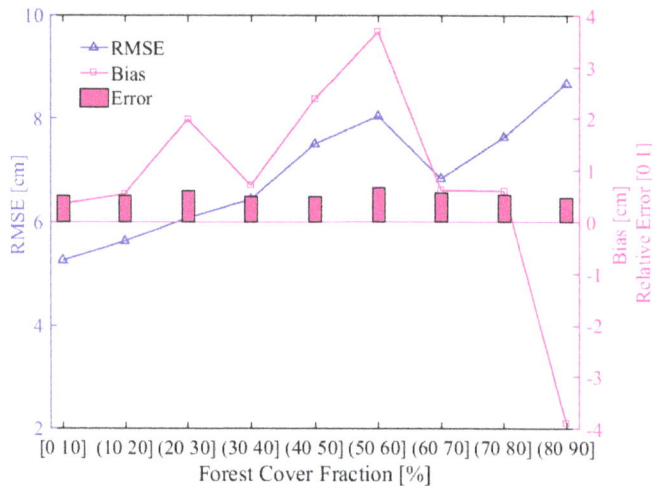

Figure 18. Influence of forest cover fraction on snow depth estimation. The blue line represents RMSE, the magenta line represents bias, and the magenta histogram represents relative error (RMSE is divided by mean snow depth, in the range of 0 to 1).

5. Conclusions

In this study, the performance of five snow depth estimation methods was evaluated using in situ snow depth measurements and satellite brightness temperatures. The results show that the WESTDC and FY-3B algorithms performed well. However, there was persistent underestimation in thick snow cover (greater than 20 cm). The purpose of the study was to develop a new algorithm that can solve the problem of underestimation of deep snow. Ideally, it should be possible to retrieve deep and shallow snow depths, as the AMSR-E algorithm does. However, PMW remote sensing still cannot distinguish deep snow from thin snow and only can detect snow cover, i.e., snow vs. snow-free. Thus, establishing regional algorithms calibrated at a local scale is a promising approach and may improve snow depth estimation. Thus, we developed regional algorithms for Xinjiang, Northeast China, and other areas. Based on our evaluation and analysis, the FY-3D algorithm performed better than other algorithms. The RMSE and bias were 6.6 cm and 0.2 cm, respectively. Based on an in-situ snow density of 0.18 g/cm^3, the RMSE of the SWE is approximately 12 mm.

Author Contributions: L.J. and S.W. conceived and designed the study; S.W. provided the FY-3C/MWRI data and contributed the analysis; J.Y. analyzed the data and wrote the paper; and G.W., J.W. and X.L. contributed analytical tools and methods.

Funding: This research was funded by the Science and Technology Basic Resources Investigation Program of China [2017FY100502] and the National Natural Science Foundation of China [41671334].

Acknowledgments: The authors would like to thank the China Meteorological Administration, National Geomatics Center of China, and National Snow and Ice Data Center for providing land-cover products and satellite data. We would also like to thank the Chinese snow survey teams for providing in situ data. We would like to thank the three anonymous reviewers for their helpful comments.

Conflicts of Interest: The authors declare no conflict of interest.

References

1. Fernandes, R.; Zhao, H.; Wang, X.; Key, J.; Qu, X.; Hall, A. Controls on Northern Hemisphere snow albedo feedback quantified using satellite Earth observations. *Geophys. Res. Lett.* **2009**, *36*. [CrossRef]
2. Hernández-Henríquez, M.A.; Déry, S.J.; Derksen, C. Polar amplification and elevation-dependence in trends of Northern Hemisphere snow cover extent, 1971–2014. *Environ. Res. Lett.* **2015**, *10*, 044010. [CrossRef]

3. Derksen, C.; Brown, R. Spring snow cover extent reductions in the 2008–2012 period exceeding climate model projections. *Geophys. Res. Lett.* **2012**, *39*. [CrossRef]

4. Safavi, H.R.; Sajjadi, S.M.; Raghibi, V. Assessment of climate change impacts on climate variables using probabilistic ensemble modeling and trend analysis. *Theor. Appl. Climatol.* **2017**, *130*, 635–653. [CrossRef]

5. De Rosnay, P.; Balsamo, G.; Albergel, C.; Muñoz-Sabater, J.; Isaksen, L. Initialisation of land surface variables for numerical weather prediction. *Surv. Geophys.* **2014**, *35*, 607–621. [CrossRef]

6. Bell, V.A.; Kay, A.L.; Davies, H.N.; Jones, R.G. An assessment of the possible impacts of climate change on snow and peak river flows across Britain. *Clim. Chang.* **2016**, *136*, 539–553. [CrossRef]

7. Barnett, T.P.; Adam, J.C.; Lettenmaier, D.P. Potential impacts of a warming climate on water availability in snow-dominated regions. *Nature* **2005**, *438*, 303–309. [CrossRef]

8. Dressler, K.A.; Leavesley, G.H.; Bales, R.C.; Fassnacht, S.R. Evaluation of gridded snow water equivalent and satellite snow cover products for mountain basins in a hydrologic model. *Hydrol. Process.* **2006**, *20*, 673–688. [CrossRef]

9. Gong, G.; Cohen, J.; Entekhabi, D.; Ge, Y. Hemispheric-scale climate response to Northern Eurasia land surface characteristics and snow anomalies. *Glob. Planet. Chang.* **2007**, *56*, 359–370. [CrossRef]

10. Lemmetyinen, J.; Schwank, M.; Rautiainen, K.; Kontu, A.; Parkkinen, T.; Mätzler, C.; Wiesmann, A.; Wegmüller, U.; Derksen, C.; Toose, P.; et al. Snow density and ground permittivity retrieved from L-band radiometry: Application to experimental data. *Remote Sens. Environ.* **2016**, *180*, 377–391. [CrossRef]

11. Zheng, X.; Li, X.; Jiang, T.; Ding, Y.; Wu, L.; Zhang, S.; Zhao, K. Retrieving soil surface temperature under snowpack using special sensor microwave/imager brightness temperature in forested areas of Heilongjiang, China: An improved method. *J. Appl. Remote Sens.* **2016**, *10*, 26016. [CrossRef]

12. Rautiainen, K.; Lemmetyinen, J.; Schwank, M.; Kontu, A.; Ménard, C.B.; Mätzler, C.; Drusch, M.; Wiesmann, A.; Ikonen, J.; Pulliainen, J. Detection of soil freezing from L-band passive microwave observations. *Remote Sens. Environ.* **2014**, *147*, 206–218. [CrossRef]

13. Chang, A.T.C.; Foster, J.L.; Hall, D.K. Nimbus-7 SMMR derived global snow cover parameters. *Ann. Glaciol.* **1987**, *9*, 39–44. [CrossRef]

14. Foster, J.L.; Chang, A.T.C.; Hall, D.K. Comparison of snow mass estimates from a prototype passive microwave snow algorithm, a revised algorithm and a snow depth climatology. *Remote Sens. Environ.* **1997**, *62*, 132–142. [CrossRef]

15. Derksen, C.; Walker, A.; Goodison, B. Evaluation of passive microwave snow water equivalent retrievals across the boreal forest/tundra transition of western Canada. *Remote Sens. Environ.* **2005**, *96*, 315–327. [CrossRef]

16. Che, T.; Li, X.; Jin, R.; Armstrong, R.; Zhang, T. Snow depth derived from passive microwave remote-sensing data in China. *Ann. Glaciol.* **2008**, *49*, 145–154. [CrossRef]

17. Jiang, L.; Wang, P.; Zhang, L.; Yang, H.; Yang, J. Improvement of snow depth retrieval for FY3B-MWRI in China. *Sci. China Earth Sci.* **2014**, *44*, 531–547. [CrossRef]

18. Santi, E.; Paloscia, S.; Pampaloni, P.; Pettinato, S.; Brogioni, M.; Xiong, C.; Crepaz, A. Analysis of Microwave Emission and Related Indices Over Snow using Experimental Data and a Multilayer Electromagnetic Model. *IEEE Trans. Geosci. Remote Sens.* **2017**, *55*, 2097–2110. [CrossRef]

19. Cai, S.; Li, D.; Durand, M.; Margulis, S.A. Examination of the impacts of vegetation on the correlation between snow water equivalent and passive microwave brightness temperature. *Remote Sens. Environ.* **2017**, *193*, 244–256. [CrossRef]

20. Li, Q.; Kelly, R.E.J. Correcting Satellite Passive Microwave Brightness Temperatures in Forested Landscapes Using Satellite Visible Reflectance Estimates of Forest Transmissivity. *IEEE J. Sel. Top. Appl. Earth Obs. Remote Sens.* **2017**, *10*, 3874–3883. [CrossRef]

21. Roy, A.; Royer, A.; Hall, R.J. Relationship Between Forest Microwave Transmissivity and Structural Parameters for the Canadian Boreal Forest. *IEEE Geosci. Remote Sens. Lett.* **2014**, *11*, 1802–1806. [CrossRef]

22. Takala, M.; Ikonen, J.; Luojus, K.; Lemmetyinen, J.; Metsämäki, S.; Cohen, J.; Arslan, A.N.; Pulliainen, J. New Snow Water Equivalent Processing System with Improved Resolution Over Europe and its Applications in Hydrology. *IEEE J. Sel. Top. Appl. Earth Obs. Remote Sens.* **2017**, *10*, 428–436. [CrossRef]

23. Shi, L.; Qiu, Y.; Shi, J.; Lemmetyinen, J.; Zhao, S. Estimation of Microwave Atmospheric Transmittance Over China. *IEEE Geosci. Remote Sens. Lett.* **2017**, *99*, 1–5. [CrossRef]

24. Ji, D.; Shi, J. Water Vapor Retrieval Over Cloud Cover Area on Land Using AMSR-E and MODIS. *IEEE J. Sel. Top. Appl. Earth Obs. Remote Sens.* **2014**, *7*, 3105–3116. [CrossRef]

25. Kelly, R. A Prototype AMSR-E Global Snow Area and Snow Depth Algorithm. *IEEE Trans. Geosci. Remote Sens.* **2003**, *41*, 230–242. [CrossRef]

26. Kelly, R. The AMSR-E snow depth algorithm: Description and initial results. *J. Remote Sens. Soc. Jpn.* **2009**, *29*, 307–317. [CrossRef]

27. Pulliainen, J.T.; Grandell, J.; Hallikainen, M.T. HUT snow emission model and its applicability to snow water equivalent retrieval. *IEEE Trans. Geosci. Remote Sens.* **1999**, *37*, 1378–1390. [CrossRef]

28. Pulliainen, J. Mapping of snow water equivalent and snow depth in boreal and sub-arctic zones by assimilating space-borne microwave radiometer data and ground-based observations. *Remote Sens. Environ.* **2006**, *101*, 257–269. [CrossRef]

29. Takala, M.; Luojus, K.; Pulliainen, J.; Derksen, C.; Lemmetyinen, J.; Kärnä, J.-P.; Koskinen, J.; Bojkov, B. Estimating northern hemisphere snow water equivalent for climate research through assimilation of space-borne radiometer data and ground-based measurements. *Remote Sens. Environ.* **2011**, *115*, 3517–3529. [CrossRef]

30. Derksen, C.; Toose, P.; Rees, A.; Wang, L.; English, M.; Walker, A.; Sturm, M. Development of a tundra-specific snow water equivalent retrieval algorithm for satellite passive microwave data. *Remote Sens. Environ.* **2010**, *114*, 1699–1709. [CrossRef]

31. Sorman, A.U.; Beser, O. Determination of snow water equivalent over the eastern part of Turkey using passive microwave data. *Hydrol. Process.* **2013**, *27*, 1945–1958. [CrossRef]

32. Jiang, L.; Shi, J.; Tjuatja, S.; Chen, K.S.; Du, J.; Zhang, L. Estimation of Snow Water Equivalence Using the Polarimetric Scanning Radiometer from the Cold Land Processes Experiments (CLPX03). *IEEE Geosci. Remote Sens. Lett.* **2011**, *8*, 359–363. [CrossRef]

33. Pan, J.; Durand, M.T.; Vander Jagt, B.J.; Liu, D. Application of a Markov Chain Monte Carlo algorithm for snow water equivalent retrieval from passive microwave measurements. *Remote Sens. Environ.* **2017**, *192*, 150–165. [CrossRef]

34. Gu, L.; Zhao, K.; Huang, B. Microwave Unmixing With Video Segmentation for Inferring Broadleaf and Needleleaf Brightness Temperatures and Abundances from Mixed Forest Observations. *IEEE Trans. Geosci. Remote Sens.* **2016**, *54*, 279–286. [CrossRef]

35. Liu, X.; Jiang, L.; Wang, G.; Hao, S.; Chen, Z. Using a Linear Unmixing Method to Improve Passive Microwave Snow Depth Retrievals. *IEEE J. Sel. Top. Appl. Earth Obs. Remote Sens.* **2018**, *11*, 4414–4429. [CrossRef]

36. Larue, F.; Royer, A.; De Sève, D.; Roy, A.; Cosme, E. Assimilation of passive microwave AMSR-2 satellite observations in a snowpack evolution model over northeastern Canada. *Hydrol. Earth Syst.* **2018**, *22*, 5711–5734. [CrossRef]

37. Dai, L.; Che, T.; Wang, J.; Zhang, P. Snow depth and snow water equivalent estimation from AMSR-E data based on a priori snow characteristics in Xinjiang, China. *Remote Sens. Environ.* **2012**, *127*, 14–29. [CrossRef]

38. Che, T.; Dai, L.; Zheng, X.; Li, X.; Zhao, K. Estimation of snow depth from passive microwave brightness temperature data in forest regions of northeast China. *Remote Sens. Environ.* **2016**, *183*, 334–349. [CrossRef]

39. Liang, J.; Liu, X.; Huang, K.; Li, X.; Shi, X.; Chen, Y.; Li, J. Improved snow depth retrieval by integrating microwave brightness temperature and visible/infrared reflectance. *Remote Sens. Environ.* **2015**, *156*, 500–509. [CrossRef]

40. Xiao, X.; Zhang, T.; Zhong, X. Support vector regression snow-depth retrieval algorithm using passive microwave remote sensing data. *Remote Sens. Environ.* **2018**, *210*, 48–64. [CrossRef]

41. Bair, E.H.; Abreu Calfa, A.; Rittger, K.; Dozier, J. Using machine learning for real-time estimates of snow water equivalent in the watersheds of Afghanistan. *Cryosphere* **2018**, *12*, 1579–1594. [CrossRef]

42. Yang, J.; Luojus, K.; Lemmetyinen, J.; Jiang, L.; Pulliainen, J. Comparison of SSMIS, AMSR-E and MWRI brightness temperature data. In Proceedings of the 2014 IEEE Geoscience and Remote Sensing Symposium, Quebec, QC, Canada, 13–18 July 2014. [CrossRef]

43. Li, X.J.; Liu, Y.J.; Zhu, X.X.; Zheng, Z.J.; Chen, A.J. Snow Cover Identification with SSM/I Data in China. *J. Appl. Meteorol. Sci.* **2007**, *18*, 12–20.

44. Liu, X.; Jiang, L.; Wu, S.; Hao, S.; Wang, G.; Yang, J. Assessment of Methods for Passive Microwave Snow Cover Mapping Using FY-3C/MWRI Data in China. *Remote Sens.* **2018**, *10*, 524. [CrossRef]

45. Durand, M.; Kim, E.J.; Margulis, S.A. Quantifying uncertainty in modeling snow microwave radiance for a mountain snowpack at the point-scale, including stratigraphic effects. *IEEE Trans. Geosci. Remote Sens.* **2008**, *46*, 1753–1767. [CrossRef]

46. Martinec, J. *Remote Sensing of Ice and Snow*; Springer: Dordrecht, The Netherlands, 1985. [CrossRef]

47. Dong, C.H.; Zhang, G.C.; Xing, F.Y. *Manual for the Interpretation of Meteorological Satellite Business Products*; China Meteorological Press: Beijing, China, 1999; pp. 81–90.

48. Chen, A.J.; Liu, Y.J.; Du, B.Y. Preliminary research on monitoring snow-cover over China with AMSU data. *J. Appl. Meteorol. Sci.* **2005**, *16*, 35–44.

49. Montpetit, B.; Royer, A.; Roy, A.; Langlois, A.; Derksen, C. Snow Microwave Emission Modeling of Ice Lenses Within a Snowpack Using the Microwave Emission Model for Layered Snowpacks. *IEEE Trans. Geosci. Remote Sens.* **2013**, *51*, 4705–4717. [CrossRef]

50. Dai, L.; Che, T.; Ding, Y.; Hao, X. Evaluation of snow cover and snow depth on the Qinghai–Tibetan Plateau derived from passive microwave remote sensing. *Cryosphere* **2017**, *11*, 1933–1948. [CrossRef]

51. Dahe, Q.; Shiyin, L.; Peiji, L. Snow cover distribution, variability, and response to climate change in western China. *J. Clim.* **2006**, *19*, 1820–1833. [CrossRef]

52. Yang, J.; Jiang, L.; Ménard, C.B.; Luojus, K.; Lemmetyinen, J.; Pulliainen, J. Evaluation of snow products over the Tibetan Plateau. *Hydrol. Process.* **2015**, *29*, 3247–3260. [CrossRef]

53. Xue, Y.; Forman, B.A. Atmospheric and Forest Decoupling of Passive Microwave Brightness Temperature Observations Over Snow-Covered Terrain in North America. *IEEE J. Sel. Top. Appl. Earth Obs. Remote Sens.* **2017**, *10*, 3172–3189. [CrossRef]

54. Vander Jagt, B.J.; Durand, M.T.; Margulis, S.A.; Kim, E.J.; Molotch, N.P. On the characterization of vegetation transmissivity using LAI for application in passive microwave remote sensing of snowpack. *Remote Sens. Environ.* **2015**, *156*, 310–321. [CrossRef]

55. Sturm, M.; Holmgren, J.; Liston, G.E. A seasonal snow cover classification system for local to global applications. *J. Clim.* **1995**, *8*, 1261–1283. [CrossRef]

56. Sturm, M.; Wagner, A.M. Using repeated patterns in snow distribution modeling: An arctic example. *Water Resour. Res.* **2010**, *46*, 65–74. [CrossRef]

remote sensing

MDPI

Article

Development of a Parameterized Model to Estimate Microwave Radiation Response Depth of Frozen Soil

Tao Zhang [1], Lingmei Jiang [2,*], Shaojie Zhao [3], Linna Chai [3], Yunqing Li [4] and Yuhao Pan [2]

[1] Land Satellite Remote Sensing Application Center, Ministry of Natural Resources of the People's Republic of China, Beijing 100048, China
[2] State Key Laboratory of Remote Sensing Science, Jointly Sponsored by Beijing Normal University and Institute of Remote Sensing and Digital Earth of Chinese Academy of Sciences, Beijing Engineering Research Center for Global Land Remote Sensing Products, Faculty of Geographical Science, Beijing Normal University, Beijing 100875, China
[3] State Key Laboratory of Earth Surface Processes and Resource Ecology, Faculty of Geographical Science, Beijing Normal University, Beijing 100875, China
[4] School of Urban Construction, Beijing City University, Beijing 100083, China
* Correspondence: jiang@bnu.edu.cn; Tel.: +86-10-5880-5042

Received: 22 July 2019; Accepted: 26 August 2019; Published: 28 August 2019

Abstract: The sensing depth of passive microwave remote sensing is a significant factor in quantitative frozen soil studies. In this paper, a microwave radiation response depth (MRRD) was proposed to describe the source of the main signals of passive microwave remote sensing. The main goal of this research was to develop a simple and accurate parameterized model for estimating the MRRD of frozen soil. A theoretical model was introduced first to describe the emission characteristics of a three-layer case, which incorporates multiple reflections at the two boundaries. Based on radiative transfer theory, the total emission of the three layers was calculated. A sensitivity analysis was then performed to demonstrate the effects of soil properties and frequency on the MRRD based on a simulation database comprising a wide range of soil characteristics and frequencies. Sensitivity analysis indicated that soil temperature, soil texture, and frequencies are three of the primary variables affecting MRRD, and a definite empirical relationship existed between the three parameters and the MRRD. Thus, a parameterized model for estimating MRRD was developed based on the sensitivity analysis results. A controlled field experiment using a truck-mounted multi-frequency microwave radiometer (TMMR) was designed and performed to validate the emission model of the soil freeze–thaw cycle and the parameterized model of MRRD developed in this work. The results indicated that the developed parameterized model offers a relatively accurate and simple way of estimating the MRRD. The total root mean square error (RMSE) between the calculated and measured MRRD of frozen loam soil was approximately 0.5 cm for the TMMR's four frequencies.

Keywords: frozen soil; microwave radiation response depth (MRRD); microwave radiometer experiment; parameterized model

1. Introduction

Permafrost and seasonally frozen soil, whose thermal and physical properties differ from unfrozen soil, are key components of the cryosphere. The regional energy and water balance are dramatically modified by the phase transition of soil water during the freeze–thaw process. The freezing–thawing of soil induced the release of decomposable organic carbon, and thus had a profound influence on the overall functioning of ecosystems [1]. However, the degradation of permafrost, which releases latent heat and carbon, has become a positive global warming feedback [2]. Previous experiments have shown that the carbon emission in soil profile varied spatially and temporally and was correlated with soil frozen depth [3]. Hence, it is necessary to determine the soil frozen depth for a given frozen area and explore the extent and distribution of frozen areas.

Satellite remote-sensing technology is an ideal tool for obtaining the spatial and temporal information of frozen areas. Passive microwave remote sensing has been used to provide qualitative information in previous studies because of less vulnerably linked to cloud cover and more frequent revisit time comparing to optical remote sensing. In some previous research, many scientists have devoted to exploring freeze/thaw discrimination methods using passive microwave remote sensing data [4–8]. Recently, multiple studies have focused on the quantitative frozen ground remote sensing, such as the amount of water released by the phase transition during the freeze/thaw process, the frost penetration velocity [9,10], and so on. It should be noted that in these investigations, land surface properties were monitored using microwave remote-sensing sensors, which provided an integrated microwave emission signal over a certain surface within a certain depth. Knowing this depth is important for this research referred to above. It is an indication of the thickness of the surface layer, within which variations in soil moisture or other soil parameters can significantly affect the emitted radiation at a certain frequency [11,12]. For different frequencies, the depth varies too. Hence, it can be an explanation for some phenomenon in freeze–thaw related research because signals from different surfaces might indicate different depth. In addition, it is useful for planning ground data collection campaigns for model and algorithm development and validation in this field.

However, what the definition of this 'depth' is and how deep the 'depth' refers to are two main issues in front of us. Many scientists attempt to answer these two questions and have contributed to some related work. Ulaby proposed a penetration depth model for active microwaves [13]. It might cause errors for the active microwave model to be used in passive microwave remote sensing applications because of the disparate working mechanisms between active and passive microwave remote sensing [14]. Wilheit proposed a thermal sampling depth and reflectivity sampling depth based on a multiple-layered and coherent radiative transfer model [15]. The thermal sampling depth is the depth at which thermal radiation upwelling originates in the soil and the reflectivity sampling depth was determined as the depth at which the isothermal and non-isothermal reflectivities were equal. The reflectivity characteristics changed over the depth of approximately 1/10–1/7 of the wavelengths in the medium, and the thermal radiation was generally larger [16]. Blinn and Paloscia gave a feasible way to study the question regarding which depth of soil is responsible for the greatest part of the emission upwelling from soil by conducting a controlled ground experiment [17,18]. The conclusion of the sensitivity of L band emission to the moisture content of a soil layer about 5 cm thick was confirmed. There are also several theoretical and experimental results that demonstrated that the sensing depths were approximately 2–20 cm for different soil conditions at L band [12,19–22] and about 2 cm at 5 GHz with a soil moisture of 0.1 cm^3/cm^3 [20].

Although, previous studies have revealed different definitions and results of the passive microwave remote-sensing depth, a feasible method to estimate a depth that determined as the source of the predominant microwave remote-sensing signals has been poorly investigated. As a result, scientists commonly use an empirical depth. For example, although the depth varies with frequency and soil characteristics, soil moisture measurements with depths of at least 0–5 cm are widely used to validate remote-sensing data at L band [23–25]. It is possible that for unfrozen soil the measurements of soil moisture are only partially representative of average values measured by a radiometer. Finally, it can be seen that most of the research referred to unfrozen soil. As the dielectric and microwave radiation characteristics of unfrozen soil are quite different with that of frozen soil [26–28], the models of estimating penetration depth for unfrozen soil usually do not work well for frozen soil.

In this research, we will try to propose a passive microwave remote-sensing response depth (MRRD) and develop a simple parameterized model to determine its value for frozen soil. A field experiment using a truck-mounted multi-frequency microwave radiometer (TMMR) will be introduced in Section 2.1. Field experiment data will be used for validation. In Section 2.2, we propose the theory used to investigate the passive microwave remote sensing response depth and develop a parameterized model to estimate the frozen soil response depth. The measurements are analyzed and discussed in Sections 3 and 4. In Section 5 we give the conclusions of this study.

2. Materials and Methods

2.1. Field Experiment

As has already been done by other investigators [17,18], we carried out a "microwave radiation response depth experiment" by putting soil samples over a metal plate. A TMMR was used to measure the microwave radiation from an aluminum sheet covered with soil samples of various thicknesses. The controlled experiment data is used to investigate the effects of frequency and soil parameters on the microwave remote sensing response depth and validated a parameterized model of the microwave radiation response depth (MRRD) in frozen soil. The detailed field experiment is presented in the following sections.

2.1.1. Experimental Setup and Material Preparation

The experiment was performed in Baoding, Hebei Province, China (38°42'10.21"N, 115°23'18.23"E) from 12 to 19 January 2012. During this period, the soil typically froze at night and thawed during the day. The most important instrument in this experiment is the TMMR. It is an eight-channel radiometer with four frequencies (6.925, 10.65, 18.7, and 36.5 GHz), with vertical and horizontal polarization at each frequency. The TMMR can collect data at multiple angles, both in the zenith (from −90° to 90°) and azimuthal directions (from 0° to 360°). The TMMR was calibrated on a four-point calibration scheme before the experiment, and the precision of measurements was also tested by measuring a calm water area. An absolute Tb accuracy of 1 K could be achieved with the accuracy of calibration target temperature sensors and the minimization of thermal gradients. Detailed calibration and test procedures were described in [29].

The radiometer was placed on a truck using a hydraulic lifting platform, which lifted the apparatus to 4.78 m above the ground. Figure 1 presents the observation field. Measurements were made toward the south to avoid the shadow of the truck. The incident angle was fixed at 45° during the whole experiment. The −10 dB footprints of the four antennas were calculated and labeled on the ground according to the radiometer configuration. Due to the low emissivity of aluminum, aluminum sheets were used as the background for the following measurements. Eight aluminum sheets (each 1 m × 2 m) were arranged on the ground in two rows to form a 4 m × 4 m mosaic, which completely covered the four antenna footprints. This can ensure that the signals received by TMMR are from the objects on the aluminum sheets. The soil was air dried and sieved to ensure a homogeneous soil moisture and texture.

For each thickness, the appropriate amount of water was mixed with the dried soil to obtain the desired soil moisture. The mixture was used to cover the aluminum sheets, and the layer was artificially smoothed. To improve soil plasticity and artificially smooth the soil surface, appropriate soil moisture content is needed. Note that covering the eight aluminum sheets with the area of 16 m² using the mixture and artificially smoothing the sample surface is a very time-consuming and labor-intensive process. In addition, a heterogeneous soil temperature profile may exist if the soil thickness is too large. Conversely, soil samples with thin thickness are more easily to be freeze uniformly. Thus, only five and relatively thin soil samples were analyzed in this experiment.

Figure 1. Viewing scene of the truck-mounted multi-frequency microwave radiometer (TMMR).

2.1.2. Measurements

The TMMR was used to measure the microwave radiation from the aluminum sheet to obtain the emissivity at four frequencies. During the experiment, the brightness temperature of the aluminum sheet, which was covered by various soil layer thicknesses, was measured using a TMMR. To eliminate the atmospheric effect on each measurement, the atmospheric downward radiations in different frequencies were measured by TMMR and then were used to calculate the emissivity values of soil samples.

In this experiment, the initial soil moisture of soil sample is 0.433 cm^3/cm^3, and the thicknesses of five soil samples were 0.18, 0.43, 0.63, 0.96, and 1.06 cm, respectively. The continuous measurements of each sample lasted for approximately 24 hours to include the typical thawing and freezing processes that occur during the day and night, respectively. Because the physical temperature remained too high for completely freezing during the first 24-hour period, the 1.06 cm soil sample was observed for about two days. To avoid soil moisture evaporation, a plastic film was used to cover the soil. A trial conducted before our experiment revealed that the effect of the plastic film was negligible. The soil temperature and moisture were automatically measured every 3 minutes by the temperature and moisture sensors and collected using an Intelligent Data Acquisition Collector (IDAC) (Figure 1). As one thickness measurement was being conducted, another sample was being prepared. The thickness of the soil sample was measured after one sample observation was performed.

A cutting ring was used to measure the soil bulk density. The mean bulk density was 1.41 g/cm^3. The soil texture is classified as loam (sand: 30.16%, silt: 48.85%, clay: 20.99%) according to the U.S. Department of Agriculture classification scheme.

2.2. Methodology

To develop a parameterized model to estimate MRRD of frozen soil, the definitions of the MRRD and the theoretical model utilized were introduced firstly. Sensitivity analysis were then performed to determine the relationship between MRRD and soil parameters using the theoretical model. The parameterized model was developed based on sensitivity analysis results. The field experiment data introduced in 2.1 were used for validation of theorical model and parameterized model. A flow chart has shown the parameterized model development and validation process (Figure 2).

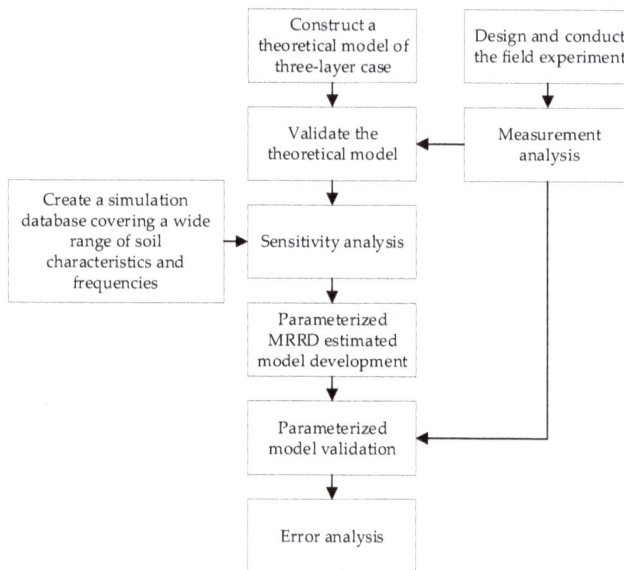

Figure 2. The flow chart of the parameterized model development and validation process.

2.2.1. Microwave Radiation Response Depth (MRRD)

To describe the characteristics observed by passive microwave remote sensing, the MRRD was defined in this paper. We used the following experiment to obtain an expression for the MRRD. A radiometer was used to measure the emissivity of aluminum sheets (theoretically, its emissivity is 0 and reflectivity is 1.0) covered with various soil thicknesses. As a result, when the soil layer was thin, the brightness temperature measured by the radiometer was very low. The observed brightness temperature increased as the soil thickness increased, and it will stabilize when the soil thickness increases. Because the emissivity can be deduced by normalizing brightness temperature to a physical temperature of the target, herein, the emissivity was used to describe the definition of MRRD. If the observed emissivity reaches e_{max}, i.e., achieves stability, the MRRD can be defined as the depth at which:

$$e_{max} - e_z = 0.001 \tag{1}$$

Where, e_z is the emissivity of soil with thickness of z. The above formula indicates that at depth z, the difference between the emissivity and the maximum emissivity (the stable value) is only 0.001 (approximately 0.1–0.3 K for frozen soil). This depth is defined as the MRRD. The MRRD provides a reference for the primary source of the observed signals.

2.2.2. Theoretical Model

A three-layer case was used to illustrate the theoretical model for frozen/thaw soil (Figure 3). We define the air, soil, and aluminum sheet as the first, second, and third layers and the air–soil and soil–aluminum sheet boundaries as the first and second boundaries, respectively.

The model was used to calculate emissions using the radiative transfer theory [13]. It is a non-coherent model, which does not include interference. The brightness temperature consists of two contributions:

$$T_B(\theta_1; p) = T_{B2}(\theta_1; p) + T_{B3}(\theta_1; p) \tag{2}$$

where T_{B2} and T_{B3} are the brightness temperature contributions due to the layer 2 and 3 emissions, respectively. θ_1 is the angle of the radiation emitted into layer 1.

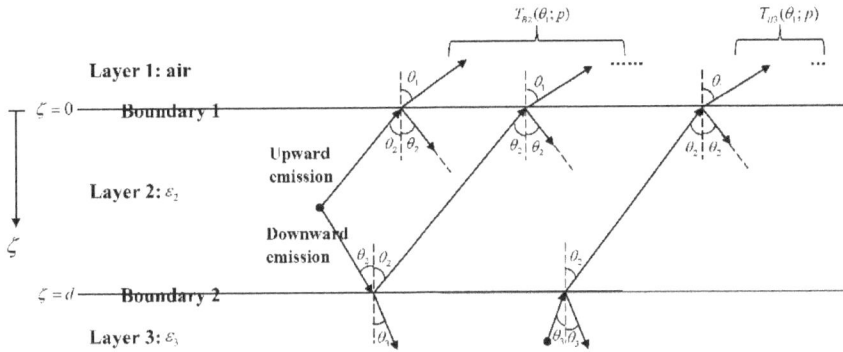

Figure 3. The emissions from layer 2 and layer 3 into layer 1 (air).

Multiple reflections at the two boundaries were incorporated into this model. Each of the two bottom layers is assumed to isotropically radiate, and the boundary roughness is ignored. We are only interested in the radiation that is emitted into layer 1 at an angle of θ_1. The only radiation of interest in layer 2 is that emitted at boundary 1 at an angle of θ_2. This consists of upward emissions and downward emissions that are reflected by boundary 2 toward boundary 1. The contributions of these two components to T_{B2} can be designated by the upward emission, T_{2U}, and downward emission, T_{2D}:

$$T_{B2}(\theta_1;p) = T_{2U}(\theta_1;p) + T_{2D}(\theta_1;p) \tag{3}$$

Consider a thin horizontal stratum in layer 2 at a depth of ζ and a thickness of $d\zeta$. In the upward direction, this stratum's emissions are first attenuated by the stratum at depth ζ and boundary 1 along the path. Then, at boundary 1, a fraction of this attenuated emission is transmitted across the boundary and the remainder is reflected. The reflected portion decreases as it travels to boundary 2. It is then partially reflected toward boundary 1 and partially transmitted into layer 3. This process will infinitely continue. In the downward direction, this stratum's emissions are also attenuated by the stratum at depth ζ and boundary 2 along the path. A fraction is transmitted across boundary 2, while the remainder is reflected. It then experiences the same reflections between boundaries 1 and 2 as the upward emission. The total energy can be found by integrating over the 0 to d depth range.

For layer 3, the upwelling emission at angle θ_3 is transmitted by boundary 2 and attenuated by layer 2, then reflected toward boundary 1, which is the only component we consider. It can be computed in a similar manner as layer 2. The theoretical model can be described as:

$$T_B(\theta_1;p) = \frac{(1-\Gamma_1)}{\left(1-\Gamma_1\Gamma_2/L_2^2\right)}\left[\left(1+\frac{\Gamma_2}{L_2}\right)(1-a)\left(1-\frac{1}{L_2}\right)T_2 + \frac{(1-\Gamma_2)T_3}{L_2}\right] \tag{4}$$

The above expression gives the total brightness temperature, T_B, of the three layers. In the model, Γ is the reflectivity of a boundary, and the subscripts 1 and 2 indicate boundaries 1 and 2. The reflectivity can be calculated using Fresnel's law at the first boundary. T is the physical temperature of each layer. Subscripts 1, 2, and 3 indicate layers 1, 2, and 3, respectively. θ $\theta$and p are the incident angle and polarization. L is the power loss factor, which can be expressed by:

$$L = \exp(\kappa_e d \sec \theta_2) \tag{5}$$

where d is the soil thickness, θ_2 is the real transmission angle, and κ_e is the extinction coefficient. If soil medium scattering is ignored, κ_e is approximately equal to the absorption coefficient, κ_a, given by:

$$\kappa_e \cong \kappa_a = \frac{4\pi}{\lambda_0}\left|Im\left(\sqrt{\varepsilon}\right)\right| \tag{6}$$

where λ_0 is the wavelength in free space and $\varepsilon = \varepsilon' - j\varepsilon''$ is the corresponding permittivity of soil.

The low emissivity of the aluminum sheet makes it an ideal background material. Additionally, it can shield signals from below the aluminum sheet. We assume that the soil properties are uniform with depth, the reflectivity of the aluminum sheet is 1.0 and the single scattering albedo is sufficiently small that diffuse scattering can be ignored (a = 0). The implications of these assumptions will be discussed in Section 4.

Because the physical characteristics of the soil layer are unknown, several empirical models were introduced. Unfrozen soil can be considered a mixture of air, solid soil, bound water, and free water. Dobson developed an empirical model to calculate the permittivity [30], which has been widely used in soil permittivity estimations.

$$\varepsilon_m^\alpha = 1 + (\rho_b/\rho_s)(\varepsilon_s^\alpha - 1) + m_v^\beta \varepsilon_{fw}^\alpha - m_v \tag{7}$$

Here, ε_m, ε_s, and ε_{fw} are the dielectric constants of the unfrozen soil, solid soil, and free water, respectively. ρ_b and ρ_s are the bulk density and specific density in g/cm^3, respectively. m_v is the volumetric soil moisture in cm^3/cm^3. α is a constant shape factor, and β is a coefficient dependent on the soil textural composition.

Due to the adsorption forces and curvature at the soil particle surfaces in frozen soil, a certain amount of water remains unfrozen, even when the soil temperature is below 0 °C. This is called the unfrozen water content. The amount of unfrozen water mainly depends on the physical properties of the soil. Typically, the specific surface, which is the ratio of total surface area to the mass of the soil, controls the binding force on the water. The higher the soil specific surface area, the greater the binding force, making the water prone to remaining unfrozen. However, the unfrozen water content overcomes the adsorption forces of soil particles and freezes as the soil temperature continues to decrease. Anderson developed an empirical model for estimating the unfrozen water content using the soil temperature and specific surface area of the soil [31] which is given by:

$$\begin{aligned} m_u &= a|T - 273.15|^{-b} \\ \ln a &= 0.5519 \ln S + 0.2618 \\ \ln b &= -0.264 \ln S + 0.3711 \end{aligned} \tag{8}$$

where m_u is the unfrozen water content (cm^3/cm^3), T is the soil temperature in Kelvin, and S is the soil specific area in m^2/g, which is determined based on the particle-size distribution of the soil. The soil specific area can be predicted using an empirical model [32] based on the sand, silt, and clay contents:

$$S = 0.042 + 4.23\text{clay}\% + 1.12\text{silt}\% - 1.16\text{sand}\% \tag{9}$$

The unfrozen water content directly affects the permittivity of the frozen soil due to the large difference in the permittivity between ice and water. Based on Dobson's work, Zhang added a term to the parameterized model to describe the ice fraction contribution to the frozen soil permittivity [26]. This model can be used for various soil types and includes soil texture, bulk density, soil moisture, and temperature inputs. It has been used in previous passive microwave remote sensing studies of frozen soil [7,9]. The simulated results were also validated using experimental data obtained with an Agilent PNA Network Analyzer E8362B. The expression can be written as:

$$\varepsilon_{mf}^\alpha = 1 + (\rho_b/\rho_s)(\varepsilon_s^\alpha - 1) + m_{vu}^\beta \varepsilon_{fw}^\alpha - m_{vu} + m_{vi}\varepsilon_i^\alpha - m_{vi} \tag{10}$$

where the subscripts s, i, fw, vu, and vi refer to solid soil, ice, free water, volumetric unfrozen water, and the volumetric ice content, respectively.

2.2.3. Sensitivity Analysis

The attenuation of electromagnetic waves in the soil is determined based on the soil's permittivity and the wavelength. Hence, the MRRD is mainly affected by the factors impact on the dielectric characteristics of the soil. A sensitivity analysis was conducted to determine the relationship between MRRD and soil parameters using the theoretical model described above. We have performed the comparison between MRRD computed using the theoretical model for horizontal and vertical polarization. It can be concluded that the MRRDs for horizontal and vertical polarization are highly correlated and have negligible difference when the incident angle equals zero. However, the MRRD for horizontal polarization was slightly higher than that for vertical polarization because different Fresnel reflectivity of polarizations exists at the air–soil boundary if the incident angle is not zero [12,33]. For simplicity, the vertical polarization brightness temperature is used in the following analysis. A simulation database covering a wide range of soil characteristics and frequencies was created. The unfrozen water content, soil temperature, frequency, bulk density, and soil specific surface area were selected as the factors to be analyzed. The key input parameters of the theoretical model are listed in Table 1. Note that the unfrozen water content was calculated by temperature and soil specific surface area using Equation (8). The incidence angle determines the microwave radiation path, which can be calculated with vertical path. Therefore, it was set to be 55° according to the commonly used Advanced Microwave Scanning Radiometer–Earth Observing System (AMSR-E) observation configuration.

Table 1. Setup of the key parameters in the simulation database.

Parameters	Value
Temperature	243.15 to 271.15 K(-30 to -2 °C)
Frequency	4 to 40 GHz
Soil specific surface area	37.442 to 253.042 m^2/g
Unfrozen water content	0.02 to 0.31 cm^3/cm^3
Bulk density	1.2 to 1.8 g/cm^3
Incident angle	55°
Initial soil moisture	0.433 cm^3/cm^3

To compare the sensitivity of soil characteristics and frequencies to MRRD, the normalized index for each parameter was used in the sensitivity analysis. The normalized index for a parameter can be expressed as

$$P_N = (P_i - P_{min}) / (P_{max} - P_{min}) \tag{11}$$

Where, P_i indicates the values of soil temperature, frequency, soil specific surface area and bulk density. P_N is the normalized value of these parameters. The subscript max and min are the maximum and minimum value of the corresponding parameters set in the simulation database.

As described above, not all of the liquid water transforms into ice when the temperature drops below 0 °C. The unfrozen water in the soil is determined based on the temperature and soil texture. Figure 4a–c show the dependence of the MRRD on temperature, soil specific area, and unfrozen soil water content, respectively. For a given frequency, the MRRD increases when the temperature varies from -2 °C to -30 °C and the soil specific surface area ranges from 253.042 to 37.442 m^2/g. This can be explained by the positive correlation between the unfrozen water and temperature. The unfrozen water content in the frozen soil decreases as the temperature decreases, leading to a lower soil permittivity and weaker electromagnetic extinction when all other soil characteristics are held constant. Another parameter affecting the unfrozen water in the soil is the soil texture, which impacts the soil particle water adsorption. The higher the specific surface area, the stronger the water binding force, leading to more unfrozen water and a higher permittivity (Figure 4b). Figure 4d illustrates the MRRD decrease with increasing frequency (decreasing the wavelength). Additionally, at -15 °C, the difference in the MRRD is 2 cm (approximately 8 cm and 6 cm) for frequencies ranging from 6.925 GHz to 10.65 GHz, but only 0.4 cm (5.1 cm and 4.7 cm) for frequencies from 18 GHz to 36.5 GHz. The MRRD is weakly dependent on the frequency above 10 GHz. This behavior is clearly shown in Figure 4d and agrees with our expectations of the microwave remote sensing penetrability.

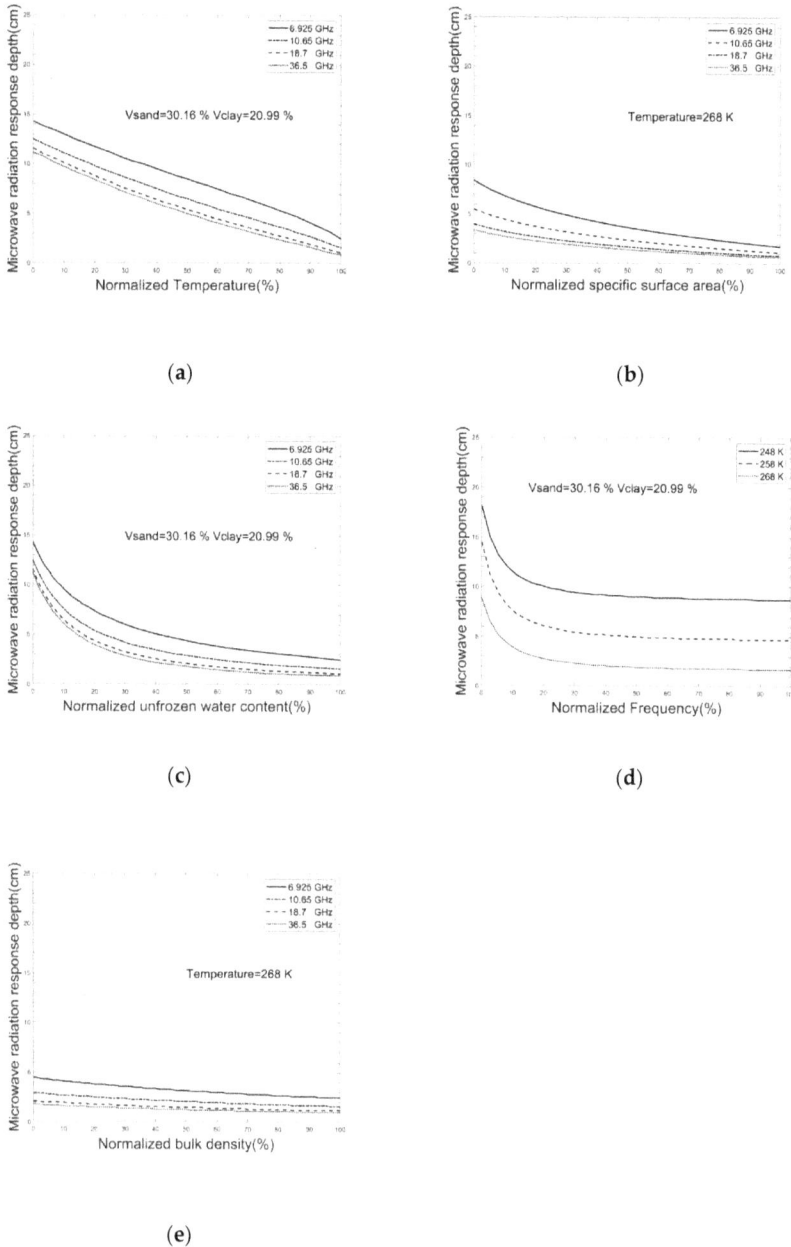

(a)

(b)

(c)

(d)

(e)

Figure 4. The microwave radiation response depth (MRRD) computed using the theoretical model as a function of normalized soil temperature (**a**), soil specific surface area (**b**), unfrozen soil water content (**c**), frequency (**d**), and bulk density (**e**).

Overall, the MRRD is negatively correlated with all parameters. The sensitivity of MRRD to unfrozen water content, temperature, frequency, and soil specific surface area is larger than that to bulk density (Figure 4e). Furthermore, the bulk density generally relates to the composition of soil particles. Hence, it is not considered in the parameterized model development.

2.2.4. Parameterized MRRD Estimated Model Development

Based on the sensitivity analysis in part 2.2.3, we concluded that the observation frequency, unfrozen water content, temperature, and specific surface area are important factors for determining the frozen soil MRRD. Note that the measurements of unfrozen water content in frozen soil is generally very complicated and difficult. Furthermore, the unfrozen water content was represented by temperature and soil specific surface area. Therefore, it was not included in the parameterized model development. Hence, three parameters, including temperature, frequency, and specific surface area, were used to develop a simple, accurate model to estimate the MRRD in this research. The relationship between the temperature and MRRD was analyzed firstly based on the simulation database as shown in Figure 5.

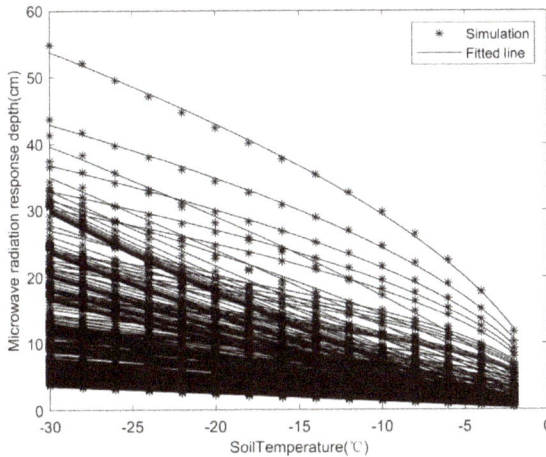

Figure 5. The fitted result of the relationship between MRRD and soil temperature.

The relationship between the MRRD and temperature can be expressed by the exponential function:

$$d = A|T - 273.15|^B \qquad (12)$$

where d is the MRRD in cm and T is the physical soil temperature in Kelvin. Because the physical temperature is lower than 273.15 K in frozen soil, the absolute value was used in the function.

The empirical coefficients A and B depend on observation frequency and specific surface area. The least squares fitting method was used to seek the relationship between the coefficients and frequency (Figure 6).

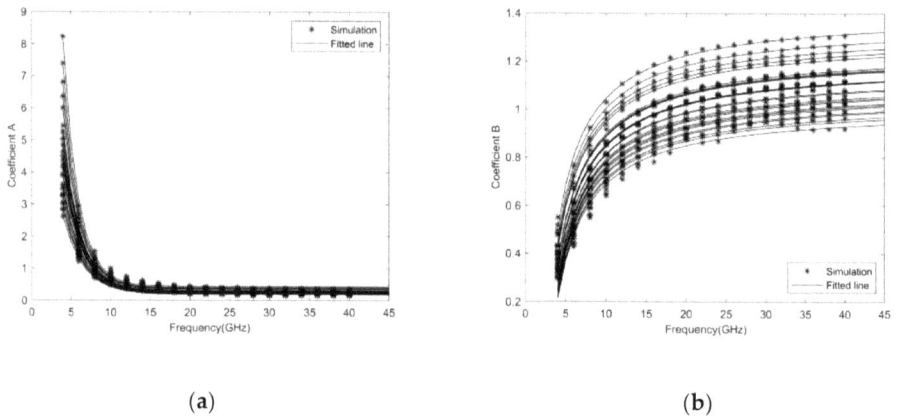

(**a**)   (**b**)

Figure 6. The fitted results of the relationship between coefficients A(**a**) and B(**b**) and frequency.

According to the fitted results, the coefficient A and B is negatively and positively correlated with frequency, respectively. They can be expressed by a function of frequency as:

$$A = a_1 e^{a_2 f} + a_3$$
$$B = b_1 + b_2 / f \tag{13}$$

Then a regression analysis was performed to explore the relationship between the coefficients a1, a2, a3, b1, and b2 with specific surface area (Figure 7). Except for coefficient b2, the coefficients a1, a2, a3, and b1 were high related to specific surface area and the R^2 was 0.96, 0.91, 0.93, and 0.84, respectively.

The relationship between the coefficients and specific surface area can be expressed as (14). It should be noted that the b2 was set to be a constant of average values because it is slightly correlated with specific surface area.

$$a_1 = -8.316 \ln(S) + 50.991$$
$$a_2 = 0.0004S - 0.368$$
$$a_3 = -0.116 \ln(S) + 0.8004$$
$$b_1 = -0.197 \ln(S) + 2.1617$$
$$b_2 = -3.97168 \tag{14}$$

Formulas (12) to (14) comprise the parameterized model for estimating the MRRD. In this model, the MRRD can be estimated using three common and easily acquired parameters: soil temperature, T, frequency, f, and soil specific surface area, S. Note that the empirical model was developed according to the AMSR-E configuration, at an incident angle of 55°. Thus, d is the soil layer vertical thickness, not the extinction path in the soil layer. Hence, the MRRD is $d \sec \theta_2$ when the wave is vertically incident upon the soil.

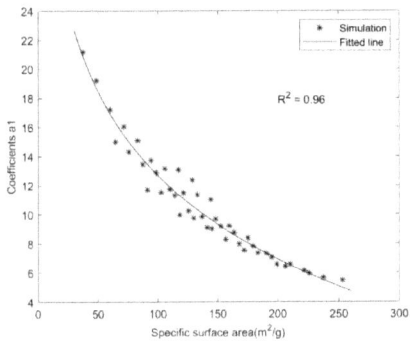

(**a**)

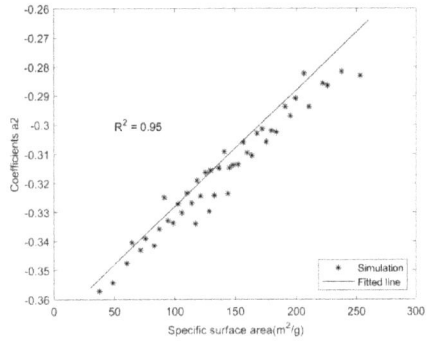

(**b**)

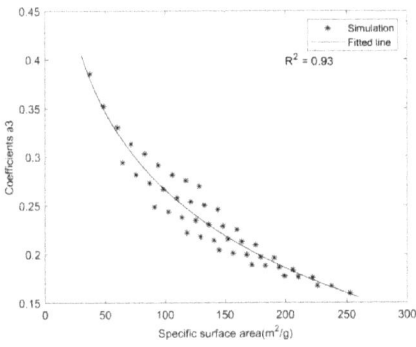

(**c**)

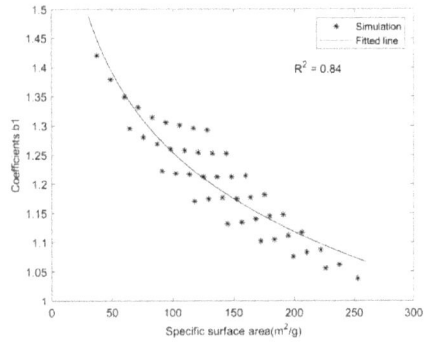

(**d**)

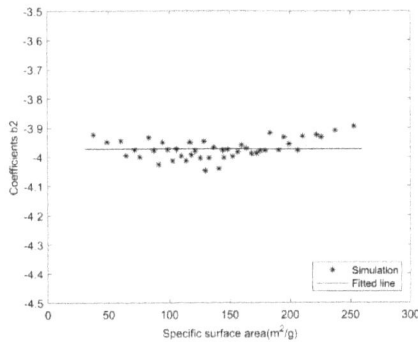

(**e**)

Figure 7. The fitted results of the relationship between coefficients a1(a), a2(b), a3(c), b1(d), and b2(e) and specific surface area.

3. Results

3.1. Measurement Results

The observation duration and the lowest/highest temperature of five soil samples were listed in Table 2. The lowest temperature of all samples is lower than 268.5 K (−4.65 °C) and the highest temperature is higher than 275.9 K (2.75 °C), ensuring all samples occurred complete thawing and freezing processes.

Table 2. The observed lowest and highest temperature of five soil samples.

Samples Thickness(cm)	Observation Duration	The Lowest and Highest Temperature(K)
0.18	Jan.12th 21:45~Jan. 13th, 15:24	260.3 ~ 294.1
0.43	Jan.13th 18:06~Jan. 14th, 16:10	263.3 ~ 288.7
0.63	Jan.14th 17:52~Jan. 15th, 16:32	265.7 ~ 283.7
0.96	Jan.15th 17:40~Jan. 16th, 15:56	267.5 ~ 275.9
1.06	Jan.17th 17:30~Jan. 19th, 09:53	268.5 ~ 277.8

Typical brightness temperature and physical temperature variations are shown in Figure 8. (using a 0.63 cm-thick sample as an example). It can be seen that the soil on the aluminum sheet experienced a thaw-freeze-thaw cycle from 17:52 on 14 January to 16:32 on 15 January. The lowest physical temperature was 265.7K (−7.45 °C), which occurred at 7:47. The highest temperature was 283.7 K (10.55 °C), which occurred at 13:41.

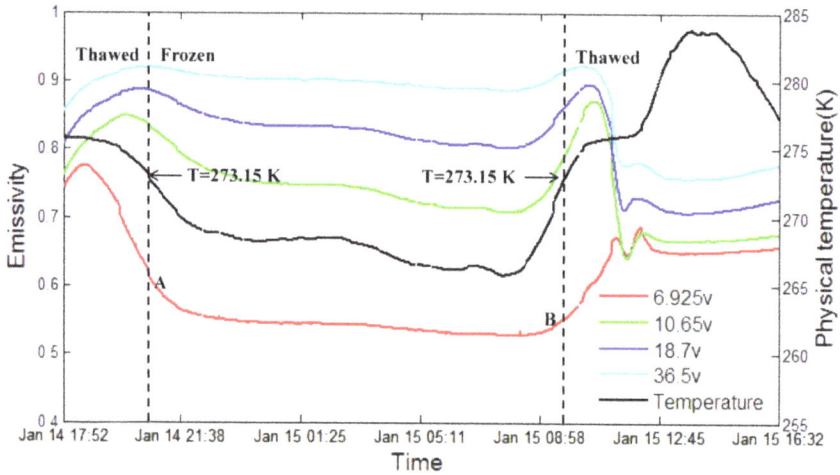

Figure 8. Emissivity and physical temperature variations measured in 0.63 cm-thick soil.

In the experiment, the aluminum sheets were the "cold" source. At the beginning of the measurements, all or part of the signal from the aluminum sheet and wet soil was measured by the radiometer. As the soil's physical temperature decreases, the liquid water in the soil freezes, which weakens the attenuation ability of the soil and increases its emissivity. Thus, the "cold" signal from the aluminum sheet can easily be observed by the radiometer and initially contribute to the total emissions. Due to the low emissivity of the aluminum sheet, the observed emissivity quickly decreased as its contribution increased. Moreover, due to the greater penetrability, the lower frequency emissivity values were smaller. Hence, the emissivity at 6.925 GHz decreased first, followed by 10.65, 18.7, and 36.5 GHz. After the wet soil was completely frozen, the emissivity values became relatively stable at all four frequencies.

During the soil-thawing process, some of the ice melted, forming liquid water, when the temperature increased on the morning of 15 January. Thus, the "cold" signal of aluminum sheet is shielded by the wet soil, and the observed emissivity increases. However, the emissivity decreases as the amount of ice transformed into liquid water increases. As expected, the emissivity again stabilizes when the soil is completely thawed.

The emissivity and physical temperature trends were similar for each soil sample, regardless of thickness. Two emissivity values exist for each frequency at a given physical temperature (e.g., points A and B on the 6.925 GHz vertical polarization line in Figure 8). At point A, the soil temperature is decreasing from 0+ to 0− °C ('+' and '−' indicate that the temperature is just above or below 0 °C). Thus, the soil was freezing. In contrast, point B (from 0− to 0+ °C) corresponds to thawing. Although the temperatures are equal, the physical processes are different, resulting in small emissivity differences. According to a previous experimental study [9,26], at the same negative soil temperature, more unfrozen water can be found in frozen soil during freezing than thawing. Therefore, the unfrozen soil in the thawing process lags behind that in the freezing process. Hence, the unfrozen water content at point B is smaller than at point A. Accounting for the "cold" signal of the aluminum sheet, the observed emissivity at point A is larger than at point B.

To further analyze the emissivity difference in the same negative temperature between freezing and thawing process, we analyzed the measured emissivity with temperature of 268 K, 269 K, 270 K, 271 K, and 272 K. The emissivity at different frequencies during freezing and thawing process was shown in Figure 9. It has shown that the emissivity at all frequencies during freezing process were higher than that during thawing process with the same soil sample temperature.

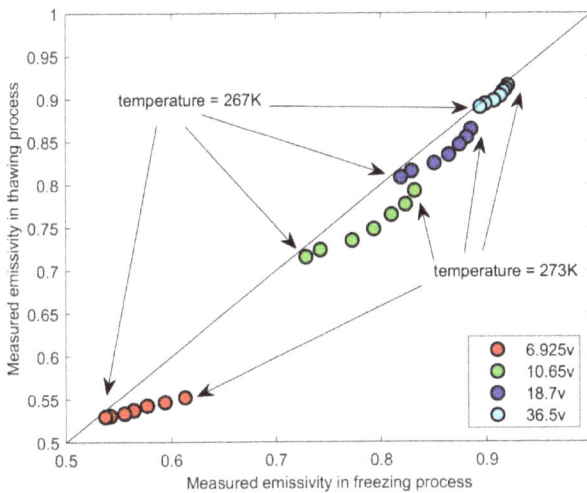

Figure 9. The measured emissivity at different frequencies during freezing and thawing process.

3.2. Theoretical Model Validation

The theoretical model was evaluated based on the field experiment radiometer measurements. Five soil samples of different thicknesses were analyzed. The measured emissivities with the soil temperature of 268 K, 269 K, 270 K, 271 K, and 272 K in the field experiment were selected to compare with the corresponding simulated emissivities. The parameters used in emissivity simulation were listed in Table 3. Figure 10 illustrates the comparison between measured and simulated emissivity values for the four frequencies: 6.925, 10.65, 18.7, and 36.5 GHz, during the freezing and thawing processes. As described above, there is little difference between the freezing and thawing soil processes.

The RMSEs between the simulated and observed emissivity values are 0.076 and 0.096 for the freezing and thawing processes, respectively.

Table 3. Setup of the key parameters in emissivity simulation.

Parameters	Value
Temperature	268 K, 269 K, 270 K, 271 K, and 272 K
Frequency	6.925, 10.65, 18.7, and 36.5 GHz
Soil texture	sand: 30.16%, silt: 48.85%, clay: 20.99%
Bulk density	1.41 g/cm^3
Initial soil moisture	0.433 cm^3/cm^3

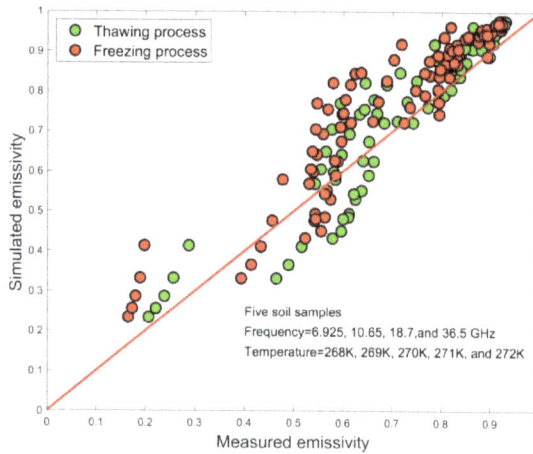

Figure 10. Comparison between measured and simulated emissivity using theoretical model during freezing and thawing processes.

3.3. Parameterized Model Validation

Five soil samples of different thicknesses were studied in the experiment. Accounting for the emissivity of the aluminum sheet as the background value (soil thickness = 0), we obtained six points on the curve expressing the relationship between emissivity and soil thickness for a given soil physical temperature. Soil thickness is the only variable that affects the observed emissivity if the physical temperature is fixed. The emissivity will increase with increasing soil thickness due to the low emissivity of the aluminum sheet, eventually reaching a stable value. However, due to the time and material limitations of this experiment, we cannot create enough soil samples with varying thicknesses to reach these stable emissivity levels for the four frequencies analyzed. Thus, the least square method was used to fit the curves using experimental data. Furthermore, considering the relationship between the measured emissivity and soil thickness introduced in the expression for the MRRD, the exponential form was used to fit the curves.

$$e_{fit} = \alpha + \beta \cdot \exp(\gamma \cdot st) \tag{15}$$

Where, e_{fit} is the fitted emissivity based on experiment measurements, st is soil thickness in cm. α, β, and γ is coefficients.

Using a physical temperature of 268 K as an example, Figure 11 shows the measured and fitted emissivity values as a function of soil thickness at 6.925, 10.65, 18.7, and 36.5 GHz. The initial soil moisture is 0.433 cm^3/cm^3. As the soil thickness increases, the stable values of emissivity at lower frequencies lags behind that at higher frequencies because lower frequencies can penetrate greater soil depths. Moreover, there is little difference between the freezing (soil temperature varying from 268+ to 268− K) and thawing processes (from 268− to 268+ K) of soil, as noted above.

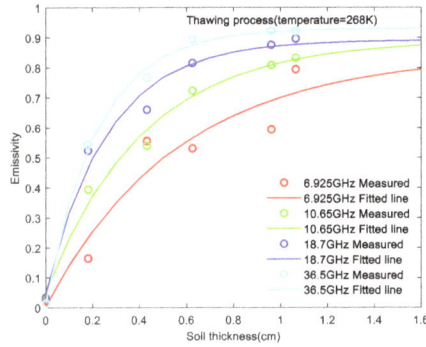

(a)

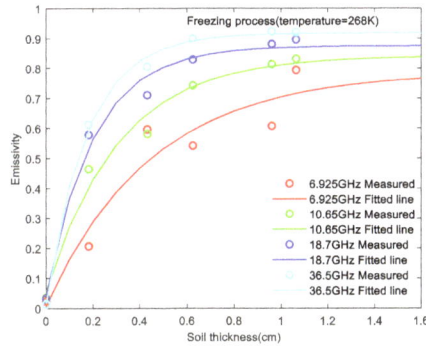

(b)

Figure 11. Measured and fitted relationships between soil thickness and emissivity at four frequencies during the thawing process (**a**) and freezing process (**b**).

Using the fitted curve obtained to describe the relationship between the emissivity and soil thickness, it is feasible to estimate the MRRD. The MRRD can be calculated based on the fitted curve for each frequency at a given physical temperature.

The parameterized model was evaluated based on the radiometer measurements from the field experiment. The emissivity values were selected based on a single physical temperature, thereby guaranteeing a consistent soil condition at different soil thicknesses. Thus, the soil thickness is the only variable affecting the observed emissivity for a given frequency. The soil physical temperatures used in this study are 268 K, 269 K, 270 K, 271 K, and 272 K. The MRRD was calculated based on the previously discussed methods at each physical temperature and all four TRMM frequencies (6.925, 10.65, 18.7 and 36.5 GHz). The MRRD can also be estimated using the parameterized model. Because the MRRD from the model and experiment were derived with different incident angles, they must be modified to the appropriate depth at 90° using the real angle of transmission [13]. The estimated MRRD is plotted with the measured values in Figure 12. A good agreement can be seen between the two parameters, except for several points at a low frequency (6.925 GHz), which display a higher MRRD. Our previous study [27] focused on the coherent effects of freezing soil, which may be significant in ground radiometric experiments, especially when using a low-frequency radiometer. This effect could be attributed to the interference caused by frozen layer thickness variations, which results in brightness temperature oscillations (Figure 11). However, the theoretical model used in this study is a non-coherent model, which does not include interference. Furthermore, due to the time limitation and avoiding a heterogeneous frozen soil sample in this experiment, the thickest soil sample was only 1.06 cm. It limited the accuracy of the fitted curves, especially the measurements at lower frequencies, i.e. 6.925 GHz. The fitting curve introduced errors to results of measured MRRD at 6.925 GHz and resulted in disagreement of measured and estimated MRRDs. Therefore, the estimated points at 6.925 GHz do not agree well with the measured MRRD.

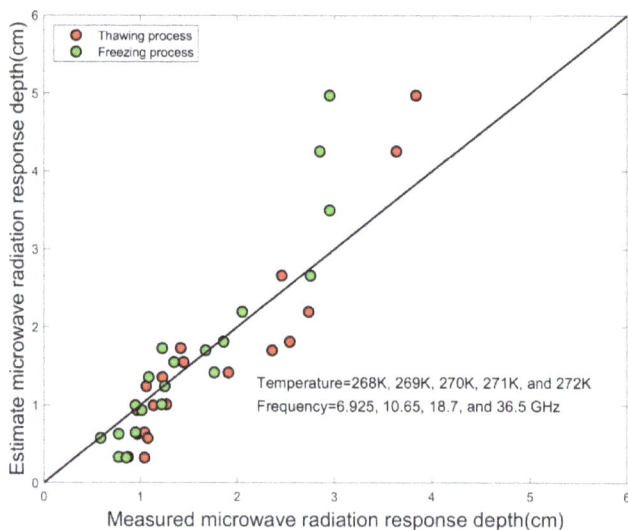

Figure 12. Comparison of measured and estimated MRRDs using the parameterized model.

The averaged root mean square errors (RMSE) of the soil at the four frequencies are 0.402 cm and 0.537 cm for the thawing and freezing processes, respectively. Note that the validation conducted based on controlled experimental data (i.e., specific soil moistures, textures, temperatures and other parameters). In addition, the parameterized model accuracy is difficult to address based on real soil data. However, the parameterized model was derived based on the radiative transfer theory and the assumptions are both reasonable and consistent with previous research. Thus, there is reason to believe that the developed parameterized model offers an acceptable and simple way for estimating the MRRD using three common parameters: temperature, frequency, and specific surface area.

4. Discussion

In this study, the theoretical model was used to describe the soil microwave radiation in the soil freezing and thawing process. It was then used to develop the parameterized model for estimating the MRRD by using multiple regression analysis based on a simulation database. However, the developed parameterized model has its limitations because the diffuse scattering is not taken into consideration. At low microwave frequencies, the wavelength in soil is in the order of several centimeters, which is significantly larger than the soil particles. At high frequencies, which correspond to shorter wavelengths, the soil particles and voids between particles may cause scattering. For the frozen soil, their emission characteristics are similar to very dry soil, due to an absence of liquid water. Thus, scattering is possible, especially at higher frequencies.

However, there are few published models to describe the complicated diffuse scattering of soil. Herein, the diffuse scattering of soil was ignored in our work and the effect of soil diffuse scattering on MRRD estimation was evaluated by analyzing MRRD at different frequencies as a function of the single scattering albedo (Figure 13). It has shown that the single scattering albedo varying from 0.01 to 0.09 resulted in about 0.43, 0.46, 0.56, and 0.75 cm variation of MRRD at 6.925 GHz, 10.65 GHz, 18.7 GHz, and 36.5 GHz, respectively. This finding suggests that the omission of diffused albedo can lead to a slight MRRD overestimation.

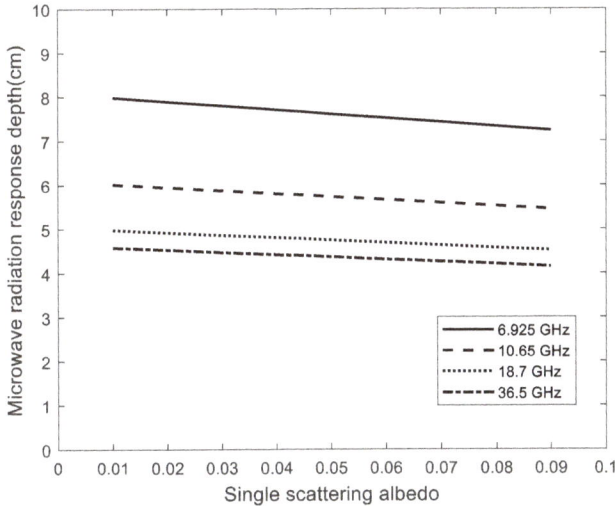

Figure 13. MRRD as a function of single scattering albedo at different frequencies.

The assumption of equal soil moisture and soil temperature throughout the soil is not strictly true. Temperature and moisture gradients exist within the frozen soil, especially when the soil depth is very thick and the soil temperature is approximately 0 °C. Because the soil sample on the aluminum sheet is not very thick and the data used in this study were below −1 °C, the assumption of uniform soil properties with depth is reasonable.

Another explanation for the disagreement between the measured and estimated MRRDs is the difference of aluminium sheet emissivity used in the theoretical model and measurements. Theoretically, aluminium is an ideal conductor and the waves transmitted to a smooth aluminium sheet will be totally reflected [34]. Therefore, the emissivity of the aluminium sheets should be zero at any observation angle or frequency. In the theoretical model, the emissivity of aluminium were set to be zero. Obviously, they are about 0.01 ~ 0.03, which may be caused by the surface oxidation of the aluminium sheets in the experiment (as shown in Figure 11, when the soil thickness is zero, the measured emissivity is not zero). These results were also found in our other field experiments [35].

5. Conclusions

The objective of this work was to develop a simple, accurate approach for calculating the MRRD of frozen soil. The concept of the MRRD was proposed to describe the source of the primary signal of passive microwave remote sensing and provide a specific expression to define the depth. A controlled truck-mounted experiment was conducted using a three-layer setup. A three-layer theoretical model was then introduced and altered to satisfy the specific experimental design of the study. A database was constructed and sensitivity analysis was conducted based on this theoretical model to study the

Remote Sens. **2019**, *11*, 2028

relationship among the MRRD, soil properties, and frequency. According to the sensitivity analysis, the sensitivity of the MRRD to soil bulk density is much less than that of the three main variables temperature, frequency, and specific surface area for the frozen soil. A parameterized model was then developed for estimating the MRRD of frozen soil using three common and easily acquired parameters: temperature, frequency, and specific surface area. Finally, the parameterized model was validated using experimental data. Some interesting conclusions can be made regarding the MRRD of frozen soil. The soil temperature, frequency, and soil texture are the three main variables affecting the MRRD. The MRRD is negatively correlated with soil temperature, frequency, and specific surface area. Additionally, the MRRD is more sensitive to temperature and frequency than specific surface area. Based on the sensitivity analysis, a parameterized model was developed for estimating the MRRD, and the validation results indicated that the parameterized model has an acceptable accuracy. The RMSEs of MRRD at TMMR's four frequencies were 0.402 cm and 0.537 cm for the thawing and freezing soil processes, respectively. However, because the validation was performed based on controlled experimental data (i.e., specific soil moistures, textures, temperature variations and other parameters), the parameterized model should be further tested in future studies prior to application. This testing could include field data collection and brightness temperature simulation using a radiative transfer model.

Author Contributions: L.J. conceived and designed the study; T.Z. analyzed the data and wrote the paper; S.Z. and Y.P. collected the experiment data; and L.C. and Y.L. contributed analytical tools and methods.

Funding: This work was funded by the National Natural Science Foundation of China under Grant 41801287, 41671334, and 41871228.

Acknowledgments: The authors would like to thank the anonymous reviewers for their valuable comments.

Conflicts of Interest: The authors declare no conflict of interest.

References

1. Kurganova, I.; Teepe, R.; Loftfield, N. Influence of freeze-thaw events on carbon dioxide emission from soils at different moisture and land use. *Carbon Balance Manag.* **2007**, *2*, 2. [CrossRef] [PubMed]
2. Yang, Z.; Ou, Y.H.; Xu, X.; Zhao, L.; Song, M.; Zhou, C. Effects of permafrost degradation on ecosystems. *Acta Ecologica Sinica* **2010**, *30*, 33–39. [CrossRef]
3. Yang, J.; Zhou, W.; Liu, J.; Hu, X. Dynamics of greenhouse gas formation in relation to freeze/thaw soil depth in a flooded peat marsh of Northeast China. *Soil Biol. Biochem.* **2014**, *75*, 202–210. [CrossRef]
4. Zuerndorfer, B.; England, A.W. Radiobrightness decision criteria for freeze/thaw boundaries. *IEEE Trans. Geosci. Remote Sens.* **1992**, *30*, 89–102. [CrossRef]
5. Judge, J.; Galantowicz, J.F.; England, A.W.; Dahl, P. Freeze/thaw classification for prairie soils using SSM/I radiobrightnesses. *IEEE Trans. Geosci. Remote Sens.* **1997**, *35*, 827–832. [CrossRef]
6. Zhang, T.; Armstrong, R.L.; Smith, J. Investigation of the near-surface soil freeze-thaw cycle in the contiguous United States: Algorithm development and validation. *J. Geophys. Res.-Atmos.* **2003**, *108*, 8860. [CrossRef]
7. Zhao, T.; Zhang, L.; Jiang, L.; Zhao, S.; Chai, L.; Jin, R. A new soil freeze/thaw discriminant algorithm using AMSR-E passive microwave imagery. *Hydrol. Process.* **2011**, *25*, 1704–1716. [CrossRef]
8. Du, J.; Kimball, J.S.; Azarderakhsh, M.; Dunbar, R.S.; Moghaddam, M.; McDonald, K.C. Classification of Alaska Spring Thaw Characteristics Using Satellite L-Band Radar Remote Sensing. *IEEE Trans. Geosci. Remote Sens.* **2015**, *53*, 542–556.
9. Zhang, L.; Zhao, T.; Jiang, L.; Zhao, S. Estimate of Phase Transition Water Content in Freeze-Thaw Process Using Microwave Radiometer. *IEEE Trans. Geosci. Remote Sens.* **2010**, *48*, 4248–4255. [CrossRef]
10. Schwank, M.; Stahli, M.; Wydler, H.; Leuenberger, J.; Matzler, C.; Fluhler, H. Microwave L-band emission of freezing soil. *IEEE Trans. Geosci. Remote Sens.* **2004**, *42*, 1252–1261. [CrossRef]
11. Wigneron, J.; Chanzy, A.; de Rosnay, P.; Rudiger, C.; Calvet, J. Estimating the Effective Soil Temperature at L-Band as a Function of Soil Properties. *IEEE Trans. Geosci. Remote Sens.* **2008**, *46*, 797–807. [CrossRef]
12. Escorihuela, M.J.; Chanzy, A.; Wigneron, J.P.; Kerr, Y.H. Effective soil moisture sampling depth of L-band radiometry: A case study. *Remote Sens. Environ.* **2010**, *114*, 995–1001. [CrossRef]

13. Ulaby, F.T.; Moore, R.K.; Fung, A.K. *Microwave Remote Sensing: Active and Passive. Volume 1—Microwave Remote Sensing Fundamentals and Radiometry*; Remote Sensing A; Addison-Wesley Publishing Company: Boston, MA, USA, 1981; Volume 1, ISBN 0-201-10759-7.

14. Zhou, F.; Song, X.; Leng, P.; Li, Z. An Effective Emission Depth Model for Passive Microwave Remote Sensing. *IEEE J. Sel. Top. Appl. Earth Observ. Remote Sens.* **2016**, *9*, 1752–1760. [CrossRef]

15. Wilheit, T.T. Radiative Transfer in a Plane Stratified Dielectric. *IEEE Trans. Geosci. Remote Sens.* **1978**, *16*, 138–143. [CrossRef]

16. Wang, J.R. Microwave emission from smooth bare fields and soil moisture sampling depth. *IEEE Trans. Geosci. Remote Sens.* **1987**, *25*, 616–622. [CrossRef]

17. Blinn, J.C.; Conel, J.E.; Quade, J.G. Microwave emission from geological materials: Observations of interference effects. *J. Geophys. Res.* **1972**, *77*, 4366–4378. [CrossRef]

18. Paloscia, S.; Pampaloni, P.; Chiarantini, L.; Coppo, P.; Gagliani, S.; Luzi, G. Multifrequency passive microwve remote sensing of soil moisture and roughness. *Int. J. Remote Sens.* **1993**, *14*, 467–483. [CrossRef]

19. Newton, R.W.; Black, Q.R.; Makanvand, S.; Blanchard, A.J.; Jean, B.R. Soil Moisture Information and Thermal Microwave Emission. *IEEE Trans. Geosci. Remote Sens.* **1982**, *20*, 275–281. [CrossRef]

20. Owe, M.; Van de Griend, A.A. Comparison of soil moisture penetration depths for several bare soils at two microwave frequencies and implications for remote sensing. *Water Resour. Res.* **1998**, *34*, 2319–2327. [CrossRef]

21. Laymon, C.A.; Crosson, W.L.; Jackson, T.J.; Manu, A.; Tsegaye, T.D. Ground-based passive microwave remote sensing observations of soil moisture at S-band and L-band with insight into measurement accuracy. *IEEE Trans. Geosci. Remote Sens.* **2001**, *39*, 1844–1858. [CrossRef]

22. Pampaloni, P.; Paloscia, S.; Chiarantini, L.; Coppo, P.; Gagliani, S.; Luzi, G. Sampling depth of soil moisture content by radiometric measurement at 21 cm wavelength: Some experimental results. *Int. J. Remote Sens.* **1990**, *11*, 1085–1092. [CrossRef]

23. Jackson, T.J.; Bindlish, R.; Cosh, M.H.; Zhao, T.; Starks, P.J.; Bosch, D.D.; Seyfried, M.; Moran, M.S.; Goodrich, D.C.; Kerr, Y.H.; et al. Validation of Soil Moisture and Ocean Salinity (SMOS) Soil Moisture Over Watershed Networks in the U.S. *IEEE Trans. Geosci. Remote Sens.* **2012**, *50*, 1530–1543. [CrossRef]

24. Al Bitar, A.; Leroux, D.; Kerr, Y.H.; Merlin, O.; Richaume, P.; Sahoo, A.; Wood, E.F. Evaluation of SMOS Soil Moisture Products Over Continental US Using the SCAN/SNOTEL Network. *IEEE Trans. Geosci. Remote Sens.* **2012**, *50*, 1572–1586. [CrossRef]

25. Dall'Amico, J.T.; Schlenz, F.; Loew, A.; Mauser, W. First Results of SMOS Soil Moisture Validation in the Upper Danube Catchment. *IEEE Trans. Geosci. Remote Sens.* **2012**, *50*, 1507–1516. [CrossRef]

26. Zhang, L.; Shi, J.; Zhang, Z.J.; Zhao, K.G. The Estimation of Dielectric Constant of Frozen Soil-Water Mixture at Microwave Bands. In Proceedings of the 2003 IEEE International Geoscience and Remote Sensing Symposium (IGARSS 2003), Toulouse, France, 21–25 July 2003; Volume 4, pp. 2903–2905.

27. Zhao, S.; Zhang, L.; Zhang, Y.; Jiang, L. Microwave emission of soil freezing and thawing observed by a truck-mounted microwave radiometer. *Int. J. Remote Sens.* **2012**, *33*, 860–871. [CrossRef]

28. Zheng, D.; Wang, X.; Rogier, V.; Zeng, Y.; Wen, J.; Wang, Z.; Mike, S.; Paolo, F.; Su, B. L-Band Microwave Emission of Soil Freeze-Thaw Process in the Third Pole Environment. *IEEE Trans. Geosci. Remote Sens.* **2017**, *9*, 5324–5338. [CrossRef]

29. Zhao, S.; Zhang, L.; Zhang, Z. Design and Test of a New Truck-Mounted Microwave Radiometer for Remote Sensing Research. In Proceedings of the 2008 IEEE International Geoscience and Remote Sensing Symposium (IGARSS 2008), Boston, MA, USA, 7–11 July 2008; Volume 2, pp. 1192–1195.

30. Dobson, M.C.; Ulaby, F.T.; Hallikainen, M.T.; El-rayes, M.A. Microwave Dielectric Behavior of Wet Soil-Part II: Dielectric Mixing Models. *IEEE Trans. Geosci. Remote Sens.* **1985**, *GE-23*, 35–46. [CrossRef]

31. Anderson, D.M.; Tice, A.R. Predicting unfrozen water contents in frozen soils from surface area measurements. *Highw. Res. Rec.* **1972**, 12–18.

32. Ersahin, S.; Gunal, H.; Kutlu, T.; Yetgin, B.; Coban, S. Estimating specific surface area and cation exchange capacity in soils using fractal dimension of particle-size distribution. *GEODERMA* **2006**, *136*, 588–597. [CrossRef]

33. Nedeltchev, N.M. Thermal microwave emission depth and soil moisture remote sensing. *Int. J. Remote Sens.* **1999**, *20*, 2183–2194. [CrossRef]

34. Guru, B.S.; Hiziroglu, H.R. *Electromagnetic Field Theory Fundamentals*; Cambridge University Press: Cambridge, UK, 2004; ISBN 0-521-83016-8.
35. Chai, L.; Zhang, L.; Shi, J.C.; Wu, F. Equivalent scattering albedo estimation of cotton and soybean. *J. Remote Sens.* **2013**, *17*, 17–33.

remote sensing

MDPI

Letter

Mapping High Mountain Lakes Using Space-Borne Near-Nadir SAR Observations

Shengyang Li *, Hong Tan, Zhiwen Liu, Zhuang Zhou, Yunfei Liu, Wanfeng Zhang, Kang Liu and Bangyong Qin

Key Laboratory of Space Utilization, Technology and Engineering Center for Space Utilization, Chinese Academy of Sciences, Beijing 100094, China; tanhong@csu.ac.cn (H.T.); zwliu@csu.ac.cn (Z.L.); zhouzhuang@csu.ac.cn (Z.Z.); liuyunfei@csu.ac.cn (Y.L.); wfzhang@csu.ac.cn (W.Z.); liukang@csu.ac.cn (K.L.); qinby@csu.ac.cn (B.Q.)
* Correspondence: shyli@csu.ac.cn; Tel.: +86-10-82178203

Received: 10 August 2018; Accepted: 5 September 2018; Published: 6 September 2018

Abstract: Near-nadir interferometric imaging SAR (Synthetic Aperture Radar) techniques are promising in measuring global water extent and surface height at fine spatial and temporal resolutions. The concept of near-nadir interferometric measurements was implemented in the experimental Interferometric Imaging Radar Altimeters (InIRA) mounted on Chinese Tian Gong 2 (TG-2) space laboratory. This study is focused on mapping the extent of high mountain lakes in the remote Qinghai–Tibet Plateau (QTP) areas using the InIRA observations. Theoretical simulations were first conducted to understand the scattering mechanisms under near-nadir observation geometry. It was found that water and surrounding land pixels are generally distinguishable depending on the degree of their difference in dielectric properties and surface roughness. The observed radar backscatter is also greatly influenced by incidence angles. A dynamic threshold method was then developed to detect water pixels based on the theoretical analysis and ancillary data. As assessed by the LandSat results, the overall classification accuracy is higher than 90%, though the classifications are affected by low backscatter possibly from very smooth water surface. The algorithms developed from this study can be extended to all InIRA land measurements and provide support for the similar space missions in the future.

Keywords: near-nadir SAR; Tian Gong 2; Qinghai–Tibet Plateau; lake

1. Introduction

Lakes are essential components in global hydraulic cycle and climate processes. The expansion and shrinkage of lake extent are strongly influenced by seasonal climate patterns as well as long term environmental changes [1,2]. Lakes provide strong feedbacks to regional and global environments, and the emergence and expansion of thaw-lakes are found having profound impacts on high-latitude ecosystems [3]. There is high priority to monitor the dynamics of lake area and water storage for assessing global change impacts and forecasting the future climate change scenarios. The areas and levels of high mountain lakes in the Qinghai–Tibet Plateau (QTP) are particularly important for monitoring the impacts of glacier melting [4], mitigating the hazards from glacier lake outbursts [5], and detecting the climate pattern changes [6] due to the region's high sensitivity and vulnerability to climate changes [7–9].

A large number of lakes are distributed in the QTP whose lake areas are more than half of the total lake area of China [1]. The unique high mountain lakes in QTP also represents one of the largest lake systems in the world [10]. However, most of the lakes in the region are located in remote areas, which make it difficult for human surveying of the lake physical and chemical parameters. Remote sensing has been successfully used for mapping global water bodies. A variety of studies applying optical remote sensing have been conducted for detecting land surface water over the QTP and the globe [11–16]. Among these studies, the research described in [16] used satellite optical images from the China–Brazil Earth Resources Satellite (CBERS) and the National Aeronautics and Space Administration (NASA)/United States Geological Survey (USGS) Landsat to provide detailed mapping of the boundaries of natural lakes in China including the QTP region. However, the CBERS and LandSat images or similar optical observations only work well for clear-sky conditions, and this restriction normally leads to retrievals with coarse temporal resolutions [17]. Satellite passive microwave radiometer is capable of all-sky sensing of land surface and has been used for deriving global surface water products. The products derived from the Special Sensor Microwave Imagers (SSMI), the Advanced Microwave Scanning Radiometer on the Earth Observing System (AMSR-E), and the Soil Moisture Active Passive (SMAP) are sensitive to short-term changes of water surface but are at resolutions about 25 km [18–20]. These products therefore cannot be used for mapping small lakes. In contrast, the next generation near-nadir interferometric imaging SAR (Synthetic Aperture Radar) such as the NASA Surface Water Ocean Topography (SWOT) mission is promising for high spatial and temporal mapping of lakes including those smaller than 1 km^2 (SWOT) [21–24]. The main payloads of SWOT include two Ka-band Radar Interferometers designed for near-nadir SAR imaging and interferometric measurement [22]. SWOT is scheduled for launch in 2021 and its potential for accurate inundation mapping has been demonstrated by SWOT airborne simulator in the field campaign [24]. Similar to SWOT in the designing concept, a pair of near-nadir Interferometric Imaging Radar Altimeters (InIRA) mounted on Chinese unmanned space laboratory Tian Gong 2 (TG-2) have collected Ku-band SAR observations over the globe since 2016. The InIRA provides a unique opportunity to evaluate the near-nadir imaging techniques in measuring water cycle components from space. In this study, we developed an algorithm to map lake bodies in Northern QTP based on TG-2 InIRA observations and theoretical simulations. The derived lake body maps were evaluated by LandSat results. Detailed descriptions of the study region and data set are in the following Sections 2.1 and 2.2, theory and algorithms are described in Sections 2.3 and 2.4, and the results are presented in Section 3 and discussed in Section 4.

2. Materials and Methods

2.1. Study Region

The study region (within the red rectangles of Figure 1) was located within Nagqu prefecture of Tibet Autonomous Region, China, covering about 3600 km^2 area. The natural environment of the region is cold and dry, typical of that of the Northern QTP where elevations are normally higher than 4000 m and annual precipitation is as low as 247.3 mm [1,25]. Major lakes within the study areas include Que'er Caka, Khongnam Tso, Dorosidong Co, Maqiao Co., and Chibzhang Co. Most of the lakes in the regions are saline lakes [26,27]. Due to the scarce of precipitation, glaciers of the region provide important water supplies to many lakes; the lake area changes are affected by factors such as glacier retreat, permafrost degradation, and climate pattern changes [1,26,27]. The two largest lakes of the region are Dorsoidong Co (center latitude 33.41°, longitude 89.88°) and Chibzhang Co (center latitude 33.47°, longitude 90.34°), both of which are glacier-fed lakes undergoing expansions and interlinked with other in recent years [26].

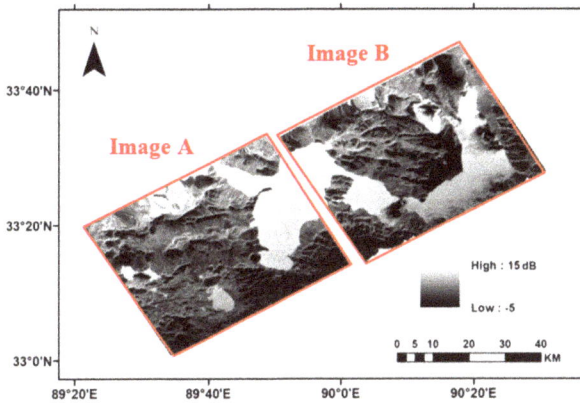

Figure 1. The InIRA observations over Nagqu area of Qinghai–Tibet Plateau (QTP).

2.2. Instrument and Data Set

The InIRA mounted on TG-2 space laboratory was developed by the China Manned Space Engineering Project and has been operational since 15 September 2016. The TG-2 space laboratory is served as a test bed for scientific research and new technologies, and carries more than 50 scientific instruments. There are also more than 10 science and application space experiments conducted in TG-2 in the fields of earth science, astronomy, microgravity physics, microgravity fluid physics, space life science, space environment, and space physics. The InIRA is an interferometric synthetic aperture radar (InSAR) system with near-nadir imaging capacity at Ku-band. As the first space-borne interferometric radar altimeters, the experimental InIRA system is designed for evaluating the near-nadir imaging and interferometric techniques in measuring sea surface height, water body extent and land elevation. The detailed parameters of the instrument for land applications are listed in Table 1.

Table 1. Parameters of Tian Gong 2 Interferometric Imaging Radar Altimeter.

Altitude	Frequency	Incidence Angles	Spatial Resolution
400 km	13.58 GHz	2.5–7.5°	40 m/200 m
Swath Width	**Look Direction**	**Baseline length**	**Cover area**
35 km	Right	2.3 m	±42 degrees of latitudes

For this study, the VV-polarized SAR backscatter images (Figure 1) acquired by InIRA of the study region on 23 September 2016 were used for mapping lake extent. The two images (Images A and B of Figure 1) sequentially acquired by InIRA were Level 2 scientific data with radiometric and geometric calibrations conducted and downloaded from Space Application Data Promoting Service Platform for China Manned Space Engineering (http://www.msadc.cn). As seen in Figure 1, the backscattering coefficients range from −5 to 15 dB and the overall bright lake areas are generally distinguishable from the surrounding darker land areas. However, geometric distortions are found over the northern edges of the images. The layover and shadow regions which are typical to the SAR observations over mountainous areas are also shown in the two images.

Two ancillary data sets were used in this study for algorithm development. The data sets include the official elevation product of InIRA and LandSat-based water occurrence dataset (WOD) [15]. The official Level 3 elevation products of InIRA (Figure 2) derived using signals with high interferometric coherence were obtained from http://www.msadc.cn. The elevation data are used to calculate terrain slopes for identifying hilly areas. Since no valid data were available over areas with strong geometric distortions, the elevation product also serves as a reference to mask out the non-retrievable pixels. The second ancillary data set WOD was generated based on 30 m Landsat images from 1984 to 2015 [15]. The WOD was produced by analyzing the Landsat 5, 7, and 8 archives. The data of WOD range from 0% (always land) to 100% (permanent water) and represent the global surface water persistence over more than three decades. For this study, about 120 pixels with water occurrence >95% were randomly selected for each InIRA image. These pixels are highly likely to be part of a permanent water body and are used for establishing the training data sets for InIRA water detection.

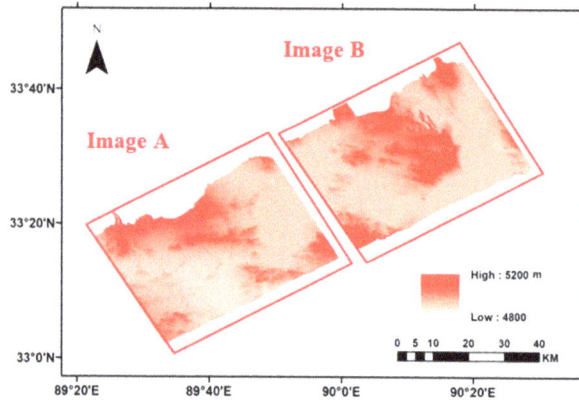

Figure 2. The elevation of the study region estimated by InIRA.

We also chose LandSat Operational Land Imager (OLI) images as the ancillary data for validation purposes. The LandSat/OLI image used for this study was acquired on 9 September 2016 over the study region, and downloaded from USGS website (https://landsat.usgs.gov/). The selected LandSat image has relatively less cloud coverage and is closest to the InIRA observations in acquisition date, though not exactly the same. For validating the InIRA water retrievals, the Landsat/OLI image was processed by Fmask algorithm [28] for generating 30-m water mask, which was then interpolated to 40-m resolution image using the nearest neighbor method. The Fmask algorithm was designed to identify cloud, cloud shadow, snow, land, and water pixels using LandSat or Sentinel 2 images, and was recently improved for its use over mountainous areas [29].

2.3. Theoretical Simulations

Similar to the conventional SAR applications, the near-nadir SAR studies requires the understanding of the interactions between microwave and land surface features through theoretical models, field experiments, or both. Radar backscattering can be simulated through numerical simulations based on rigid electromagnetic theories or their simplified forms. The Integral Equation Method (IEM) developed in the literature [30–32] has a relatively simple algebraic form with physically justified assumptions while retains high accuracy under a wide range of surface roughness and dielectric conditions. The IEM model bridges the validity gaps between the traditional small perturbation (SPM) and geometrical optics (GO) models [30,31], and has been successfully applied to microwave remote sensing including forward model simulations and inversion algorithm development [33–36].

For investigating the possibility of identifying water and land through InIRA observations, we simulated the InIRA signals under a set of soil and water conditions using IEM model. Typical input parameters used by IEM for describing surface roughness include root mean square (RMS) height, correlation length, and correlation function. The model inputs for this study are listed in Table 2. The dielectric properties of soil are calculated by the Dobson model [37]. Considering little vegetation presence in the study region, only bare soil conditions were simulated. The QTP lakes are brackish or saline with salinity ranging from <3‰ to hypersaline conditions [38]. For simulating the backscattering from saline lake surfaces, we assumed the salinity 35.0‰, which is the average salinity level of sea water. The dielectric properties of salty water are calculated by the Stogryn dielectric model [39].

Table 2. Input parameters of theoretical model.

Frequency	Incidence Angles	RMS Height	Correlation Length	Soil Moisture	Correlation Function
13.58 GHz	2, 5, and 8°	0.125–3 cm interval 0.125 cm	10 cm	10–40% interval 15% (cm^3/cm^3)	Exponential function

2.4. SAR Water Detection Method

We developed a dynamic threshold algorithm for InIRA aiming at detecting water bodies from bare or sparsely vegetated soil of the focused region using limited pixels selected from WOD as prior knowledge. The first step is to select pixels representing permanent water within a InIRA image. The water pixels are randomly selected from the WOD data whose historical water occurrence is higher than 95%, meaning likely part of a permanent water body. Similarly, the same number of land pixels is determined using the WOD data if 0% water occurrence is documented. The backscattering coefficients of the selected land and water pixels are then obtained from the InIRA images and constitute a training data set. For this study, the training data set for each InIRA image contains about 120 water pixels and 120 land pixels. The portion of training pixels over the number of total retrievable pixels is very small (less than 0.1%) but provide important reference information of land and water scattering characteristics at different incidence angles and across the study regions since the training points were randomly selected and distributed over the image. As the second step, the terrain slope derived from Tian Gong 2 elevation products was calculated for each pixel and used for additional constraints to the water and land classifications. For a given pixel, its slope is determined by the rates of elevation changes in both X and Y directions of an InIRA image. To minimize possible impacts on the classifications caused by geometry distortions detected on the image edges and mountainous regions, we use TG-2 elevation product to mask out the pixels without valid elevation retrievals. The last step is to loop through each pixel and determine its land or water attribute. A pixel is assigned as water if its observed backscattering coefficient is closer to the value of nearest water pixel than that of the nearest land pixel pre-defined in the training data set, and its terrain slope is lower than 5°, indicating a relatively flat area.

The method was applied to the InIRA images and evaluated by the Fmask results derived from LandSat OLI observations. The classification results are assessed by the following metrics including overall accuracy, producer's accuracy, and user's accuracy. The overall accuracy is the ratio between the number of correctly classified pixels and the total number of pixels. The producer's accuracy represents the proportion of reference land or water features being correctly classified. The user's accuracy is referred to classification reliability and represents the proportion of classified water or land pixels being consistent with the reference pixels [40].

3. Results

3.1. Theoretical Simulations

The simulation results (Figure 3) show the backscatter patterns of soil and water over different surface roughness parameters, dielectric properties, and incidence angles. The water backscattering under small incidence angle conditions is typically stronger than that of soil for a given surface roughness. This is caused by the higher dielectric value of water than those of soil at normal wetness levels (≤ 0.4 cm^3/cm^3). Comparing with the dielectric properties, the surface roughness is a more dominant factor that determines the backscatter magnitudes. For very smooth surface (e.g., still water), the backscatter signals from soil and water are generally low and similar to each other (e.g., Figure 3B when RMS height less than 0.25 cm) since most of the scattered signals are distributed in a narrow specular direction. For smooth surface (e.g., RMS height 0.5 cm) which can be seen in water bodies with wind-driven ripples and waves, and natural (e.g., large rocks) or artificial land features (e.g., airplane runway), the near-nadir backscatter is generally very strong due to increased diffuse scattering into the observed direction. The diffuse scattering tends to be more uniform in all directions as the surface roughness further increases (e.g., RMS height 3.0 cm), therefore the intensity of the received signals at near-nadir directions decrease. In summary, the surface roughness difference between normally smooth water surface and relatively rough soil surface is the key to distinguish the two features in InIRA images. For evaluating the impacts of incidence angles, the simulations were conducted for 2, 5, and 8°, which cover the incidence angle range of InIRA. As illustrated by Figure 3A–C, the pattern of backscattering coefficients changing with surface roughness is significantly affected by the sensor viewing geometry. The backscatter peaks earlier as the RMS height increases but then drops faster for smaller incidence angles. For larger incidence angles, the magnitudes of backscattering coefficients from soil and water surfaces with RMS height smaller than 0.5 cm (Figure 3C) appear to be similarly small.

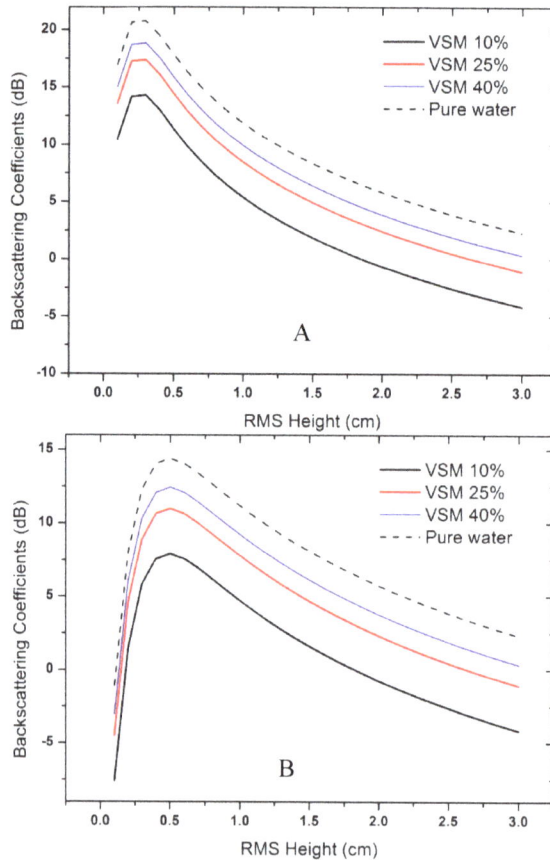

Figure 3. *Cont.*

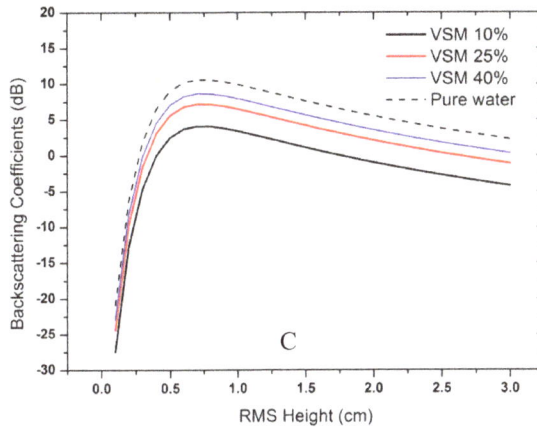

Figure 3. Radar backscattering coefficients changes with surface roughness for incidence angle (**A**) 2°; (**B**) 5° and (**C**) 8°. VSM, volumetric soil moisture.

The above simulations suggest: (a) water and surrounding soil can be generally separated through their differences in backscatter intensity except for the very calm/rough water or very smooth soil. Prior knowledge of the land feature information may help avoid the misclassifications of smooth water and land; (b) the backscatter intensity of water and soil varies with their roughness and dielectric values; and (c) the incidence angle increases from InIRA near-range to far-range, and different incidence angles significantly affects the radar backscatter observations. Therefore, for the applications over a large region with a variety of land surface and incidence angle conditions, a dynamic threshold classification method is needed.

3.2. Land and Water Classifications

The dynamic threshold algorithm was applied to InIRA images (Figure 1) for the regions with valid TG-2 elevation data (Figure 2). The mean threshold values are 9.8 dB and 7.1 dB for Images A and B, respectively. The classified lake water result (Figure 4) was compared to the LandSat/OLI water classifications by Fmask method (Figure 5). The detailed error matrices were listed in Tables 3 and 4 for the respective Images A and B. The Fmask is proved accurate in detecting land features; however, its water and land classifications are also affected by the presence of cloud and its shadow. The comparisons with Fmask result were only made when land and water classifications are available. As such, the total pixel number of Image B (Table 4) is smaller than that of Image A (Table 3) due to a larger portion of cloud coverage over the Image B area in the LandSat/OLI observations.

Table 3. Error matrix for the classification results of Image A.

	OLI Land	OLI Water	Total
InIRA Land	728,795	11,780	740,575
InIRA Water	24,822	179,899	204,721
Total	753,617	191,679	945,296

Table 4. Error matrix for the classification results of Image B.

	OLI Land	OLI Water	Total
InIRA Land	439,097	51,689	490,786
InIRA Water	26,354	270,300	296,654
Total	465,451	321,989	787,440

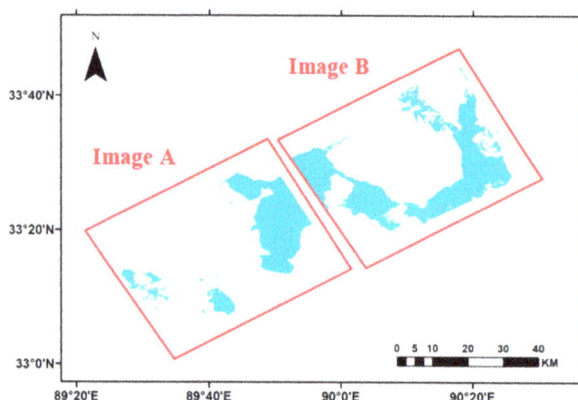

Figure 4. Lake water detected by Tian Gong 2 InIRA images acquired on 23 September 2016 using a dynamic threshold algorithm (blue: water; white: land or no retrievals).

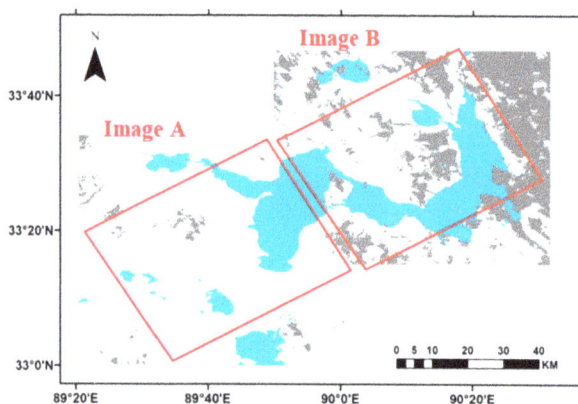

Figure 5. Lake water detected using Landsat OLI image acquired on 9 September 2016 (blue: water; grey: cloud; white: land or no retrievals).

The accuracy metrics were calculated from the error matrices. For both images, most water and land pixels as shown in the reference LandSat/OLI results are correctly detected by InIRA with the overall classification accuracy 96.13% for image A and 90.09% for Image B. For land pixels in Image A, the producer's accuracy is 96.71% and user's accuracy 98.41%; for water pixels, the producer's accuracy is 93.85% and user's accuracy 87.88%. For land pixels in Image B, the producer's accuracy is 94.34% and user's accuracy is 89.47%; for water pixels, the producer's accuracy is 83.95% and user's accuracy is 91.12%. Though the overall accuracy is higher than 90%, the producer's accuracy for water classifications using Image B decreases for about 10% as compared with that of Image A, suggesting increased misclassifications of the referenced LandSat/OLI water pixels. In Image B, the major water areas not detected by InIRA were coincident in locations with the dark regions of InIRA backscattering coefficient image (Figure 1). As analyzed by theoretical simulations (Section 3.1), low backscatter from water surface is possibly caused by very smooth water surface when there is no/little wind.

4. Discussion

The future NASA SWOT mission is promising in estimating global water budget at high accuracy, fine spatial and favorable temporal resolutions. The near-nadir SAR imaging and interferometric measuring techniques adopted by SWOT represent a novel way to realize the goal of global water measurements. The experimental InIRA mounted on TG-2 space laboratory has provided unique Ku-band observations since 2016 and these data contain rich information that needs to be interpreted for understanding the advantages and drawbacks of near-nadir SAR imaging techniques for water cycle observations. For this study, we applied the images to the classifications of high mountain waters with the supports of theoretical simulations. Both the theoretical predictions and actual SAR images prove the potential of using SWOT-like instruments in fine-scale detection of water bodies. Compared with LandSat/OLI (Figure 5), the InIRA Ku-band observations (Figure 1) and classifications (Figure 4) are less affected by the atmospheric conditions and have similarly high classification accuracy.

The main issue of the near-nadir imaging technique in water body detection is its strong dependence on surface roughness. In extreme cases when the water body is perfectly still, low backscatter from water surface or dark-water observations can be confused with soil signals. Referenced data sets from other sensors or InIRA time-series observations may help mitigate the problem. Similar issue was independently confirmed by airborne observations from AirSWOT measurements, though the flat surface problem can be partially overcome by increasing the observation frequency (e.g., to the SWOT Ka-band) [21,24]. Considering limited near-nadir microwave observations from satellites, the studies on InIRA may help to mitigate the dark-water issue for future SWOT mission. We also noticed the geometric distortions in InIRA images which need to be corrected by improved data processing. Despite the above issues, the contrasting land and water backscattering behaviors were observed by the space-borne InIRA and utilized in high accurate water body mapping at 40-m resolution by the dynamic threshold method. The findings of this work help to improve the designing of instruments, data processing flow, and algorithm performance in the future studies.

5. Conclusions

The lakes in QTP are highly sensitive to global environment changes while are less studied due to harsh natural conditions. The expansion and shrinking of these high mountain lakes contain the useful information of the surrounding environment changes. This study represents the first lake mapping efforts using the novel space-borne near-nadir SAR imaging techniques. The resulting water maps show high consistency with the alternative LandSAT/OLI results while less affected by cloud coverage and weather conditions. The algorithm was also found not applicable to the lakes with very smooth surface and needs to be improved in the future studies by introducing ancillary data sets or time series information. Besides the direct applications of InIRA in high mountain studies, the accumulated observations from InIRA and its algorithms developed in this study also provide support for future missions and their global applications.

Author Contributions: S.L. designed the whole study and supervised the data analysis. H.T., Z.L., Z.Z., and Y.L. developed the algorithm, conducted the data analysis, and crafted the manuscript. W.Z. significantly edited the manuscript. K.L. and B.Q. revised the manuscript.

Funding: This research was funded by China Manned Space Engineering Program Tiangong-2 data Platform grant number [Y3140231WN] and National basic science data sharing service platform grant number [Y7031511WY].

Acknowledgments: The LandSat data are available from the U.S. Geological Survey.

Conflicts of Interest: The authors declare no conflict of interest. The funders had no role in the design of the study; in the collection, analyses, or interpretation of data; in the writing of the manuscript, and in the decision to publish the results.

References

1. Bianduo; Bianbaciren; Li, L.; Wang, W.; Zhaxiyangzong. The response of lake change to climate fluctuation in north Qinghai-Tibet Plateau in last 30 years. *J. Geogr. Sci.* **2009**, *19*, 131–142. [CrossRef]
2. Ma, R.; Duan, H.; Hu, C.; Feng, X.; Li, A.; Ju, W.; Jiang, J.; Yang, G. A half-century of changes in China's lakes: Global warming or human influence? *Geophys. Res. Lett.* **2010**, *37*. [CrossRef]

3. Van Huissteden, J.; Berrittella, C.; Parmentier, F.J.W.; Mi, Y.; Maximov, T.C.; Dolman, A.J. Methane emissions from permafrost thaw lakes limited by lake drainage. *Nat. Clim. Chang.* **2011**, *1*, 119–123. [CrossRef]

4. Zhang, G.; Xie, H.; Kang, S.; Yi, D.; Ackley, S.F. Monitoring lake level changes on the Tibetan Plateau using ICESat altimetry data (2003–2009). *Remote Sens. Environ.* **2011**, *115*, 1733–1742. [CrossRef]

5. Yao, T.; Thompson, L.; Yang, W.; Yu, W.; Gao, Y.; Guo, X.; Yang, X.; Duan, K.; Zhao, H.; Xu, B.; et al. Different glacier status with atmospheric circulations in Tibetan Plateau and surroundings. *Nat. Clim. Chang.* **2012**, *2*, 663–667. [CrossRef]

6. Du, J.; Kimball, J.S.; Jones, L.A.; Watts, J.D. Implementation of satellite based fractional water cover indices in the pan-Arctic region using AMSR-E and MODIS. *Remote Sens. Environ.* **2016**, *184*, 469–481. [CrossRef]

7. Cheng, G.; Wu, T. Responses of permafrost to climate change and their environmental significance, Qinghai-Tibet Plateau. *J. Geophys. Res. Earth Surf.* **2007**, *112*, F02S03. [CrossRef]

8. Yang, K.; Wu, H.; Qin, J.; Lin, C.; Tang, W.; Chen, Y. Recent climate changes over the Tibetan Plateau and their impacts on energy and water cycle: A review. *Glob. Planet. Chang.* **2014**, *112*, 79–91. [CrossRef]

9. Liu, Q.; Du, J.; Shi, J.; Jiang, L. Analysis of spatial distribution and multi-year trend of the remotely sensed soil moisture on the Tibetan Plateau. *Sci. China Earth Sci.* **2013**, *56*, 2173–2185. [CrossRef]

10. Jiang, Q.; Fang, H.; Zhang, J. Dynamic changes of lakes and the geo-mechanism in Tibet based on RS and GIS technology. In *Remote Sensing of the Environment: 16th National Symposium on Remote Sensing of China*; Tong, Q., Ed.; International Society for Optics and Photonics: Bellingham, WA, USA, 2008; Volume 7123, p. 71230S.

11. Carroll, M.L.; Townshend, J.R.; DiMiceli, C.M.; Noojipady, P.; Sohlberg, R.A. A new global raster water mask at 250 m resolution. *Int. J. Digit. Earth* **2009**, *2*, 291–308. [CrossRef]

12. Zhang, G.; Yao, T.; Xie, H.; Zhang, K.; Zhu, F. Lakes' state and abundance across the Tibetan Plateau. *Chin. Sci. Bull.* **2014**, *59*, 3010–3021. [CrossRef]

13. Wan, W.; Xiao, P.; Feng, X.; Li, H.; Ma, R.; Duan, H.; Zhao, L. Monitoring lake changes of Qinghai-Tibetan Plateau over the past 30 years using satellite remote sensing data. *Chin. Sci. Bull.* **2014**, *59*, 1021–1035. [CrossRef]

14. Chen, J.; Chen, J.; Liao, A.; Cao, X.; Chen, L.; Chen, X.; He, C.; Han, G.; Peng, S.; Lu, M.; et al. Global land cover mapping at 30 m resolution: A POK-based operational approach. *ISPRS J. Photogramm. Remote Sens.* **2015**, *103*, 7–27. [CrossRef]

15. Pekel, J.F.; Cottam, A.; Gorelick, N.; Belward, A.S. High-resolution mapping of global surface water and its long-term changes. *Nature* **2016**, *540*, 418–422. [CrossRef] [PubMed]

16. Ma, R.; Yang, G.; Duan, H.; Jiang, J.; Wang, S.; Feng, X.; Li, A.; Kong, F.; Xue, B.; Wu, J.; et al. China's lakes at present: Number, area and spatial distribution. *Sci. China Earth Sci.* **2011**, *54*, 283–289. [CrossRef]

17. Joshi, N.; Baumann, M.; Ehammer, A.; Fensholt, R.; Grogan, K.; Hostert, P.; Jepsen, M.R.; Kuemmerle, T.; Meyfroidt, P.; Mitchard, E.T.; et al. A review of the application of optical and radar remote sensing data fusion to land use mapping and monitoring. *Remote Sens.* **2016**, *8*, 70. [CrossRef]

18. Prigent, C.; Papa, F.; Aires, F.; Rossow, W.B.; Matthews, E. Global inundation dynamics inferred from multiple satellite observations, 1993–2000. *J. Geophys. Res. Atmos.* **2007**, *112*, D12107. [CrossRef]

19. Schroeder, R.; McDonald, K.C.; Chapman, B.D.; Jensen, K.; Podest, E.; Tessler, Z.D.; Bohn, T.J.; Zimmermann, R. Development and evaluation of a multi-year fractional surface water data set derived from active/passive microwave remote sensing data. *Remote Sens.* **2015**, *7*, 16688–16732. [CrossRef]

20. Du, J.; Kimball, J.S.; Galantowicz, J.; Kim, S.B.; Chan, S.K.; Reichle, R.; Jones, L.A.; Watts, J.D. Assessing global surface water inundation dynamics using combined satellite information from SMAP, AMSR2 and Landsat. *Remote Sens. Environ.* **2018**, *213*, 1–17. [CrossRef] [PubMed]

21. Solander, K.C.; Reager, J.T.; Famiglietti, J.S. How well will the Surface Water and Ocean Topography (SWOT) mission observe global reservoirs? *Water Resour. Res.* **2016**, *52*, 2123–2140. [CrossRef]

22. Fu, L.L.; Alsdorf, D.; Rodriguez, E.; Morrow, R.; Mognard, N.; Lambin, J.; Vaze, P.; Lafon, T. The SWOT (Surface Water and Ocean Topography) Mission: Spaceborne radar interferometry for oceanographic and hydrological applications. In Proceedings of the OCEANOBS'09 Conference, Venice, Italy, 21–25 September 2009.

23. Durand, M.; Fu, L.L.; Lettenmaier, D.P.; Alsdorf, D.E.; Rodriguez, E.; Esteban-Fernandez, D. The surface water and ocean topography mission: Observing terrestrial surface water and oceanic submesoscale eddies. *Proc. IEEE* **2010**, *98*, 766–779. [CrossRef]

24. Fjortoft, R.; Gaudin, J.M.; Pourthie, N.; Lalaurie, J.C.; Mallet, A.; Nouvel, J.F.; Martinot-Lagarde, J.; Oriot, H.; Borderies, P.; Ruiz, C.; et al. KaRIn on SWOT: Characteristics of near-nadir Ka-band interferometric SAR imagery. *IEEE Trans. Geosci. Remote Sens.* **2014**, *52*, 2172–2185. [CrossRef]

25. Gao, Q.; Li, Y.; Wan, Y.; Qin, X.; Jiangcun, W.; Liu, Y. Dynamics of alpine grassland NPP and its response to climate change in Northern Tibet. *Clim. Chang.* **2009**, *97*, 515. [CrossRef]

26. Song, C.; Sheng, Y. Contrasting evolution patterns between glacier-fed and non-glacier-fed lakes in the Tanggula Mountains and climate cause analysis. *Clim. Chang.* **2016**, *135*, 493–507. [CrossRef]

27. Mao, D.; Wang, Z.; Yang, H.; Li, H.; Thompson, J.R.; Li, L.; Song, K.; Chen, B.; Gao, H.; Wu, J. Impacts of climate change on Tibetan lakes: Patterns and processes. *Remote Sens.* **2018**, *10*, 358. [CrossRef]

28. Zhu, Z.; Wang, S.; Woodcock, C.E. Improvement and expansion of the Fmask algorithm: Cloud, cloud shadow, and snow detection for Landsats 4–7, 8, and Sentinel 2 images. *Remote Sens. Environ.* **2015**, *159*, 269–277. [CrossRef]

29. Qiu, S.; He, B.; Zhu, Z.; Liao, Z.; Quan, X. Improving Fmask cloud and cloud shadow detection in mountainous area for Landsats 4–8 images. *Remote Sens. Environ.* **2017**, *199*, 107–119. [CrossRef]

30. Wu, T.D.; Chen, K.S. A reappraisal of the validity of the IEM model for backscattering from rough surfaces. *IEEE Trans. Geosci. Remote Sens.* **2004**, *42*, 743–753.

31. Fung, A.K.; Chen, K.S. An update on the IEM surface backscattering model. *IEEE Geosci. Remote Sens. Lett.* **2004**, *1*, 75–77. [CrossRef]

32. Chen, K.S.; Wu, T.D.; Tsay, M.K.; Fung, A.K. Note on the multiple scattering in an IEM model. *IEEE Trans. Geosci. Remote Sens.* **2000**, *38*, 249–256. [CrossRef]

33. Bindlish, R.; Barros, A.P. Multifrequency soil moisture inversion from SAR measurements with the use of IEM. *Remote Sens. Environ.* **2000**, *71*, 67–88. [CrossRef]

34. Jiang, L.; Shi, J.; Tjuatja, S.; Dozier, J.; Chen, K.; Zhang, L. A parameterized multiple-scattering model for microwave emission from dry snow. *Remote Sens. Environ.* **2007**, *111*, 357–366. [CrossRef]

35. Du, J.; Shi, J.; Sun, R. The development of HJ SAR soil moisture retrieval algorithm. *Int. J. Remote Sens.* **2010**, *31*, 3691–3705. [CrossRef]

36. Baghdadi, N.; Gherboudj, I.; Zribi, M.; Sahebi, M.; King, C.; Bonn, F. Semi-empirical calibration of the IEM backscattering model using radar images and moisture and roughness field measurements. *Int. J. Remote Sens.* **2004**, *25*, 3593–3623. [CrossRef]

37. Dobson, M.C.; Ulaby, F.T.; Hallikainen, M.T.; El-Rayes, M.A. Microwave dielectric behavior of wet soil—Part II: Dielectric mixing models. *IEEE Trans. Geosci. Remote Sens.* **1985**, *1*, 35–46. [CrossRef]

38. Kropáček, J.; Maussion, F.; Chen, F.; Hoerz, S.; Hochschild, V. Analysis of ice phenology of lakes on the Tibetan Plateau from MODIS data. *Cryosphere* **2013**, *7*, 287–301. [CrossRef]

39. Stogryn, A. Equations for calculating the dielectric constant of saline water (correspondence). *IEEE Trans. Microw. Theory Tech.* **1971**, *19*, 733–736. [CrossRef]

40. Congalton, R.G.; Green, K. *Assessing the Accuracy of Remotely Sensed Data: Principles and Practices*; CRC Press: Boca Raton, FL, USA, 2008.

remote sensing

MDPI

Article

Development of Supraglacial Ponds in the Everest Region, Nepal, between 1989 and 2018

Mohan Bahadur Chand [1,*] and Teiji Watanabe [2]

[1] Graduate School of Environmental Science, Hokkaido University, Sapporo, Hokkaido 060-0810, Japan

[2] Faculty of Environmental Earth Science, Hokkaido University, Sapporo, Hokkaido 060-0810, Japan; teiwata@mac.com

* Correspondence: mohanchand06@gmail.com; Tel.: +81-11-706-2213

Received: 25 March 2019; Accepted: 2 May 2019; Published: 5 May 2019

Abstract: Several supraglacial ponds are developing and increasing in size and number in the Himalayan region. They are the precursors of large glacial lakes and may become potential for glacial lake outburst floods (GLOFs). Recently, GLOF events originating from supraglacial ponds were recorded; however, the spatial, temporal, and seasonal distributions of these ponds are not well documented. We chose 23 debris-covered glaciers in the Everest region, Nepal, to study the development of supraglacial ponds. We used historical Landsat images (30-m resolution) from 1989 to 2017, and Sentinel-2 (10-m resolution) images from 2016 to 2018 to understand the long-term development and seasonal variations of these ponds. We also used fine-resolution (0.5–2 m) WorldView and GeoEye imageries to reveal the high-resolution inventory of these features and these images were also used as references for accuracy assessments. We observed a continuous increase in the area and number of ponds from 1989–2017, with minor fluctuations. Similarly, seasonal variations were observed at the highest ponded area in the pre- and postmonsoon seasons, and lowest ponded area in the winter season. Substantial variations of the ponds were also observed among glaciers corresponding to their size, slope, width, moraine height, and elevation. The persistency and densities of the ponds with sizes >0.005 km^2 were found near the glacier terminuses. Furthermore, spillway lakes on the Ngozompa, Bhote Koshi, Khumbu, and Lumsamba glaciers were expanding at a faster rate, indicating a trajectory towards large lake development. Our analysis also found that Sentinel-2 (10-m resolution) has good potential to study the seasonal changes of supraglacial ponds, while fine-resolution (<2 m) imagery is able to map the supraglacial ponds with high accuracy and can help in understanding the surrounding morphology of the glacier.

Keywords: glacial lake; supraglacial pond; Himalaya; Everest; remote sensing

1. Introduction

High Mountain Asian glaciers are the perennial sources of water for approximately 1.4 billion people [1]. Glaciers in this region are losing their mass and volume [2–5] at a significant rate due to a warming climate. The increased storage of meltwater from glaciers and snow in the form of supraglacial and proglacial lakes is also an indication of volumetric loss of glacier ice and snow [6,7]. The number of glacial lakes and supraglacial ponds have been increasing in size and number [8,9] in the region. Supraglacial ponds are common features on the surfaces of relatively slow-moving, debris-covered glaciers [10] in comparison to clean glaciers. These features grow by the coalescence of small ponds [11]. About 13–36% of the Himalayan region's glacierized area exhibits debris cover [12], which shows very slow movement rates at their tongues [13,14]. The debris-covered glaciers have heterogeneous surfaces with debris thickness ranging from a few centimeters to meters, and they have different thermal properties [15]. The thermal properties of the debris play an important role in the heat conduction from the surface to debris–ice interface. Setting of the debris-covered glaciers favors the formation and expansion of supraglacial ponds, whose hydrological buffering roles remain unconstrained [12]. Supraglacial ponds are known for meltwater storage [16], progressively buffering the runoff regimes of the glacier-originated river in increased projections of debris cover [12]. They play an important role in the ablation of debris-covered glaciers [17,18] through absorbing atmospheric energy [19,20]. The majority of absorbed atmospheric energy leaves the pond system through englacial conduits [16,20,21], and hydraulic connection of pond to englacial water level exerts a key control on whether the pond contributes to longer-term terminus disintegration [10]. This process enlarges the englacial conduits which can collapse the roof of the conduits, leading to the formation of ice cliffs and new ponds [20,22,23]. However, the majority of ponds occupy closed basins with no perennial connection to the englacial system and can undergo rapid growth until they find the connection [24]. Ponds are highly recurrent and persistent with high interannual variability [19], but small ponds have the potential to expand rapidly [25].

Previous studies on supraglacial ponds have shown that the ponded areas change from year-to-year [11,16,19,26], which may be due to the downwasting of glaciers [4–6,27]. These features also show substantial seasonal variations in response to draining and freeze–thaw activities in different seasons [19], and seasonal differences in the ice melt [28]. The condition of pond formation according to the glacier's characteristics, including slope and surface velocity has also been demonstrated [18,19,29,30].

The use of multitemporal satellite imagery is a common technique for monitoring large glacial lakes [9,11,31–34]. Several previous studies were conducted in the Nepal Himalaya, focusing on the development of such lakes (e.g., [31,34–36]), hazard assessments (e.g., [13,30,34,37,38]), and community involvement in glacial lake research (e.g., [39,40]). Most of these studies demonstrated the development of glacial lakes usually on a decadal basis [8,41,42], and were regionally aggregated [11]. Furthermore, these studies were glacier or lake specific [13,24,43,44] or used one time satellite imagery [45]. Remote sensing techniques are also used for monitoring supraglacial ponds [16,19,26]. Spatial, seasonal, and interannual patterns of the ponds for five debris-covered glaciers in the Langtang Valley, Nepal were studied [19] using Landsat images of 30-m resolution, which found high variability in the emergence of ponds among glaciers and also pronounced seasonal variations.

The Everest region in the Nepal Himalaya can be considered a hotspot of glacial lakes and supraglacial ponds [16,37,45]. However, there is no research being conducted on the spatial and seasonal variations and long-term development of the ponds on an annual basis, despite their importance in studies on the impact of recent climate change [45,46]. Such studies are also important in understanding the evolution of ponds into large glacial lakes in the future. Efforts for documenting the development of the ponds and their variations were made in the region [16] by using satellite imageries of 0.5–2 m resolution. However, this study incorporated only eight glaciers and used historical imageries only from 2000 to 2015.

To address this shortcoming, we assessed the development of supraglacial ponds on all of the 23 debris-covered glaciers in the region. Our first aim was to present the historical development of the supraglacial ponds on annual basis from 1989 to 2017, to understand the year-to-year variations and long-term evolution in the Everest region of Nepal. We used atmospherically corrected surface reflectance Level-2 science products of Landsat images for this purpose. Secondly, we aimed, for the first time in this area of research, to understand the seasonal variations of the ponds by analyzing Sentinel-2 images of 10-m resolution in combination with long-term development. Our third aim was to prepare a high-resolution inventory (2-m spatial resolution) of the supraglacial ponds by using WorldView-2, WorldView-3 and GeoEye-1 images for 2015–2016. Finally, we evaluated the relationship between the ponded areas and the morphometric characteristics of the glaciers.

2. Materials and Methods

2.1. Study Area

The Everest region is located in the Solukhumbu District in the northeastern part of Nepal (Figure 1). This area includes Sagarmatha (Mt. Everest) National Park (SNP), a World Heritage site that is the highest mountainous area in the world. This region includes the upper catchment of the Dudh Koshi River (DKR) basin, which is one of the most widely glaciated regions in the Nepal Himalaya. This river is one of the seven major tributaries of the Koshi River. The SNP covers the northern part of the DKR basin, and encompasses an area of 1148 km^2, with elevations ranging from 2845 m a.s.l. at Jorsalle to 8848 m a.s.l. at the peak of Mt. Everest. More than 60% of the park area has an elevation higher than 5000 m. The total number of glaciers in the whole DKR basin is 287 and cover an area of 391.1 km^2 which was 9.62% of the basin in 2010 [47]. The glaciers in this region are characterized by the presence of debris in their lower reaches. The debris area covers approximately 28% (110 km^2) of the total glacier area in the DKR basin [47]. The SNP includes 132 glaciers covering an area of 262 km^2 which is 23% of the park area.

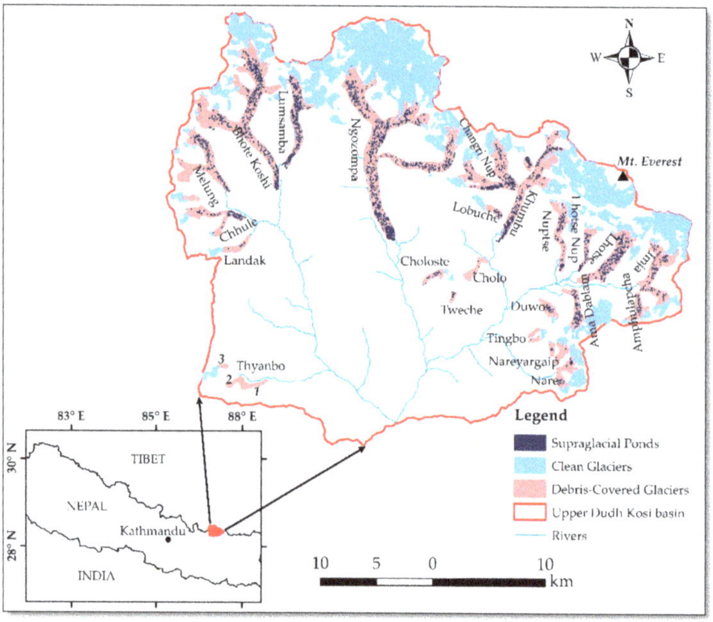

Figure 1. The glaciers studied in the Sagarmatha National Park (Upper Dudh Koshi basin), Nepal. Glacier outlines and supraglacial ponds were delineated using 2-m resolution images of WorldView and GeoEye except Thyanbo Glacier, for which Sentinel-2 of 10-m resolution image was used.

We chose 23 debris-covered glaciers in the SNP covering approximately 88% and 230.7 km^2 of the total glacier area in the park. The debris portion of the selected glaciers occupy an area of 103.4 km^2, which is 45% of the total area of the studied glaciers. Debris-covered glaciers slope relatively gently in comparison to clean glaciers and have the potential to form large glacial lakes. All supraglacial ponds that were plotted by Salerno [45] in 2008 covered 18% of the total lake area in the park. These supraglacial ponds are precursors to large glacier lakes. They are vulnerable to increasing temperatures. The Nare Drangka and Dig Tsho glacial lakes experienced glacier lake outburst floods (GLOFs) in 1977 and 1985, respectively. Some of the supraglacial ponds on the Lhotse and Changri Shar Glaciers also experienced GLOFs in 2015, 2016 [48], and in 2017 [49], respectively. Inter- and intra-annual changes in glacier-scale ponded areas of up to 17% and 52% respectively, have been observed, which indicates drainage events, pond expansion and coalescence, and melt season pond expansion [16].

2.2. Datasets and Preprocessing

To assess the evolution and variation of the supraglacial ponds from 1989 to 2018 we used data from multiple platforms and sensors with medium to high resolution. These were Landsat (30-m resolution), Sentinel-2 (10-m resolution); WorldView-2, WorldView-3 and GeoEye-1 (2-m and 0.5-m resolution).

2.2.1. Landsat

We used surface reflectance Level-2 science products of the Landsat 5 Thematic Mapper (TM), Landsat 7 Enhanced Thematic Mapper (ETM+), and Landsat 8 operational land imager (OLI) to study the long-term development of the supraglacial ponds. These products were available after 1987 for our study site. Images were downloaded from the USGS website for each year from 1987 to 2017. Unfortunately, no suitable scenes were available from the same month for each year, which would minimize error from monthly variation of the ponds due to the presence of significant cloud during the summer monsoon season and snow during the winter and premonsoon season. Seasonal changes in glaciers and glacial lakes are relatively minor from September to December [50]. Therefore, images that lie within the three-month period of October to December (Figure 2) were selected for the whole period except for 1990, 2013, and 2014. We obtained the images from January for 1990 and September for 2013 and 2014. Most utilized scenes were without snow or cloud cover on the debris portions of the glaciers and were suitable for pond identification. However, no suitable scenes were available for 1987, 1988, 1991, 1997, 1999, 2006, 2007, 2011, and 2012 due to extensive snow or cloud cover and data gaps caused by a scan line error. We used two scenes for 2014, one from September, and the other from November, to minimize the effect of clouds. In total, 23 scenes (Table S1) were used for the 22 different years for our study. Surface reflectance products are atmospherically corrected products using a radiative transfer model, which are the Second Simulation of Satellite Signal in the Solar Spectrum (6S) for the Landsat 5 and 7 and an internal algorithm for Landsat 8. In these models, the effects of water vapor, aerosol, and ozone were removed to obtain accurate surface reflectance. Landsat images were also radiometrically calibrated and orthorectified using ground control points and a digital elevation model (DEM).

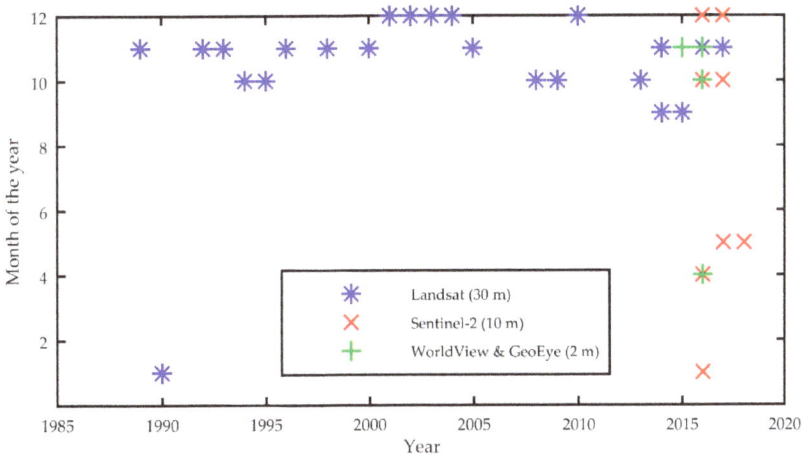

Figure 2. Temporal and seasonal distributions of scenes used in the study. Blue marks are for Landsat, red marks for Sentinel-2, and green marks for WorldView and GeoEye scenes.

2.2.2. Sentinel-2

We obtained Sentinel-2A and 2B images for 2016–2018 from the European Space Agency's (ESA) Sentinel Scientific Data Hub, incorporating the postmonsoon (October), winter (December–January), and premonsoon (April–May) seasons (Figure 2). Two scenes of Sentinel-2 were acquired to cover the entire study area for each period. We acquired images from two time periods of the same month for the premonsoon season of 2018 to minimize the error due to freezing of the ponds. We used 14 Sentinel 2A and four Sentinel 2B images of the level 1C (Table S2) covering the entire study area for the eight different time periods. The ESA sen2cor plugin, which is available on the Sentinel Application Platform (SNAP) was used for atmospheric and terrain correction of the Level 1C images and to produce an atmospherically corrected Level 2A bottom of atmosphere (BOA) reflectance product. The Level 2A product is similar to that of the Landsat surface reflectance product. We chose only 10-m resolution bands, being Blue, Green, Red, and Infrared bands for our study's purposes. A mosaic of two scenes of the same period was created to incorporate the entire area.

2.2.3. WorldView and GeoEye

High-resolution WorldView and GeoEye images from 2015 and 2016 were obtained from the DigitalGlobe Foundation. We obtained the 0.5-m (panchromatic) and 2-m (multispectral) resolution Basic 1B imagery products (Level 1) of WorldView-2, WorldView-3 and GeoEye-1 images (Table S3). These images were radiometrically and sensor corrected but not projected to a plane using a map projection or datum. Therefore, we orthorectified each scene in ERDAS Imagine using rational polynomial coefficients and the 30-m Shuttle Radar Topography Mission (SRTM) DEM. These images covered 22 of the 23 debris-covered glaciers studied here and were used to prepare the high resolution inventory map of the supraglacial ponds in the SNP.

2.2.4. Digital Elevation Model (DEM)

We obtained High Mountain Asia 8-m DEMs derived from along-track optical imagery, version 1 for the period 2015 to 2016 from the Earth Data website [51] for an analysis of glacier characteristics. These DEMs were generated from very-high-resolution (VHR) stereoscopic imagery from DigitalGlobe satellites. We filled the gap that existed in each of the individual DEM tiles using the focal statistics tool in ArcMap and the filled tiles were mosaicked to cover the entire glacier area. However, these DEMs have significant data gaps in accumulation zones of the glaciers so we used the SRTM DEM based on data collected in 2000 to identify the accumulation and ablation areas.

2.3. Methods

2.3.1. Glacier Characteristics

We manually identified and digitized the boundaries of the 22 debris-covered glaciers using 2-m resolution images of WorldView and GeoEye from 2015 to 2016. A Sentinel-2 image from 2016 was used for the Thyanbo Glacier, for which high-resolution images were not available. In our study, the terminus of the Changri Nup and Changri Shar Glaciers were merged together and considered a single glacier, the Changri Nup Glacier. The Imja and Lhotse Shar Glaciers were also merged together and considered a single glacier, the Imja Glacier. Clean-type glaciers were ignored from the inventory in our study.

We evaluated eight descriptive metrics for the debris-covered area of the glaciers and compared them with ponded areas using Spearman's rank-order correlation [19]. We produced correlation coefficients (r_s) ranging from a perfect negative correlation (−1) to a perfect positive correlation (+1).

The total glacier area, debris-covered area, and width of the glaciers were determined by using the outlines of the glaciers. We averaged the width of the glaciers based on 3–13 transects, depending on the sizes of the glaciers. The minimum and mean elevations of the debris-covered glaciers were computed based on the 8-m DEM, assuming that air temperature has a strong control on the surface mass balance of the glaciers. Similarly, the accumulation-area ratio (AAR) was calculated based on the equilibrium-line altitude (ELA) of 5477 based on a previous study in the Everest region [52]. We approximated the average height of the moraine from the glacier surface (DGM) based on the 3–13 transects to understand the cumulative downwasting of the glacier surface [19,29]. Similarly, we estimated the mean gradients of the glaciers using a recent 8-m DEM, which has a strong control on pond formation and distribution [19,30,45]. The aspect of the debris-covered part of the glaciers was computed, however, r_s was not estimated for this metric, and we used this metric solely to understand the dominance of the ponds on certain aspects of the glaciers.

The boundaries of the glaciers for 1990 was adopted from the glacier inventory of the ICIMOD [47] and were modified to obtain the extent of the debris-covered area using Landsat images of the corresponding period. The boundaries of the debris-covered glaciers from the 1990s were used as references to map the supraglacial ponds for the entire study period.

2.3.2. Supraglacial Pond Mapping

Automatic lake mapping methods for glacial lakes and supraglacial ponds have been well discussed [16,19,53]. However, the possibility of misclassification and omission of ponds increases significantly with moderate resolution of the dataset [16]. Manual editing is recommended to increase the accuracy of the mapping [54,55], and therefore, to improve the results, we adopted postediting after applying the water index and band ratios. Preprocessed scenes of each Landsat, Sentinel, and WorldView and GeoEye images were clipped to the boundary of the debris-covered glaciers. Several combination of bands for normalized difference water index (NDWI) have been proposed by several previous studies [11,19,53,56]. Modified NDWI (MNDWI) that uses the SWIR and Green band is useful in built-up or urbanized area to minimize the noise [56]. The bands used in MNDWI are similar to normalized difference snow index (NDSI) and it omits the ponded area especially when pond is frozen. The NDWI proposed by [53] uses the Blue band in combination with NIR band, which misclassify the ponded area as shadow area [11], especially in high mountain areas with significant shadow. Therefore, here we used *NDWI* by using NIR and Green band (Equation (1)) for each scene as used by previous studies (e.g., [11,19,53,57,58]) to delineate the boundaries of the supraglacial ponds.

$$NDWI = \frac{B_{Green} - B_{NIR}}{B_{Green} + B_{NIR}} \tag{1}$$

Band ratios (*BR*1) of green-to-near-infrared (Equation (2)) were applied, which is useful for differentiating between moisture and nonmoisture [19].

$$BR1 = \frac{B_{Green}}{B_{NIR}} \tag{2}$$

The presence of shadow leads to misclassified ponds [53] due to similar reflectance with water bodies. Therefore, we used the quality assessment (QA) band available in the Landsat surface reflectance product and scene classification algorithm in sen2cor for Sentinel-2 images to remove the effect of the shadow. The above-mentioned metrics did not work efficiently for the images with snow cover and frozen ponds. Prior efforts to identify snow have used *NDSI* [59] which is similar to the *MNDWI* (Equation (3)) [56], but cannot differentiate between the presence of snow and water bodies.

$$NDSI \; and \; MNDWI = \frac{B_{Green} - B_{SWIR}}{B_{Green} + B_{SWIR}} \tag{3}$$

The mask obtained from the conventional *NDSI* leads to the significant removal of ponds. Therefore, considering that snow had the highest reflectance in the mountains, we applied the spectral metric (*BR2*) (Equation (4)).

$$BR2 = \frac{B_{Blue} + B_{Green} + B_{Red}}{3} \tag{4}$$

Supraglacial lakes are evident where the surface gradient of a glacier is less than 2° [18,19,30], while discrete and small isolated ponds are evident where slopes are between 2 and 10° [18]. This suggests that a glacial lake can expand in a debris-covered glacier which has low inclination and little ice flux from upstream [60]. We used a higher surface slope threshold of 30° [19] to eliminate steep avalanche fans or icefall from the debris-covered area in which ponds can form.

In our work, initially, we applied several thresholds of NDWI that range from 0.0 to 0.50 and checked the results from each threshold. The results obtained from threshold 0.3 was better than the other results, which was crosschecked with the histogram of NDWI. Similar approach was applied to detect threshold for $BR2$ and threshold of 1.2 for $BR1$ used by [19] was adopted. Finally, ponds that met the slope threshold as well as $NDWI > 0.3$ or $BR1 > 1.2$ and $BR2 > 0.45$ were delineated, following an approach similar to [19] and [11]. However, the threshold values can vary with time and may lead to overestimation and underestimation of ponds. Therefore, all delineated ponds were checked for accuracy and edited manually to minimize the error due to variations in thresholds values and reflectance among turbid and blue ponds. During manual editing, we edited the boundaries of ponds by including all pure pixels of water body and about half of the pixels that surround the pure pixels. Different area thresholds have been used for glacial lake mapping, ranging from 0.003 to 0.1 km^2 [11,37,41,61,62]. The possibility of an overestimation of a pond area can increase at smaller thresholds (4 pixels or less), particularly for coarse resolution images. Therefore, we used the minimum of 5 pixels [50] for mapping supraglacial ponds, an area of 0.005, 00005, 0.00002 km^2 for the Landsat, Sentinel, and WorldView and GeoEye images, respectively. Polygons smaller than these thresholds were removed. We also generated the Sentinel-2 and WorldView and GeoEye ponds with an area threshold of 0.005 km^2 to compare with the Landsat ponds.

3. Results

3.1. Glacier Distribution and Characteristics

The morphometric characteristics of the debris portions of the 23 glaciers studied for 2015–2016 are presented in Table 1. They exhibited a wide range of geometric conditions. The smallest glacier was Tweche Glacier (0.31 ± 0.003 km^2) in which no accumulation zone was observed, while the largest was Ngozompa Glacier with a total area of 77.71 ± 0.24 km^2, with debris-covering area of 25.99 ± 0.09 km^2. The proportion of debris-covered glaciers to the total glacier area ranged from 30% (Khumbu) to 100% (Tweche). The average glacier width was 465.8 m. The two extremes were Changri Nup Glacier, with the largest average width of 923 m, and Thaynbo Glacier, with the smallest average width of 206 m. The mean DGM, that is, the elevation difference between the lowest elevation of the glacier and the dominant outermost lateral moraine peak elevation, was 63.2 m. The Thyanbo Glacier had the lowest mean DGM (15 m), and the Imja Glacier had the highest mean DGM (129 m). The minimum elevation of all glaciers (debris-covered area) exceeded 4600 m a.s.l., with the exception of Cholo and Thyanbo glaciers that extended below 4500 m a.s.l. Only six glaciers had a mean elevation below 5000 m a.s.l.

The mean slope of the glaciers ranged from 6.8° (Lumsamba) to 20.3° (Tingbo), and 11 glaciers had mean surface gradients below 10°. Only four glaciers, Ngozompa, Imja, Nareyargaip, and Khumbu glaciers, had an accumulation area greater than 50% and the remainder of the glaciers were dominated by the ablation part with debris-covered portions. The Cholo and Thyanbo glaciers were oriented towards the east, with mean average azimuth of 88.8°, while the remainder of the glaciers were oriented towards the south (south, southwest, and southeast) with an average azimuth of 188.7°.

3.2. High-Resolution Inventory of Supraglacial Ponds in the Everest Region

Our mapping of supraglacial ponds in the Everest region using 2-m resolution imagery and an area threshold of 0.00002 km^2 has identified 3009 ponds (Figure 1) with a total area of 2.04 ± 0.32 km^2 and a mean size of 0.0007 km^2 in the years 2015 and 2016. The Shapiro–Wilk distribution test at 95% confidence interval reveals that the distribution of ponds was not normal (Figure 3a,b) and skewed positively with a factor of 20.9. The probability distribution of ponded area reveals that ~98% (n = 2949) of ponds had an area of <0.005 km^2, which contributes approximately 45% of total ponded area (Figure 3b). Only ~2% (n = 60) of the supraglacial ponds of sizes >0.005 km^2 contribute to 55% of the total ponded area. The three largest studied glaciers, Ngozompa, Bhote Koshi, and Khumbu glaciers feature 60% of the total ponded area among 22 glaciers, and the maximum number (896) and area (0.61 ± 0.1 km^2) of supraglacial ponds were observed in the Ngozompa Glacier. The majority of the glaciers (n = 14) that have debris-covered areas <5 km^2 (Table 1) exhibited only 13% of the total ponded area.

Table 1. Morphometric characteristics of the glaciers and supraglacial ponds and lakes of the Sagarmatha National Park in 2015 and 2016.

Glacier	Area (km^2) Clean + Debris	Debris (%)	Width (m)	DGM (m)	Elevation[1] (m a.s.l.) Min.	Mean	Slope (°)	AAR (%)	Aspect	Pond/Lake Cover (%)
Landak	1.6	1.0 (60)	312	41	4857	5030	12.0	30	SE	0.21
Chhule	4.9	3.4 (69)	408	22	4794	4980	10.5	14	SE	1.67
Melung	7.2	6.3 (88)	443	58	4967	5184	9.7	11	SE	0.69
Bhote Koshi	30.3	17.9 (59)	510	63	4756	5104	9.6	38	S	1.55
Lumsamba	10.9	5.1 (47)	463	61	4900	5166	6.8	45	S	2.98
Ngozompa	77.7	26.0 (33)	904	75	4669	5022	7.0	57	S	2.33
Changri Nup	12.3	7.4 (60)	923	98	5094	5257	9.6	38	SE	3.11
Nuptse	5.3	3.3 (63)	419	49	4938	5237	9.2	44	S	1.22
Lhotse Nup	2.3	1.6 (69)	297	39	4930	5075	8.9	18	SW	1.53
Lhotse	10.5	5.9 (56)	740	42	4813	5051	7.1	33	SW	1.54
Amphu	2.2	1.3 (60)	380	113	5021	5166	14.5	12	SW	0.65
Imja	15.3	5.5 (36)	718	129	4980	5145	8.7	53	SW	0.52
Ama Dablam	7.7	2.4 (31)	441	63	4753	4911	8.8	37	S	2.06
Duwo	1.5	1.2 (81)	616	57	4714	4809	13.4	1	SW	1.30
Lobuche	1.4	0.6 (44)	364	44	4943	5018	15.8	48	SE	3.24
Cholotse	1.2	0.8 (72)	344	70	4859	4967	13.2	21	SW	0.81
Tweche	0.3	0.3 (100)	268	64	4967	5035	13.7	0	SW	1.62
Cholo	1.0	1.0 (95)	253	39	4427	4732	16.5	5	E	0.06
Nareyargaip	5.4	2.1 (39)	375	107	5042	5268	15.5	61	S	2.17
Nare	1.6	0.7 (42)	526	108	4983	5112	12.5	24	S	0.17
Thyanbo[2]	2.2	1.4 (62)	206	15	4347	4653	13.9	26	E	
Tingbo	0.9	0.5 (56)	235	26	4855	5051	20.3	19	SW	0.03
Khumbu	27.2	8.0 (30)	568	70	4885	5132	7.7	66	SW	3.89
r$_s$	0.90	0.90	0.70	0.21	−0.11	0.72	−0.75	0.61		

[1] Elevation data are based on the debris-covered portions of the glaciers. [2] High-resolution pond cover was not available for this glacier.

We observed significant variability in pond cover among the studied glaciers, ranging from 0.03% (Tingbo Glacier) to 3.89% (Khumbu Glacier) of the debris-covered area in 2015 and 2016. The rank-order correlation coefficient between pond area and different morphometric characteristics of the glaciers was estimated and it exhibited a very strong rank-order correlation with the total glacier area (r$_s$ = 0.90) and debris area (r$_s$ = 0.90), a strong correlation with the mean slope (r$_s$ = −0.75), mean elevation (r$_s$ = 0.72), and glacier width (r$_s$ = 0.70), and a moderate correlation with the AAR (r$_s$ = 0.61) which is statistically significant at 99% confidence level. However, no significant rank-order correlation was found for glacier minimum elevation (r$_s$ = −0.11) and DGM (r$_s$ = 0.21). The altitudinal area distribution of the ponds shows that supraglacial ponds can be found as high as ~5560 m a.s.l., ~200 m lower than the upper extent of debris portion of the glacier (Figure 4a). Most ponds were concentrated at lower reaches of the glacier, below 5200 m a.s.l. and the highest area (~20%) of the pond was observed between 5100 and 5200 m a.s.l. About 87% of the ponded area was observed in the glaciers with slopes of <10°, of which 55% of the ponded area was observed on slopes from 2–6°, 17% on slopes <2° and 15% on slopes from 6–10° (Figure 4b).

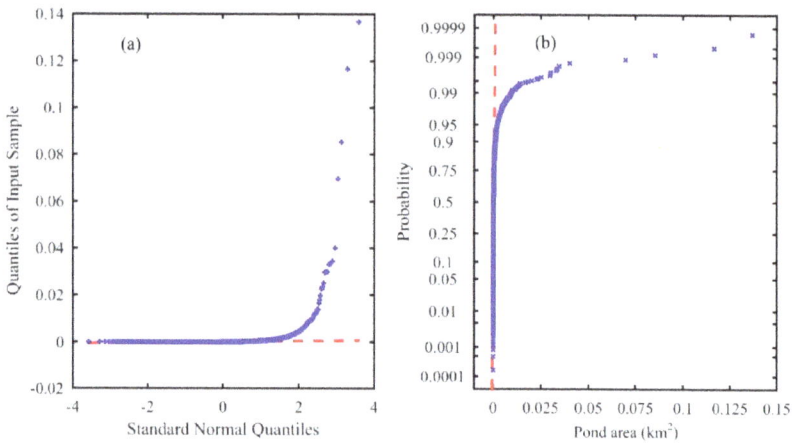

Figure 3. Distribution of ponded areas (1989–2017) in the Everest region using (**a**) normal Q-Q plot, and (**b**) normal probability plot.

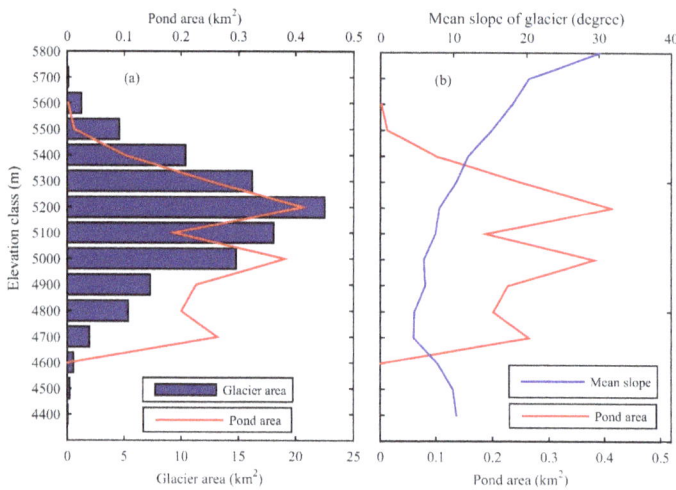

Figure 4. The relationship between ponded area and glacier characteristics with (**a**) elevation and (**b**) slope. The altitudinal area distribution of the debris portion of the glacier and area of all supraglacial ponds were mapped using 2-m resolution imagery in the Everest region. The elevation class value on the *y*-axis indicates the uppermost value.

3.3. Long-Term Evolution of the Ponds

Between 1989 and 2017 we mapped a total of 1026 supraglacial ponds (>0.005 km^2) in the Everest region on the surfaces of 23 debris-covered glaciers. We excluded the large pond at the Changri Nup Glacier for an analysis. This was observed in only a few images due to the presence of shadow. Approximately 59% (n = 594) of the supraglacial ponds were <0.01 km^2 in size, which accounts for only one-third of the total ponded area over the period studied. Of the total ponded area studied during the period, ponds with sizes of 0.01–0.02 km^2, and >0.02 km^2 had ponded areas of 32% (n = 294), and 35% (n = 120), respectively.

The Shapiro–Wilk distribution test statistics at 95% confidence interval reveals that the distribution of ponds was not normal (Figure 5a) and skewed positively with a factor of 4.23. The skewness factor was reduced to 3.71 when we ignored the large ponds at one of the tributaries of the Khumbu Glacier. The probability distribution of ponded area also reveals that ~90% (n = 923) of ponds have an area <0.025 km^2, and approximately 96% (n = 983) of the ponds have an area <0.05 km^2, which contributes approximately 60% and 73% of total ponded area between 1989 and 2017, respectively (Figure 5b,c). Only 4% (n = 43) of the ponds comprise approximately 27% of the total pond coverage. It is notable that the distribution of supraglacial ponds according to dimensional size was also far from normal in each year, with an abundance of small ponds and few large ponds. The frequency of ponds in the 0.02 to 0.04 km^2 dimensional class has increased significantly in recent years, which was also indicated by the increase in the median size of the ponds.

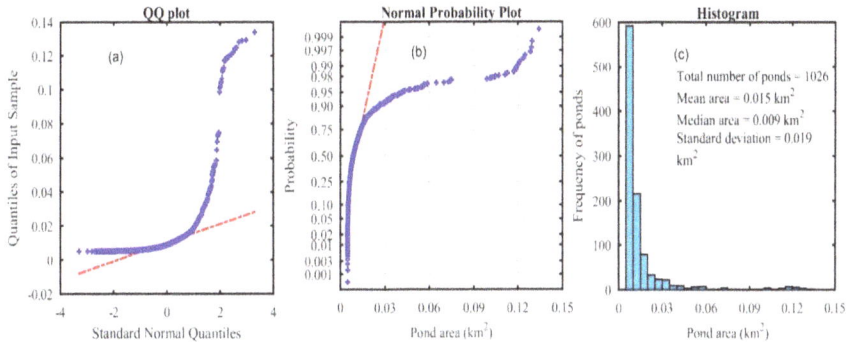

Figure 5. The distribution of pond areas (1989–2017) in the Everest region with (**a**) normal Q-Q plot of pond area, (**b**) normal probability plot of pond area, and (**c**) frequency distribution of ponds.

We observed an overall increase in area (Figure 6a,d) and number (Figure 6b) of ponds from 1989 to 2017, with minor fluctuations (Tables S4–S6). The overall area of the ponds has increased from 0.378 ± 0.19 km^2 in 1989 to approximately 1.324 ± 0.727 km^2 in 2017, representing an overall growth of 350%. The size of the ponds and lakes varied from 0.005 to 0.13 km^2, with a mean size of 0.015 ± 0.019 km^2 (Figure 6c). The rate of increase was comparatively slower between 1980 and 2005 (0.01 km^2/yr) than between 2008 and 2017 (0.07 km^2/yr). In the 1989 imagery, 25 supraglacial ponds were identified and this number increased to 85 (340%) in 2017, with the highest number of ponds in 2015 (88). The number of ponds almost doubles between 1989 and 2002, slightly decreases between 2003 and 2005, and increases rapidly from 2009. The year-by-year variations in number and total area for different dimensional classes of supraglacial ponds show that the frequency of small-sized ponds (<0.01 km^2) was higher than the larger ponds (>0.02 km^2) in each studied year, however, the area covered by larger ponds contributes significantly more than the small ponds in the majority of years (Figure 6b,d). We observed the highest increase in the number of ponds that have dimensional classes of <0.01 km^2 in the recent period. However, the highest increase in pond areas was observed in the dimensional class of >0.02 km^2. The increase in pond areas with the class >0.02 km^2 ranges from 45 to 54% after 2008. Ponds were observed in 16 of the 23 studied glaciers in the region, whereas ponded areas were not observed in the Cholo, Choloste, Landak, Nare, Tingbo, and Tweche glaciers (<1 km^2). However, a significant ponded area was observed in frequent years from 1989 to 2017 in the smaller Lobuche Glacier (0.45 km^2).

3.3.1. Glacier Wise Trends of Pond Cover

The glaciers studied here demonstrate the significant variability in pond cover over time. Of the 23 glaciers we investigated, supraglacial ponds were observed in nine glaciers in 1989 and 16 glaciers in 2014 and 2015, while no ponds were observed in 13 and 7 glaciers in 1989 and 2017, respectively. All glaciers except for Lobuche Glacier experienced either increments in area of ponds or the appearance of new ponds. The increments of pond cover vary among the glaciers, and the largest increase in number, from 8 to 27 and in area, from 0.063 ± 0.048 to 0.424 ± 0.237 km^2 was observed on the Ngozompa Glacier during the period studied of 28 years (Figure 6a). Similarly, the Bhote Koshi and Khumbu glaciers contributed significant increases in ponded area. The Ngozompa, Bhote Koshi, and Khumbu glaciers exhibited a total ponded area of 81% in 1989 and 67% in 2017. The Ngozompa Glacier alone contributed approximately 32% of total ponded area in 2017. The Nareyargaip and Lumsamba glaciers record increased pond coverage by approximately ten times than that in 1989, although the actual increase in pond area was 0.053 km^2.

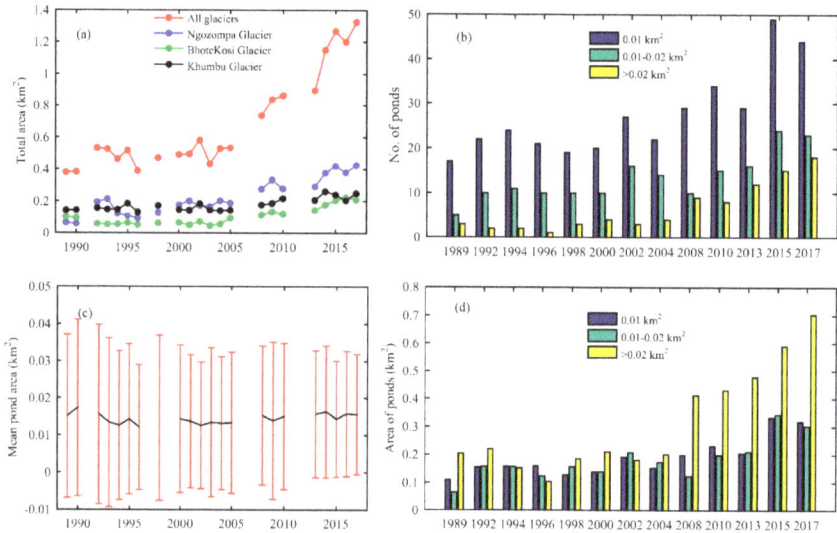

Figure 6. Long-term development of the supraglacial ponds in the Everest region from 1989–2017 with (**a**) total ponded area for all study glaciers and three selected glaciers, Ngozompa, Bhote Koshi, and Khumbu glaciers, (**b**) changes in number for three dimensional classes, (**c**) variation in the mean area, and (**d**) changes in the area of three-dimensional classes of supraglacial ponds.

Substantial interannual variation in the ponded area was observed due to draining of the ponds within the year. The ponded area on the Ngozompa Glacier increased by approximately 370% from 1990–1993 and then decreased until 1996. This variation in pond area was contributed to by the formation and development of one large pond (0.1 km^2) 1.5 km from the current outlet of the glacier. It was almost completely drained (0.007 km^2) by 1994. Similarly, a large pond on the Bhote Koshi Glacier also lost its size by 530% within a period of 1–2 years, from 1990–1992 and contributed to a decrease in ponded area on this glacier within this period.

The Thyanbo, Chhule, Melung, Bhote Koshi, Lumsamba, Ngozompa, Khumbu, and Nuptse glaciers exhibited the presence of either spillway lakes or ponds near their terminuses. However, the area of the spillways on the Ngozompa, Bhote Koshi, Khumbu, and Lumsamba glaciers increased significantly. Most of these ponds were larger than the ponds in the upstream region and exhibited ~40% of the total ponded area in 2017.

3.3.2. Pond Persistency

Pond persistency is described in terms of the frequency of the ponds. Most ponds at the glacier terminuses, especially on Ngozompa, Khumbu, and Bhote Koshi glaciers were very persistent (Figure 7). One of the tributaries of the Khumbu Glacier had a very persistent pond over 22 scenes over the studied 28-year period, with very little expansion in size. However, ponds in the smaller glaciers with relatively higher slopes tend to be less persistent for a longer time period (Figure S1). Pond frequency maps for the periods 1989–1998, 1999–2008, and 2009–2017 were computed to highlight the expansion, distribution, and persistence between different shorter periods. The distribution of ponds has expanded and shows more persistence in the later period (2009–2017) than in the early and middle periods.

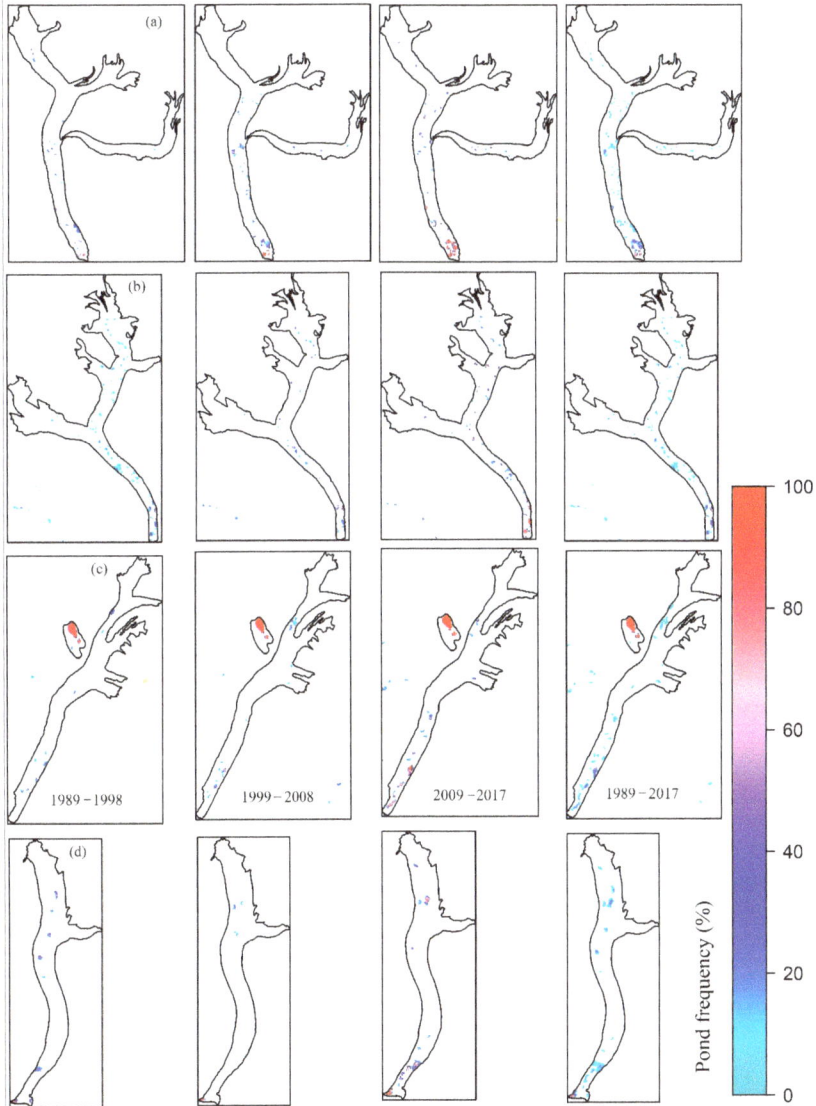

Figure 7. Distribution of supraglacial ponds in (**a**) Ngozompa, (**b**) Khumbu, (**c**) Bhote Koshi, and (**d**) Lumsamba glaciers in different time periods from 1989 and 2017, highlighting the persistence of individual ponds. The pond persistency of the remainder of the glaciers is provided in the supplementary file (Figure S1).

3.4. Seasonal Pond Cover

We mapped a total of 3027 supraglacial ponds (>0.0005 km²) for eight different seasons between January 2016 and May 2018. The mean size was 0.0046 ± 0.0014 km². Supraglacial pond cover in the Everest region shows no clear trend among seasons; however, the smallest number and area of the ponds were recorded in the winter season (Figure 8). Ponded areas in the pre- and postmonsoon seasons of 2016 were very similar at 1.7 ± 0.55 km²; however the highest number and area of the ponds were observed in the premonsoon season of 2018.

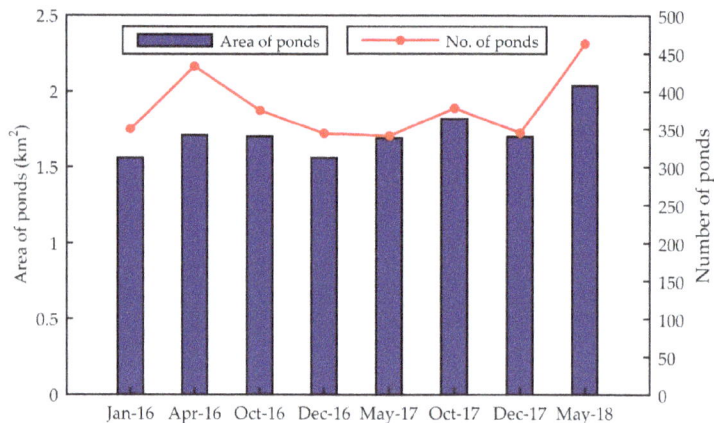

Figure 8. Seasonal changes in the number and area of the supraglacial ponds (>0.0005 km²) between January 2016 and May 2018 obtained from Sentinel-2 images of 10-m resolution.

4. Discussion

4.1. Supraglacial Pond Inventory Using Remote Sensing

We used the multitemporal and multiresolution satellite imageries for monitoring the long-term development and short-term variations of the supraglacial ponds in the Everest region and updated the glacier inventory of the ICIMOD [47] for the 23 debris-covered glaciers of the region using 2-m resolution imageries.

We were able to prepare the historical inventory for 28 years from 1989 to 2017 by using Landsat images of medium resolution with area thresholds of >0.005 km² (5 pixels). The supraglacial ponds smaller than this size were excluded from our historical inventory. The 10-m Sentinel-2 images were used for seasonal variations of the ponds and 2-m WorldView and GeoEye imageries were used to prepare the high-resolution inventory of the supraglacial ponds. The areal uncertainty of the delineated features for ponds and glaciers was generated by multiplying a perimeter by half of a cell resolution [36,50] and was varied among different resolutions. The values of uncertainties estimated using this method is higher. However, we used manual editing technique after applying automatic mapping. Therefore, highly accurate boundaries of ponds are expected from our study, which was crosschecked by using 0.5-m resolution panchromatic band of WorldView imageries. Higher-resolution panchromatic images (0.5-m) were available for 25 and 29 October 2016, which was almost the same time period as that of Sentinel-2 (30 October 2016) and Landsat (10 November 2016), and these were used for an accuracy assessment of the ponds. These images covered the Khumbu, Nuptse, Lhotse Nup, and Lhotse glaciers. Ponds (>0.005 km²) on these glaciers were manually digitized using panchromatic images. The mean uncertainty of ponds for the 0.5-m resolution images was 1.25%, which can be considered negligible. We analyzed the aerial error by comparing the area of the ponds obtained from 0.5-m resolution images with 2-, 10-, and 30-m resolution images. We found 12 supraglacial ponds covering an area of 0.3 km² on the surface of the four glaciers (>0.005 km²) and found total areal difference of 5.7, 9.2, and 14.9% with 2-, 10-, and 30-m resolutions, respectively, for the total ponded area. However, the maximum aerial error for individual ponds ranged from 20 to 85% for different resolution imageries. High error among individual ponds was mainly caused by the inability of the semiautomatic method to detect narrow channels that connect the ponds, which can be manually mapped with high-resolution images. Nevertheless, the majority of the ponds (65–75%) had an aerial error of less than 7, 14, and 22% for the 2-, 10-, and 30-m resolution imageries, respectively. Few cases of the ponds which were mapped using three different sensors are available in supplementary file (Figure S2).

The inventory using 2-m resolution with thresholds of 5 pixels for each image suggested that Landsat and Sentinel-2 images were unable to map 45% and 13%, respectively, of the total ponded area. These statistics were 98% and 83% of the total number of ponds mapped by 2-m imagery for the Landsat and Sentinel-2 imagery, respectively. The inventory also revealed that ponds <1 pixel of Landsat (900 m^2) accounted for 19% of the total ponded area and those <1 pixel of Sentinel-2 accounted for 4% of total ponded area. These estimations were upper bounds and are comparable with other findings [16]. These results suggest that Landsat images are suitable for understanding the long-term development of supraglacial ponds with sizes of $>0.005 \text{ km}^2$. Our estimations also suggest that Sentinel-2 images have the potential to study seasonal variations, and that WorldView images can be used for higher accuracy, detailed inventories of the ponds. The fluctuation of shallow and small size ponds can be large and require images with <10-m resolution [16,63]. The amount of sediments in the ponds and when frozen also hindered the pond mapping with coarse resolution imagery [16,64]. The capability of 15-m ASTER images to detect and monitor the supraglacial lakes in comparison to much coarser resolution has been highlighted [64]. Here, we demonstrated the application of the 10-m Sentinel-2 images to detect and monitor supraglacial ponds on a seasonal basis. Sentinel-2 imageries with 10-m spatial resolution and 5-day temporal resolution can potentially be used for mapping supraglacial ponds of sizes $>0.0005 \text{ km}^2$, which may help to understand, with higher accuracy, the short-term variations of the ponds.

4.2. Spatial, Temporal, and Seasonal Trends in Supraglacial Pond Development

We used satellite imagery with different resolutions ranging from 2-m to 30-m to study the long-term development and short-term variation of the ponds as well as to prepare the high-resolution inventory. Previous studies on supraglacial ponds in the Everest region were conducted by using single image or by using imageries which cover part of the Everest region or by decadal timespan studies [16,41,45]. However, year-to-year variations of the ponds from glacier-to-glacier are required to understand the pond dynamics of a studied glacier. Here, we present the year-to-year, season-to-season, and glacier-to-glacier variations of the supraglacial ponds in the SNP, Nepal. The results from this historical study reveal an increase in the number and area of the ponds with substantial temporal, spatial, and seasonal variations. The detection of substantial increases in area of the supraglacial ponds suggests that ice melt is increasing at a higher rate in recent time periods [4,5], and that ice melt is much higher at ice cliffs with supraglacial ponds [18,65]. Ponds that occupy a closed basin with no perennial connections can undergo rapid growth [24], and development of new ponds increase heat absorption, which increases ice melt through under- and side-cutting [20]. The development of new ponds also enhances the growth of the ponds.

The increase in the area and number of ponds can be attributed to the increase in temperature in this region. We used the Asian Precipitation–Highly-Resolved Observation Data Integration Towards Evaluation of the Water Resources (APHRODITE) dataset [66] to understand the general trend of temperature for the region. Point data at an elevation of 5000 m a.s.l. were extracted from the gridded dataset. Temperature shows a decreasing trend (-0.04 °C/yr) from 1961 to 1988 (Figure 9a) and an increasing trend (0.02 °C/yr) from 1989 to 2015 (Figure 9b). Significant surface lowering of glaciers in many parts of the Himalayas has been observed [2,3,5] with increase in temperature which resulted in lowering surface gradient and glacier velocity [14], which favors the development of supraglacial ponds and glacial lakes [24,67,68]. Wastage of glacier in recent period provides the sufficient melt water to develop the supraglacial ponds and helps in expanding their size [10,17,43], which may likely grow monotonically if glaciers continue losing their mass [24]. The expansion in ponded area contributes substantially to ablation of the glacier due to undercutting, calving, and melting imposed by ponded water.

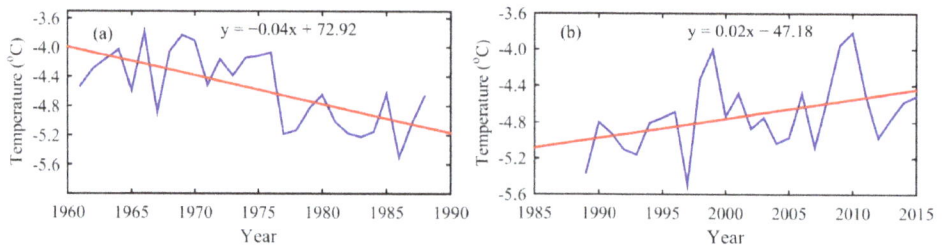

Figure 9. The temperature trend in the Everest region for the time period (**a**) 1961–1988 and (**b**) 1989–2015 using APHRODITE gridded dataset.

The results of seasonal variation of the ponds obtained from the 10-m Sentinel-2 imageries show the intra-annual dynamics for three seasons: premonsoon, postmonsoon, and winter, corresponding to seasonal ice melt [15,28]. Here, we excluded the monsoon season because satellite observations during monsoon periods are severely limited by sporadic cloud cover. Considering the three seasons for analysis, supraglacial pond cover in the Everest region showed the least during the winter season and ponded area was comparable in the premonsoon and postmonsoon seasons. Seasonality of the ponds by previous studies due to the ablation processes of the glaciers in different seasons has been reported [19,26]. The main reason for less pond cover during the winter season is the unavailability of melt water and the presence of a frozen surface, which makes accurate mapping challenging.

4.3. Glacier Characteristics and Pond Cover

We utilized the 2-m imagery to map the glacier boundaries of the 22 debris-covered glaciers and pond covers on them. The debris-covered areas of the studied glaciers showed significant variability in pond cover in 2015 and 2016, ranging from 0.03% (Tingbo Glacier) to 3.89% (Khumbu Glacier) of the debris-covered area. The pond cover on each glacier was correlated with the glacier's characteristics and showed very strong rank-order correlation with the total area of the glacier and debris-covered area [19] and strong correlation with the slope and width of the glacier. This also suggests higher pond cover for the larger glaciers, which generally have surface gradients of <10° [18]. In the Everest region, approximately 6, 45, and 69% of the glacier areas have slopes of less than 2, 6, and 10°, respectively, suggesting that all the glaciers studied have the potential to form supraglacial ponds.

The lowest percentage of pond cover on the Tingbo Glacier is correlated with its smallest size and steepest mean slope (20.3°) of all the glaciers studied. Similarly, the highest pond cover on the Khumbu Glacier is the result of a low mean gradient (7.6°) and stagnant tongue [14]. A series of several interconnected ponds at the terminus (0.12 km²) and a large pond of the same size at a tributary of the glacier in 2017 contributed to the significant pond cover on the Khumbu Glacier. The highest area of pond (0.61 km²) in the Ngozompa Glacier was also highly correlated with its largest area, low mean slope, high DGM [29,49], and glacier width, with southern aspect (Table 1). Similarly, Bhote Koshi, Changri Nup, and Lumsamba glaciers exhibited high pond cover of 1.6, 3.1, and 3% of the debris-covered area and covered an area of 0.28, 0.23, and 0.16 km², respectively. The higher ponded area on these glaciers can be explained by the large debris-covered area, low mean slope, southern aspect, and higher DGM and width of the glaciers (Table 1). The ponded area is also strongly correlated with mean elevation of the glaciers, corresponding to large area of the glaciers at higher elevations (5000–5300 m a.s.l.) and less area of the glaciers at lower elevations (4400–4900 m a.s.l.).

4.4. Future Development of the Lakes and Associated Risk

A significant increase in the area and number of ponds in the Everest region was clearly observed. The ponds which are located above the level of outlet channel of the glacier can grow until they are intercepted by an englacial conduit, and may fully or partially drain [44]. However, nearly half of the total ponded area is contributed to by the ponds near the terminus (up to ~2 km) of the glaciers where the slope was 0–4°. The highest densities of the ponds with sizes >0.005 km^2 at the terminuses of the four larger glaciers (Figure 10) suggest that these glaciers have the potential to form a large lake. A series of ponds may evolve into a large glacial lake [31,33,34,44], corresponding to warming temperatures and a trend of negative glacier mass balance [4,6,27] if the level of outlet channel remains at the same elevation. An increasing temperature can reduce the snow extent, and reducing albedo also reduces glacier extent [67], which provides more melt water to the ponds. Furthermore, we found an increase in area by 16% from the 1990s to 2016 in the debris-covered portion of the glaciers in the Everest region. We expect a potential increase in the ponded area in the future. Lakes with sizes <0.1 km^2 have been considered as less hazardous [38]; however, drainage of supraglacial ponds with sizes <0.1 km^2 also have the potential for GLOFs [48,49] by coalescing several ponds and contributing water from subglacial storage. Therefore, an estimation of the volume of these features is required [16] to understand the potential flood volume, although area–volume relationships can be used [16,49,62,63].

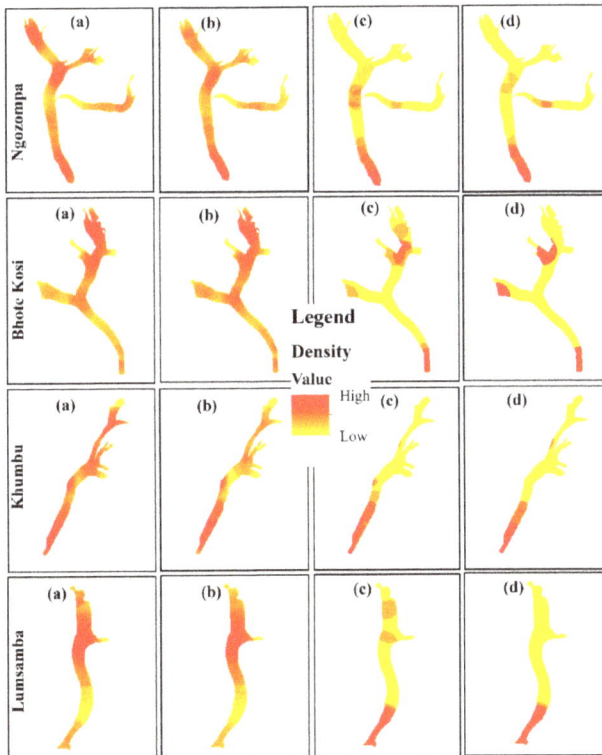

Figure 10. The densities of supraglacial ponds at the surface of the four selected glaciers with different area thresholds of the ponds, being greater than (**a**) 20 m^2, (**b**) 100 m^2, (**c**) 3600 m^2, and (**d**) 5000 m^2.

The spillway lakes on the Ngozompa, Bhote Koshi, Khumbu, and Lumsamba glaciers (Figure 11) appeared to have the greatest potential for developing into glacial lakes. This corresponds to the dominance of very gentle slopes (<2°) [18,29,30], stagnant glacier terminuses [14], and higher mean

DGM (>60 m) [29]. Furthermore, these ponds are associated with islands of ice with cliffs, and well-built terminal moraines, which also favor a trajectory towards a large glacial lake [16]. Additionally, the lake-terminating glaciers are retreating and showed maximum thinning towards their termini [69], also indicates the sign of lake expansion. It is possible that the lowering of the glacier's surface leads to a reduction in the gradients and may enhance the possibility of the development of glacial lakes [29].

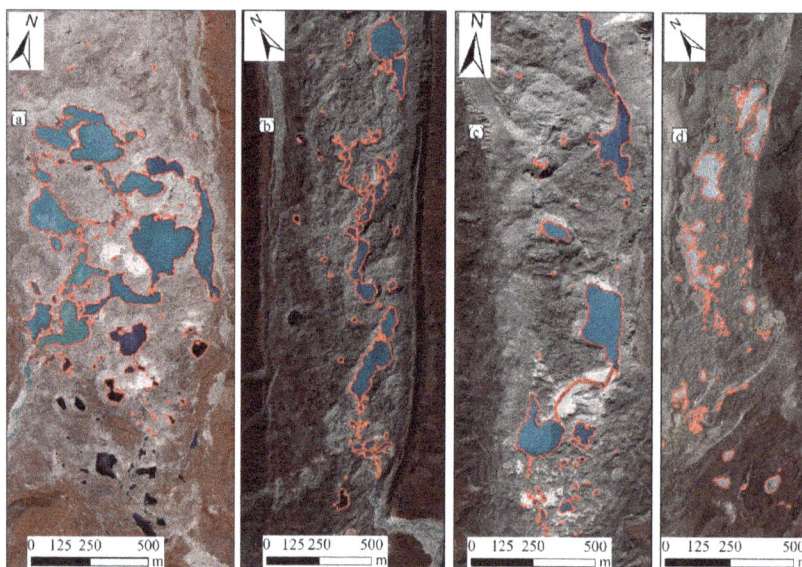

Figure 11. Spillway ponds and associated supraglacial ponds at the termini of (**a**) Ngozompa Glacier from November 2016, (**b**) Khumbu Glacier from October 2016, (**c**) Bhote Koshi Glacier from November 2015, and (**d**) Lumsamba Glacier from April 2016.

Nepal has experienced greatest national-level economic consequences (22% of global total) due to glacier floods [70]. The total number of GLOF events in Nepal documented by [71] from different sources were 24 GLOFs and out of which 14 were originated within Nepal and 10 originated in Tibet which caused floods in Nepal. Further, recent GLOFs in Upper Barun Valley in 2017 [72], Bhote Koshi and Sun Koshi River in 2016 [73] has caused geomorphic and infrastructure damage, and fatalities. Additionally, floods that originated from supraglacial ponds were also recorded in Lhotse Glacier in 2015 and 2016 [48], and Changri Shar Glacier in 2017 [49]. The potential flood volume from the supraglacial ponds assumed to be smaller and hazard associated with this could be minor due to their small size in comparison to the large glacial lakes. However, the peak discharge of GLOF event that originated from supraglacial ponds in Lhotse Glacier was estimated to be 210 ± 43 m^{3}s^{-1} with 2.65×10^6 m^3 of total maximum flood volume [48]. This event was supplemented by the stored water within englacial conduits through hydraulically efficient pathways and catastrophic glacier buoyancy. Therefore, catastrophic GLOFs might occur where series of supraglacial ponds are already developed with the presence of large spillway lakes specifically on large glacier, e.g., Ngozompa, Bhote Koshi, and Khumbu glaciers. Besides, several floods that originated from glaciers and caused serious floods in Nepal are not documented. The results obtained from our study through remote sensing techniques, can be used for "first-pass" hazard assessment in regions where field access is difficult [30]. Further detailed studies of the morphological characteristics of glaciers and the regular monitoring of lakes are required to understand the risk of GLOFs and glacier-related hazard management.

4.5. Limitations of the Study

We used Landsat images of 30-m resolution and mapped ponds with sizes >0.005 km^2 (5 pixels) to understand the historical development, and this led to the omission of 45% of the total ponded area. Images from the monsoon seasons were not used in our study due to the presence of significant cloud cover in the images. The use of radar images with a similar resolution, which can penetrate cloud, is recommended to understand the dynamics of ponds in the monsoon season. Our study was limited in the areal estimation of ponds and volumetric estimations are suggested for hazard assessment. Furthermore, pond bathymetry is required [16] as well as an understanding of englacial connectivity [22]. The exponential expansion of spillway lakes on Ngozompa, Bhote Koshi, Khumbu, and Lumsamba glaciers suggests that they require detailed investigations to understand their trajectory toward large glacial lakes. Such knowledge is lacking in this area of research. Although we used 0.5-m resolution images for accuracy assessment, field-based studies are essential for more accurate mapping and to gain a better understanding of the surrounding morphology, particularly at the locations of spillway lakes. Recent new technology, the unmanned aerial vehicle (UAV), can be used for this purpose.

5. Conclusions

We presented an extensive application of multiresolution satellite imageries to study the historical development and seasonal variations in the Everest region. We also developed a high-resolution inventory of the supraglacial ponds in the region. We used atmospherically corrected images spanning 28 years for a long-term study and three years for understanding seasonal variations of the supraglacial ponds at the surface of all debris-covered glaciers in the region. The use of Sentinel-2 images with 10-m resolution to study the seasonal variations, and the use of historical Landsat imagery from 1989–2017 to study the year-to-year variations for 28 years is novel. Also novel is the high resolution inventory of the supraglacial ponds for 2015 and 2016 presented here.

Our results show that supraglacial ponds are widely distributed on glaciers in the Everest region, and also show the rapid increase in their area and number from 1989–2017. We mapped a total of 1026 supraglacial ponds with sizes >0.005 km^2 for the entire period studied from 1989 to 2017, and only 25 in 1989 and 85 in 2017. We found a net increase in area of ponds of 350% in 28 years, from 1989–2017. High-persistence ponds were evident at the terminus of the glaciers and the persistency increased in the recent period (2009–2017) more than in the earlier period (1989–1998). It was also found that the densities of the ponds with sizes >0.005 km^2 were highest around the terminus of the glaciers. Spillway lakes and associated ponds were observed at the tongues of Ngozompa, Khumbu, Bhote Koshi, and Lumsamba glaciers, where exponential expansion was found. This is suggestive of a large lake developing in the future. The seasonal analysis showed the lowest ponded areas to be in the winter season and comparable ponded areas in the pre- and postmonsoon seasons. The temperature trend showed that annual mean temperature increased at the rate of 0.02 °C/yr from 1989–2015 in the region, which may contribute to an increase in ponded cover due to increasing ice melt. However, the characteristics of the glaciers and roles of already developed supraglacial ponds have also had a significant influence on the increase in the area of the ponds.

We have presented the results of a high-resolution inventory, in which we mapped 3009 supraglacial ponds with sizes >0.00002 km^2 on the surface of 22 debris-covered glaciers in the region, with the highest number and area of ponds on the Ngozompa Glacier, the largest glacier in the region. The ponded area was strongly correlated with area, slope, width, and mean elevation of the glacier.

We also presented the results from different resolution imageries. Landsat imagery (30 m) has the potential to map the supraglacial ponds (>0.005 km^2) to help understand their historical evolution. However, it has led to the omission of ponded areas smaller than 0.005 km^2 (45%). Sentinel-2 (10 m) with high temporal resolution has high potential to map small features and it will help to understand the seasonal variations of supraglacial ponds.

Supplementary Materials: The following materials are available online at http://www.mdpi.com/2072-4292/11/9/1058/s1, Table S1: List of Landsat imagery used in the study, Table S2: List of Sentinel-2 imageries used in the study, Table S3: List of WorldView and GeoEye imageries used in the study, Table S4: Ponded area mapped using Landsat images from 1989–2017, Table S5: Number of ponds mapped using Landsat images from 1989–2017, Table S6: Fractional pond cover correspondence to debris-covered area obtained from Landsat images, Figure S1: Distribution of the supraglacial ponds on the 12 debris-covered glaciers showing the persistency of ponds from 1989–2017, and Figure S2: Examples of ponds obtained with the Landsat (a, d, & g), Sentinel-2 (b, e, & h), WorldView-2 (c & i), and GeoEye-1 (f) at the surface of the selected glaciers i.e., terminuses of Nuptse Glacier (a, b, & c), near terminus of the Ngozompa Glacier (d, e, & f), and terminus of the Melung Glacier (g, h, & i).

Author Contributions: Conceptualization, M.B.C. and T.W.; Investigation, methodology, and formal analysis, M.B.C.; Project administration and supervision, T.W.; Writing—original draft, M.B.C.; Writing—review and editing, T.W.

Funding: The Ministry of Education, Culture, Sports, Science and Technology of Japan (MEXT) and JSPS Grant-in-aid for Scientific Research (B) Grant Number JP16H05641 (Teiji Watanabe) supported our study for the promotion of science.

Acknowledgments: We express thanks to DigitalGlobe Foundation for providing the WorldView and GeoEye imagery and Hexagon Geospatial for providing license for the ERDAS Imagine. We also thank Masami Kaneko of Rakuno Gakuen University, Japan, for providing satellite imageries.

Conflicts of Interest: The authors declare no conflicts of interest.

References

1. Immerzeel, W.W.; van Beek, L.P.H.; Bierkens, M.F.P. Climate change will affect the Asian water towers. *Science* **2010**, *328*, 1382–1385. [CrossRef] [PubMed]

2. Bolch, T.; Kulkarni, A.; Kaab, A.; Huggel, C.; Paul, F.; Cogley, J.G.; Frey, H.; Kargel, J.S.; Fujita, K.; Scheel, M.; et al. The State and Fate of Himalayan Glaciers. *Science* **2012**, *336*, 310–314. [CrossRef] [PubMed]

3. Kääb, A.; Berthier, E.; Nuth, C.; Gardelle, J.; Arnaud, Y. Contrasting patterns of early twenty-first-century glacier mass change in the Himalayas. *Nature* **2012**. [CrossRef] [PubMed]

4. Sherpa, S.F.; Wagnon, P.; Brun, F.; Berthier, E.; Vincent, C.; Lejeune, Y.; Arnaud, Y.; Kayastha, R.B.; Sinisalo, A. Contrasted surface mass balances of debris-free glaciers observed between the southern and the inner parts of the Everest region (2007–2015). *J. Glaciol.* **2017**, *63*, 637–651. [CrossRef]

5. Acharya, A.; Kayastha, R.B. Mass and Energy Balance Estimation of Yala Glacier (2011–2017), Langtang Valley, Nepal. *Water* **2018**, *11*, 6. [CrossRef]

6. Shea, J.M.; Immerzeel, W.W.; Wagnon, P.; Vincent, C.; Bajracharya, S. Modelling glacier change in the Everest region, Nepal Himalaya. *Cryosphere* **2015**, *9*, 1105–1128. [CrossRef]

7. Nie, Y.; Liu, Q.; Liu, S. Glacial Lake Expansion in the Central Himalayas by Landsat Images, 1990–2010. *PLoS ONE* **2013**, *8*, e83973. [CrossRef] [PubMed]

8. Khadka, N.; Zhang, G.; Thakuri, S. Glacial Lakes in the Nepal Himalaya: Inventory and Decadal Dynamics (1977–2017). *Remote Sens.* **2018**, *10*, 1913. [CrossRef]

9. Zhang, G.; Yao, T.; Xie, H.; Wang, W.; Yang, W. An inventory of glacial lakes in the Third Pole region and their changes in response to global warming. *Glob. Planet. Chang.* **2015**, *131*, 148–157. [CrossRef]

10. Röhl, K. Characteristics and evolution of supraglacial ponds on debris-covered Tasman Glacier, New Zealand. *J. Glaciol.* **2008**, *54*, 867–880. [CrossRef]

11. Gardelle, J.; Arnaud, Y.; Berthier, E. Contrasted evolution of glacial lakes along the Hindu Kush Himalaya mountain range between 1990 and 2009. *Glob. Planet. Chang.* **2011**, *75*, 47–55. [CrossRef]

12. Irvine-Fynn, T.D.L.; Porter, P.R.; Rowan, A.V.; Quincey, D.J.; Gibson, M.J.; Bridge, J.W.; Watson, C.S.; Hubbard, A.; Glasser, N.F. Supraglacial Ponds Regulate Runoff From Himalayan Debris-Covered Glaciers. *Geophys. Res. Lett.* **2017**, *44*, 11,894–11,904. [CrossRef]

13. Bolch, T.; Buchroithner, M.F.; Peters, J.; Baessler, M.; Bajracharya, S. Identification of glacier motion and potentially dangerous glacial lakes in the Mt. Everest region/Nepal using spaceborne imagery. *Nat. Hazards Earth Syst. Sci.* **2008**, *8*, 1329–1340. [CrossRef]

14. Quincey, D.J.; Luckman, A.; Benn, D. Quantification of Everest region glacier velocities between 1992 and 2002, using satellite radar interferometry and feature tracking. *J. Glaciol.* **2009**, *55*, 596–606. [CrossRef]

15. Chand, M.B.; Kayastha, R.B. Study of thermal properties of supraglacial debris and degree-day factors on Lirung Glacier, Nepal. *Sci. Cold Arid Reg.* **2018**, *10*, 357–368. [CrossRef]

16. Watson, C.S.; Quincey, D.J.; Carrivick, J.L.; Smith, M.W. The dynamics of supraglacial ponds in the Everest region, central Himalaya. *Glob. Planet. Chang.* **2016**, *142*, 14–27. [CrossRef]
17. Benn, D.I.; Bolch, T.; Hands, K.; Gulley, J.; Luckman, A.; Nicholson, L.I.; Quincey, D.; Thompson, S.; Toumi, R.; Wiseman, S. Response of debris-covered glaciers in the Mount Everest region to recent warming, and implications for outburst flood hazards. *Earth Sci. Rev.* **2012**, *114*, 156–174. [CrossRef]
18. Reynolds, J.M. On the Formation of Supraglacial Lakes on Debris-Covered Glaciers. Available online: http://hydrologie.org/redbooks/a264/iahs_264_0153.pdf (accessed on 1 May 2019).
19. Miles, E.S.; Willis, I.C.; Arnold, N.S.; Steiner, J.; Pellicciotti, F. Spatial, seasonal and interannual variability of supraglacial ponds in the Langtang Valley of Nepal, 1999–2013. *J. Glaciol.* **2017**, *63*, 88–105. [CrossRef]
20. Sakai, A.; Takeuchi, N.; Fujita, K.; Nakawo, M. Role of Supraglacial Ponds in the Ablation Process of a Debris-Covered Glacier in the Nepal Himalayas. Available online: http://hydrologie.org/redbooks/a264/iahs_264_0119.pdf (accessed on 1 May 2019).
21. Miles, E.S.; Pellicciotti, F.; Willis, I.C.; Steiner, J.F.; Buri, P.; Arnold, N.S. Refined energy-balance modelling of a supraglacial pond, Langtang Khola, Nepal. *Ann. Glaciol.* **2016**, *57*, 29–40. [CrossRef]
22. Miles, E.S.; Steiner, J.; Willis, I.; Buri, P.; Immerzeel, W.W.; Chesnokova, A.; Pellicciotti, F. Pond Dynamics and Supraglacial-Englacial Connectivity on Debris-Covered Lirung Glacier, Nepal. Available online: https://www.frontiersin.org/articles/10.3389/feart.2017.00069/full (accessed on 1 May 2019).
23. Watson, C.S.; Quincey, D.J.; Carrivick, J.L.; Smith, M.W.; Rowan, A.V.; Richardson, R. Heterogeneous water storage and thermal regime of supraglacial ponds on debris-covered glaciers. *Earth Surf. Process. Landf.* **2018**, *43*, 229–241. [CrossRef]
24. Benn, D.I.; Wiseman, S.; Hands, K.A. Growth and drainage of supraglacial lakes on debris-mantled Ngozumpa Glacier, Khumbu Himal, Nepal. *J. Glaciol.* **2001**, *47*, 626–638. [CrossRef]
25. Sakai, A.; Nishimura, K.; Kadota, T.; Takeuchi, N. Onset of calving at supraglacial lakes on debris-covered glaciers of the Nepal Himalaya. *J. Glaciol.* **2009**, *55*, 909–917. [CrossRef]
26. Qiao, L.; Mayer, C.; Liu, S. Distribution and interannual variability of supraglacial lakes on debris-covered glaciers in the Khan Tengri-Tumor Mountains, Central Asia. *Environ. Res. Lett.* **2015**, *10*. [CrossRef]
27. Gardelle, J.; Berthier, E.; Arnaud, Y.; Kääb, A. Region-wide glacier mass balances over the Pamir-Karakoram-Himalaya during 1999–2011. *Cryosphere* **2013**, *7*, 1263–1286. [CrossRef]
28. Chand, M.B.; Kayastha, R.B.; Parajuli, A.; Mool, P.K. Seasonal variation of ice melting on varying layers of debris of Lirung Glacier, Langtang Valley, Nepal. *Proc. Int. Assoc. Hydrol. Sci.* **2015**, *368*, 21–26. [CrossRef]
29. Sakai, A.; Fujita, K. Formation conditions of supraglacial lakes on debris-covered glaciers in the Himalaya. *J. Glaciol.* **2010**, *56*, 177–181. [CrossRef]
30. Quincey, D.J.; Richardson, S.D.; Luckman, A.; Lucas, R.M.; Reynolds, J.M.; Hambrey, M.J.; Glasser, N.F. Early recognition of glacial lake hazards in the Himalaya using remote sensing datasets. *Glob. Planet. Chang.* **2007**, *56*, 137–152. [CrossRef]
31. Lamsal, D.; Sawagaki, T.; Watanabe, T.; Byers, A.C. Assessment of glacial lake development and prospects of outburst susceptibility: Chamlang South Glacier, eastern Nepal Himalaya. *Geomat. Nat. Hazards Risk* **2016**, *7*, 403–423. [CrossRef]
32. Somos-Valenzuela, M.A.; McKinney, D.C.; Rounce, D.R.; Byers, A.C. Changes in Imja Tsho in the Mount Everest region of Nepal. *Cryosphere* **2014**, *8*, 1661–1671. [CrossRef]
33. Watanabe, T.; Ives, J.D.; Hammond, J.E. Rapid Growth of a Glacial Lake in Khumbu Himal, Himalaya: Prospects for a Catastrophic Flood. *Mt. Res. Dev.* **1994**, *14*, 329. [CrossRef]
34. Watanabe, T.; Lamsal, D.; Ives, J.D. Evaluating the growth characteristics of a glacial lake and its degree of danger of outburst flooding: Imja Glacier, Khumbu Himal, Nepal. *Nor. Geogr. Tidsskr.* **2009**, *63*, 255–267. [CrossRef]
35. Byers, A.C.; McKinney, D.C.; Somos-Valenzuela, M.; Watanabe, T.; Lamsal, D. Glacial lakes of the Hinku and Hongu valleys, Makalu Barun National Park and Buffer Zone, Nepal. *Nat. Hazards* **2013**, *69*, 115–139. [CrossRef]
36. Fujita, K.; Sakai, A.; Nuimura, T.; Yamaguchi, S.; Sharma, R.R. Recent changes in Imja Glacial Lake and its damming moraine in the Nepal Himalaya revealed by in situ surveys and multi-temporal ASTER imagery. *Environ. Res. Lett.* **2009**, *4*, 045205. [CrossRef]
37. Rounce, D.R.; McKinney, D.C.; Lala, J.M.; Byers, A.C.; Watson, C.S. A new remote hazard and risk assessment framework for glacial lakes in the Nepal Himalaya. *Hydrol. Earth Syst. Sci.* **2016**, *20*, 3455–3475. [CrossRef]

38. Aggarwal, A.; Jain, S.K.; Lohani, A.K.; Jain, N. Glacial lake outburst flood risk assessment using combined approaches of remote sensing, GIS and dam break modelling. *Geomat. Nat. Hazards Risk* **2016**, *7*, 18–36. [CrossRef]

39. Singh, R.B.; Schickhoff, U.; Mal, S. Climate change, glacier response, and vegetation dynamics in the Himalaya: Contributions toward future earth initiatives. In *Climate Change, Glacier Response, and Vegetation Dynamics in the Himalaya*; Springer International Publishing: New York, NY, USA, 2016; pp. 1–399.

40. Watanabe, T.; Byers, A.C.; Somos-Valenzuela, M.A.; McKinney, D.C. The Need for Community Involvement in Glacial Lake Field Research: The Case of Imja Glacial Lake, Khumbu, Nepal Himalaya. In *Climate Change, Glacier Response, and Vegetation Dynamics in the Himalaya*; Springer International Publishing: New York, NY, USA, 2016; pp. 235–250.

41. Shrestha, F.; Gao, X.; Khanal, N.R.; Maharjan, S.B.; Shrestha, R.B.; Wu, L.; Mool, P.K.; Bajracharya, S.R. Decadal glacial lake changes in the Koshi basin, central Himalaya, from 1977 to 2010, derived from Landsat satellite images. *J. Mt. Sci.* **2017**, *14*, 1969–1984. [CrossRef]

42. Bajracharya, S.R.; Mool, P. Glaciers, glacial lakes and glacial lake outburst floods in the Mount Everest region, Nepal. *Ann. Glaciol.* **2009**, *50*, 81–86. [CrossRef]

43. Thompson, S.S.; Benn, D.I.; Dennis, K.; Luckman, A. A rapidly growing moraine-dammed glacial lake on Ngozumpa Glacier, Nepal. *Geomorphology* **2012**, *145*, 1–11. [CrossRef]

44. Benn, D.I.; Wiseman, S.; Warren, C.R. Rapid Growth of a Supraglacial Lake, Ngozumpa Glacier, Khumbu Himal, Nepal. Available online: http://hydrologie.org/redbooks/a264/iahs_264_0177.pdf (accessed on 1 May 2019).

45. Salerno, F.; Thakuri, S.; D'Agata, C.; Smiraglia, C.; Manfredi, E.C.; Viviano, G.; Tartari, G. Glacial lake distribution in the Mount Everest region: Uncertainty of measurement and conditions of formation. *Glob. Planet. Chang.* **2012**, *92*, 30–39. [CrossRef]

46. Richardson, S.D.; Reynolds, J.M. An overview of glacial hazards in the Himalayas. *Quat. Int.* **2000**, *65–66*, 31–47. [CrossRef]

47. Bajracharya, S.; Maharjan, S.; Shrestha, F.; Bajracharya, O.; Baidya, S. *Glacier Status in Nepal and Decadal Change from 1980 to 2010 Based on Landsat Data*; International Centre for Integrated Mountain Development: Patan, Nepal, 2014.

48. Rounce, D.R.; Byers, A.C.; Byers, E.A.; Mckinney, D.C. Brief Communications: Observations of a Glacier Outburst Flood from Lhotse Glacier, Everest Area, Nepal. *Cryosphere* **2016**, *11*, 443–449. [CrossRef]

49. Miles, E.S.; Watson, C.S.; Brun, F.; Berthier, E.; Esteves, M.; Quincey, D.J.; Miles, K.E.; Hubbard, B.; Wagnon, P. Glacial and geomorphic effects of a supraglacial lake drainage and outburst event, Everest region, Nepal Himalaya. *Cryosphere* **2018**, *12*, 3891–3905. [CrossRef]

50. Jiang, S.; Nie, Y.; Liu, Q.; Wang, J.; Liu, L.; Hassan, J.; Liu, X.; Xu, X. Glacier Change, Supraglacial Debris Expansion and Glacial Lake Evolution in the Gyirong River Basin, Central Himalayas, between 1988 and 2015. *Remote Sens.* **2018**, *10*, 986. [CrossRef]

51. Shean, D. *High Mountain Asia 8-Meter Dems Derived from Along-Track Optical Imagery*; Version 1; NASA National Snow and Ice Data Center Distributed Active Archive Center: Boulder, CO, USA.

52. King, O.; Quincey, D.J.; Carrivick, J.L.; Rowan, A.V. Spatial variability in mass loss of glaciers in the Everest region, central Himalayas, between 2000 and 2015. *Cryosphere* **2017**, *11*, 407–426. [CrossRef]

53. Huggel, C.; Kääb, A.; Haeberli, W.; Teysseire, P.; Paul, F. Remote sensing based assessment of hazards from glacier lake outbursts: A case study in the Swiss Alps. *Can. Geotech. J.* **2002**, *39*, 316–330. [CrossRef]

54. Shukla, A.; Garg, P.K.; Srivastava, S. Evolution of Glacial and High-Altitude Lakes in the Sikkim, Eastern Himalaya Over the Past Four Decades (1975–2017). *Front. Environ. Sci.* **2018**, *6*, 81. [CrossRef]

55. Mergili, M.; Müller, J.P.; Schneider, J.F. Spatio-temporal development of high-mountain lakes in the headwaters of the Amu Darya River (Central Asia). *Glob. Planet. Chang.* **2013**, *107*, 13–24. [CrossRef]

56. Xu, H. Modification of normalised difference water index (NDWI) to enhance open water features in remotely sensed imagery. *Int. J. Remote Sens.* **2006**, *27*, 3025–3033. [CrossRef]

57. Jha, L.K.; Khare, D. Detection and delineation of glacial lakes and identification of potentially dangerous lakes of Dhauliganga basin in the Himalaya by remote sensing techniques. *Nat. Hazards* **2017**, *85*, 301–327. [CrossRef]

58. Bolch, T.; Peters, J.; Yegorov, A.; Pradhan, B.; Buchroithner, M.; Blagoveshchensky, V. Identification of potentially dangerous glacial lakes in the northern Tien Shan. *Nat. Hazards* **2011**, *59*, 1691–1714. [CrossRef]

59. Hall, D.K.; Riggs, G.A.; Salomonson, V.V. Development of methods for mapping global snow cover using moderate resolution imaging spectroradiometer data. *Remote Sens. Environ.* **1995**, *54*, 127–140. [CrossRef]

60. Sakai, A. Glacial Lakes in the Himalayas: A Review on Formation and Expansion Processes. *Glob. Environ. Res.* **2012**, *16*, 23–30.

61. Zhang, Y.; Fujita, K.; Liu, S.; Liu, Q.; Nuimura, T. Distribution of debris thickness and its effect on ice melt at Hailuogou glacier, southeastern Tibetan Plateau, using in situ surveys and ASTER imagery. *J. Glaciol.* **2011**, *57*, 1147–1157. [CrossRef]

62. Fujita, K.; Sakai, A.; Takenaka, S.; Nuimura, T.; Surazakov, A.B.; Sawagaki, T.; Yamanokuchi, T. Potential flood volume of Himalayan glacial lakes. *Nat. Hazards Earth Syst. Sci.* **2013**, *13*, 1827–1839. [CrossRef]

63. Cook, S.J.; Quincey, D.J. Estimating the volume of Alpine glacial lakes. *Earth Surf. Dynam* **2015**, *3*, 559–575. [CrossRef]

64. Wessels, R.L.; Kargel, J.S.; Kieffer, H.H. ASTER measurement of supraglacial lakes in the Mount Everest region of the Himalaya. *Ann. Glaciol.* **2002**, *34*, 399–408. [CrossRef]

65. Sakai, A.; Nakawo, M.; Fujita, K. Melt rate of ice cliffs on the Lirung Glacier, Nepal Himalayas, 1996. *Bull. Glacier Res.* **1998**, *16*, 57–66.

66. APHRODITE's Water Resources. Available online: http://aphrodite.st.hirosaki-u.ac.jp/ (accessed on 25 February 2019).

67. Li, Z.; Fan, K.; Tian, L.; Shi, B.; Zhang, S.; Zhang, J. Response of Glacier and Lake Dynamics in Four Inland Basins to Climate Change at the Transition Zone between the Karakorum And Himalayas. *PLoS ONE* **2015**, *10*, e0144696. [CrossRef]

68. Shrestha, A.B.; Eriksson, M.; Mool, P.; Ghimire, P.; Mishra, B.; Khanal, N.R. Glacial lake outburst flood risk assessment of Sun Koshi basin, Nepal. *Geomat. Nat. Hazards Risk* **2010**, *1*, 157–169. [CrossRef]

69. King, O.; Dehecq, A.; Quincey, D.; Carrivick, J. Contrasting geometric and dynamic evolution of lake and land-terminating glaciers in the central Himalaya. *Glob. Planet. Chang.* **2018**, *167*, 46–60. [CrossRef]

70. Carrivick, J.L.; Tweed, F.S. A global assessment of the societal impacts of glacier outburst floods. *Glob. Planet. Chang.* **2016**, *144*, 1–16. [CrossRef]

71. ICIMOD. *Glacial Lakes and Glacial Lake Outburst Floods in Nepal*; International Centre for Integrated Mountain Development: Kathmandu, Nepal, 2011.

72. Byers, A.C.; Rounce, D.R.; Shugar, D.H.; Lala, J.M.; Byers, E.A.; Regmi, D. A rockfall-induced glacial lake outburst flood, Upper Barun Valley, Nepal. *Landslides* **2019**, *16*, 533–549. [CrossRef]

73. Cook, K.L.; Andermann, C.; Gimbert, F.; Adhikari, B.R.; Hovius, N. Glacial lake outburst floods as drivers of fluvial erosion in the Himalaya. *Science* **2018**, *362*, 53–57. [CrossRef] [PubMed]

remote sensing

MDPI

Article

Impacts of Climate Change and Intensive Lesser Snow Goose (*Chen caerulescens caerulescens*) Activity on Surface Water in High Arctic Pond Complexes

T. Kiyo F. Campbell [1], Trevor C. Lantz [1,*] and Robert H. Fraser [2]

[1] School of Environmental Studies, University of Victoria, Victoria, BC V8P 5C2, Canada; tkcampbell@uvic.ca
[2] Canada Centre for Mapping and Earth Observation, Natural Resources Canada, Ottawa, ON K1A 0E4, Canada; robert.fraser@canada.ca
*   Correspondence: tlantz@uvic.ca; Tel.: +1-250-853-3566

Received: 20 October 2018; Accepted: 18 November 2018; Published: 27 November 2018

Abstract: Rapid increases in air temperature in Arctic and subarctic regions are driving significant changes to surface waters. These changes and their impacts are not well understood in sensitive high-Arctic ecosystems. This study explores changes in surface water in the high Arctic pond complexes of western Banks Island, Northwest Territories. Landsat imagery (1985–2015) was used to detect sub-pixel trends in surface water. Comparison of higher resolution aerial photographs (1958) and satellite imagery (2014) quantified changes in the size and distribution of waterbodies. Field sampling investigated factors contributing to the observed changes. The impact of expanding lesser snow goose populations and other biotic or abiotic factors on observed changes in surface water were also investigated using an information theoretic model selection approach. Our analyses show that the pond complexes of western Banks Island lost 7.9% of the surface water that existed in 1985. Drying disproportionately impacted smaller sized waterbodies, indicating that climate is the main driver. Model selection showed that intensive occupation by lesser snow geese was associated with more extensive drying and draining of waterbodies and suggests this intensive habitat use may reduce the resilience of pond complexes to climate warming. Changes in surface water are likely altering permafrost, vegetation, and the utility of these areas for animals and local land-users, and should be investigated further.

Keywords: tundra ponds; Arctic wetlands; desiccation; Landsat; aerial photographs; global change; protected areas

1. Introduction

Recent temperature increases in Arctic regions have been twice the average global change [1,2] and have triggered significant changes to regional hydrological systems, including surface water dynamics [3–7]. Changes in surface waters are concerning, because lakes, ponds, and wetlands strongly influence a range of physical, geochemical, and biological processes [8–11]. Arctic freshwater systems are also tied to the global climate system through their effects on permafrost thaw and greenhouse gas emissions from thawed ground [12–14].

Changes in the abundance and surface area of lakes and ponds in the Arctic have been attributed to increasing evaporation [15], fluctuations in precipitation [16], permafrost degradation leading to lateral and subsurface drainage [3,6,17,18], and thermokarst lake expansion [19]. The vulnerability of waterbodies to these processes depends on both the waterbody dimensions [20,21] and catchment characteristics [7,22,23]. Regional differences in these factors have resulted in considerable variation in surface water dynamics across the Arctic [7,18,23,24]. Several recent studies suggest that permafrost extent is a major determinant of change in surface water [7,23]. Most studies in discontinuous permafrost zones have reported decreases in surface water, while most studies in continuous permafrost zones have shown increases in surface water [17,19,23,25]. However, these studies have been restricted to subarctic and low Arctic regions, and trends in the high Arctic remain largely unstudied.

High Arctic pond complexes may be particularly vulnerable to the effects of increasing air and ground temperatures [26,27]. Small and shallow waterbodies, with high surface area to volume ratios, are disproportionately impacted by fluctuations in evaporation [15,21]. In addition, the shallow active layer common in the high Arctic restricts groundwater storage capacity, reducing resilience during unusually dry periods [28]. Furthermore, the high ground-ice content typically found in regularly saturated soils [28] makes areas more susceptible to thermokarst induced changes to hydrology [10,29,30].

Changes in the extent of surface water in high Arctic pond complexes will likely impact surrounding vegetation and herbivore populations, particularly migratory bird species that use these areas as breeding habitat [11,31]. Pond complexes fill an important ecological niche in the otherwise arid polar deserts of the high Arctic. However, high grazing pressures from expanding herbivore populations could also be contributing to climate-driven changes in surface water. Lesser snow goose (*Chen caerulescens caerulescens*) nesting colonies across the Arctic have seen rapid expansions in recent decades, largely due to intensified agricultural land-use providing abundant forage in their southern wintering areas and a warming climate in Arctic nesting areas [32]. These expanding nesting colonies have caused significant and lasting degradation to northern wetlands [31,33–35]. Intensive and recurring foraging can alter microtopography [36] and increase near-surface ground temperatures and evaporation [34,37], which likely decreases water retention in nearby waterbodies and may increase the risk of lateral drainage [38]. Park [38] found that ephemeral ponds surrounded by high levels of lesser snow goose grubbing had significantly shorter hydroperiods than ponds not associated with grubbing.

To improve our ability to predict the long-term impacts of climate change on high Arctic freshwater systems, additional case studies are required to understand the processes controlling surface water dynamics. The objectives of this study are to (1) explore the extent of changing waterbodies within the pond complexes of western Banks Island, Northwest Territories; and (2) to investigate the causes of this change. Landsat imagery (1985–2015) was used to detect long-term surface water trends, while higher resolution aerial photographs (1958) and satellite imagery (2014) were used to explore changes in the size and distribution of waterbodies, and field sampling investigated potential causes and contributing factors.

We tested three specific hypotheses: (1) The number and size of waterbodies on western Banks Island is decreasing; (2) the loss of small waterbodies is widespread; and (3) changes in number and size of waterbodies are following different trajectories in heavily overgrazed snow goose nesting areas, compared to areas less impacted by overgrazing.

2. Materials and Methods

2.1. Study Area

Banks Island is the westernmost island in the Canadian Arctic Archipelago and part of the Inuvialuit Settlement Region in the Northwest Territories. The community of Sachs Harbour is the only permanent settlement on the Island and has a population of approximately 100 residents. Located within the high Arctic, this area has a harsh climate with a mean annual temperature of −12.8 °C at Sachs Harbour. Summers are short with average daily temperatures rising above freezing for only 3 months of the year, peaking at 6.6 °C in July. Average annual precipitation is 151.5 mm, with only 38% falling as rain (June to September). Mean annual temperatures have shown a 3.5 °C increase since 1956, while summer precipitation and maximum snow water equivalent before spring melt have changed minimally [30,39].

The western side of Banks Island is underlain by unconsolidated Miocene-Pliocene sands and gravels and is characterized by gently rolling uplands, intersected by numerous west-flowing rivers with wide floodplains [40,41]. Alluvial terraces in these river valleys are dotted with thousands of shallow ponds and have nearly continuous vegetation cover, dominated by sedges, grasses, and mosses [31,41]. Decomposition of this vegetation has produced limited organic deposits over predominantly Gleysolic Turbic Cryosols [41]. Permafrost is continuous is this region, and ice-wedge polygons, non-sorted circles and stripes, and turf hummocks are widespread [41]. In this study, we focused on the alluvial terraces of the west-flowing rivers valleys (Figure 1).

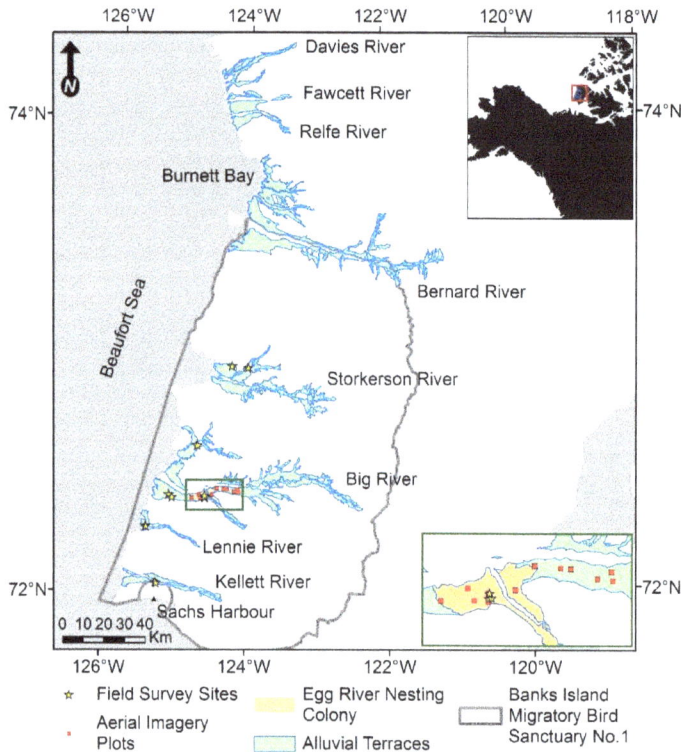

Figure 1. A map of the study area on Banks Island, Northwest Territories, showing field survey sites and areas where fine-scale imagery was analyzed. The inset map in the upper-right corner shows Banks Island as the westernmost island in the Canadian Arctic Archipelago. The inset map in the bottom-right corner is an enlarged map of the nesting colony area, within the Big River valley.

The river valleys of western Banks Island are important breeding habitat for many migratory bird species, including the lesser snow goose. This habitat supports over 95% of the western Arctic lesser snow goose population. The main nesting colony of this population is located at the confluence of the Egg and Big rivers [31,41] (Figure 1). The Banks Island Migratory Bird Sanctuary No. 1 is the second largest bird sanctuary in Canada at 20,517 km², and was created to protect this colony [31,41].

2.2. Sub-Pixel Water Fraction

To measure persistent changes in surface water, sub-pixel water fraction (SWF) was calculated for 94 30 m resolution images captured by the Landsat 5 TM, Landsat 7 ETM+, and Landsat 8 OLI sensors, between 1985 and 2015. Depending on cloud cover, 1 to 7 images were used per year, balanced over the time-series so that no significant linear relationship existed with the number of images per year. To minimize the influence of phenology and the spring freshet, all collected images fell within the period of 5 July to 10 August and were balanced across Julian Days so that no significant linear relationship existed over the time-series. Images were calibrated to top-of-atmosphere reflectance using USGS coefficients, and scan lines, clouds, and cloud shadows were masked out [42].

SWF was calculated using the Tasseled Cap wetness (TCW) index, derived from each of the 94 Landsat images, and a histogram-breakpoint method [19]. The TCW index is a transformation that contrasts shortwave infrared with visible and near-infrared bands using established Tasseled Cap (TC) coefficients [43,44]. The use of shortwave infrared bands in the TCW index makes it sensitive to water surfaces, soil moisture, and plant moisture [43,45].

Following TCW transformation, breakpoint regression was applied to the frequency distribution of pixel values in each image to identify the land limit (LL), the threshold value separating pure land pixels from pixels of mixed land-water cover, and the water limit (WL), separating pure water pixels from pixels of mixed land-water cover. Breakpoint regression was applied to each scene to reduce variabilities caused by different Landsat sensors, atmospheric conditions, and phenology states. Candidate breakpoints were determined using the 'strucchange' package [46] in R software version 3.3.2 [47] and the breakpoint algorithm for estimating multiple possible breakpoints [48]. Once range limits were obtained, the following equation was used to calculate the SWF of each mixed pixel in an image, where TCW is the Tasseled Cap Wetness value of the pixel being estimated. TCW values outside of the threshold LL and WL values were assigned 0% or 100% SWF, respectively.

$$\text{SWF} = \frac{(\text{TCW} - \text{LL}) \times 100\%}{(\text{WL} - \text{LL})} \tag{1}$$

The accuracy of the histogram-breakpoint method in this terrain type was assessed by comparing SWF estimates to manually digitized estimates of surface water within 60 (0.25 km^2) plots in the Big River valley. Manually digitized estimates of surface water were derived from WorldView-2 (WV02) satellite imagery (0.5 m resolution), acquired on 9 July 2014. Because of cloud cover and Landsat 7 scan line errors, the digitized surface water was compared against a multi-year SWF composite, calculated as the mean of SWF images from July 2013 and 2015.

To identify pixels that exhibited persistent changes in surface water, we used Theil-Sen regression and the rank-based Mann-Kendall test to determine SWF trends and significance over time (1985–2015) [19,49]. Theil-Sen regression is a nonparametric alternative to ordinary least-squares regression that uses the median of all possible pairwise slopes instead of the mean. The rank-based Mann-Kendall test of significance is calculated through comparison to all possible pairwise slopes [50]. The change in surface water for pixels with significant trends was estimated by multiplying the slope coefficient by the length of the time-series and the area of a single pixel (900 m^2). These changes were then summed within each river valley to estimate regional surface water changes.

This analysis was restricted to the alluvial terraces of major river valleys (an area of ~2335 km^2), which were manually-delineated as areas of lowland terrain within 25 km of the main river channel (<80 m above-sea-level) [41]. Lowland terrain was visually identified using 10 m resolution false-colour near-infrared Sentinel-2 satellite imagery acquired on 19 July 2017, and confirmed using a 5 m resolution digital elevation model (ArcticDEM) created by the Polar Geospatial Center from DigitalGlobe, Inc. imagery [51].

2.3. Fine-Scale Surface Water Change Detection

To explore and corroborate surface water dynamics at a finer scale, the historical and current extent of lakes and ponds was mapped within 12 (1 km^2) plots using greyscale aerial photographs and WV02 satellite images in the Big River valley. Six (1 km^2) plots were established in areas impacted by severe drying, and six (1 km^2) plots were established in stable areas that were minimally impacted by drying. Severe drying and stable plots were classified based on the composition of significant Local Indicators of Spatial Association (LISA) clusters [52] of SWF trends within the Big River valley. Severe drying plots primarily consisted of negative SWF trend clusters (mean Δ of −271.5 m^2) and stable plots primarily consisted of low positive SWF trend clusters (mean Δ of +55.6 m²). Clusters of negative and positive SWF trends were present in opposing plots; however, they did not exceed 5% of the plot area. LISA clusters were generated using GeoDa software (1.8.16.4) and an order 2 Queen contiguity weights matrix, including lower orders [52,53].

Historical waterbodies greater than 50 m² were delineated using 1:60,000 scale aerial photographs acquired on 14 July 1958, with an effective pixel size of 1.5 m. Aerial photographs were georeferenced in ArcMap (10.4.1) using a first-order polynomial transformation and 6–11 control points. The current extent of the waterbodies in these plots was delineated using WV02 satellite imagery (0.5 m resolution) acquired on 9 July 2014. Summer precipitation was similar in 1958 and 2014, reducing the likelihood that interannual variation in precipitation could influence differences in surface water extent. All waterbodies were digitized on-screen while viewing images at a 1:500 scale. If new waterbodies appeared in the 2014 imagery, their historical areas were recorded as zero. A chi-square test was used to determine if the size class distribution of waterbodies in 1958 in severe drying and stable plot types deviated from their expected distribution. Expected values were calculated by multiplying the total number of waterbodies in each size class with the total number in each plot type and dividing by the sample size. Waterbodies were tallied within eight size classes, which progressively doubled in size to account for the lower frequencies of larger waterbodies.

To explore potential drivers of surface water change in the Big River valley, we used an information-theoretic approach to compare models based on four *a priori* hypotheses regarding the cause of change in the area of individual waterbodies from 1958–2014 [54]. Hypotheses were informed by the literature (Table 1) and models were constructed using the linear models procedure in R software (3.3.2) [47]. To account for the greater potential change in surface area of larger sized waterbodies, an interaction term for pond size was also added to several models. The 2015 Tasseled Cap Greenness (TCG) parameter used in models 2 and 3 (Table 1) was calculated using the same methods as for the TCW index [43,44]. The distance from the colony parameter used in models 4 and 5 was log transformed because visual inspection of the data suggested the relationship was non-linear. The flow accumulation parameter used in models 6 and 7 was calculated using the ArcticDEM [51] and the Fill, Flow Direction, and Flow Accumulation tools on ArcMap (10.4.1). Prior to model selection, all model parameters were examined for outliers using Cleveland dot plots [55] and collinearity using Pearson correlation coefficient matrices. To keep variance inflation factors below 3.0, variable pairs with correlation values greater than 0.7 were not included in the same model [55,56].

Following model selection, we performed an additional analysis using categorical intervals of waterbody size and distance from the nesting colony to better understand how snow goose occupation may be influencing change proportional to the waterbody area. This was conducted using the GLIMMIX procedure in SAS (9.3) to construct a linear mixed effects model of proportional area change versus categorical groupings of waterbody size and distance from the colony [57]. Pairwise comparisons among categories were made using the least-squares procedure, estimated using the restricted-maximum likelihood method. The model included the 12 aerial imagery plots as a random effect. Degrees of freedom were determined using the Kenward-Roger method [57].

Table 1. Descriptions of the four *a priori* hypotheses, parameters included, and model statements. The impact column describes the hypothesized direction of the relationship between the listed parameter and waterbody area change.

Hypothesis	Parameter	Description	Impact (+/-)	Model Number and Statement	Citation
Size: Smaller ponds are more vulnerable to change and climatic variability, compared to larger ponds.	Pond size	Waterbody extent from the 1958 aerial photographs.	+	(1) Area change ~ Pond size	[15,20,21]
Vegetation: Vegetation cover insulates near-surface ground temperatures and increases soil water retention, sustaining subsurface hydrological connectivity and reducing system water loss through evaporation.	2015 TCG	Average TCG value within 0–25 m of the waterbody, taken from a 18 July 2015 Landsat scene.	+	(2) Area change ~ 2015 TCG (3) Area change ~ 2015 TCG + Pond size (4) Area change ~ 2015 TCG + Pond size + 2015 TCG × Pond size	[28,59–62]
Herbivore intensity: Intensive and recurring goose foraging can alter microtopography, and increase soil temperature and evaporation levels, which may reduce hydroperiod in nearby waterbodies or increase risk of drainage.	Colony distance	Distance from the nesting colony, which was delineated as areas of high snow goose density, using data from Samelius et al. [58].	+	(5) Area change ~ log(Colony distance) (6) Area change ~ log(Colony distance) + Pond size (7) Area change ~ log(Colony distance) + Pond size + log(Colony distance) × Pond size	[34,36–38]
Surface water connectivity: Drainage areas will have higher levels of thermal erosion gullying and are more likely to experience lateral drainage.	Flow accumulation	Maximum flow level within 25 m of the waterbody, based on the number of upslope pixels.	-	(8) Area change ~ Flow accumulation (9) Area change ~ Flow accumulation + Pond size (10) Area change ~ Flow accumulation + Pond size + Flow accumulation × Pond size	[63]

2.4. Field Surveys

To characterize field conditions in areas that showed declines in SWF and explore potential causes of these changes, surveys were conducted at 13 sites within five river valleys in July 2017 (Figure 1). Field sites were selected using LISA clusters of SWF trends and included drying sites, within clusters of negative SWF trends, and control sites, within clusters of low positive SWF trends and areas outside of significant clusters. We also visited several colony sites that were located within clusters of negative SWF trends, within 1 km of the densest parts of the nesting colony [58]. Colony sites were sampled to differentiate drying patterns in highly-used snow goose habitat areas from areas not intensively used by snow geese. All sampling locations were selected within the alluvial terraces, between 0–20 m in elevation.

At each site, 11 measurement points were established at 10 m intervals along a north-south oriented 100 m transect. At each point, measurements were made of thaw depth, soil moisture, vegetation cover, and goose grubbing. Thaw depth was measured using an active layer probe, which was pushed into the ground until the depth of refusal. Soil moisture was measured using a handheld moisture probe (HH2 Moisture Meter with a Theta Probe soil moisture sensor-ML2x, from Delta-T Devices Ltd., Cambridge, UK). Vegetation cover was measured by visually estimating the percent cover of vascular plants within a 50 cm^2 quadrat, aligned with the bottom-left corner at the measurement point. Goose grubbing was measured by counting the number of grub holes within the same 50 cm^2 quadrat. Grubbing is a particularly destructive form of foraging which targets below-ground roots and rhizomes before above-ground vegetation is available. Along each transect, we also noted if points were located within a former pond basin. Former pond basins were identified based on the absence of organic material within a pond-shaped topographic depression.

To test for significant differences in thaw depth, soil moisture, vegetation cover, and goose grubbing among drying, control, and colony site types, we constructed linear mixed effects models using the GLIMMIX procedure in SAS (9.3) [57]. Pairwise comparisons among site types were made using the least-squares procedure, estimated using the restricted-maximum likelihood method. All models included site and river valley as random effects. Degrees of freedom were determined using the Kenward-Roger method [57]. Measurements that landed within former pond basins were excluded from statistical comparisons of these variables, as they represented landcover with a different origin and substrate.

3. Results

3.1. Sub-Pixel Water Fraction

3.1.1. SWF Trends

Between 1985 and 2015, the alluvial terraces of western Banks Island lost 33.3 km^2 of surface water and gained 3.9 km^2 of surface water, resulting in a net loss of 29.3 km^2, or 7.9% of the original surface water (Figures A1–A3). Regional SWF trends indicate that surface water has declined in all river valleys except the Kellett (Figure 2). The Bernard River valley lost the largest absolute area of surface water (8.83 km^2) and the Relfe-Fawcett River valley lost the highest proportion of original surface water (17.1%).

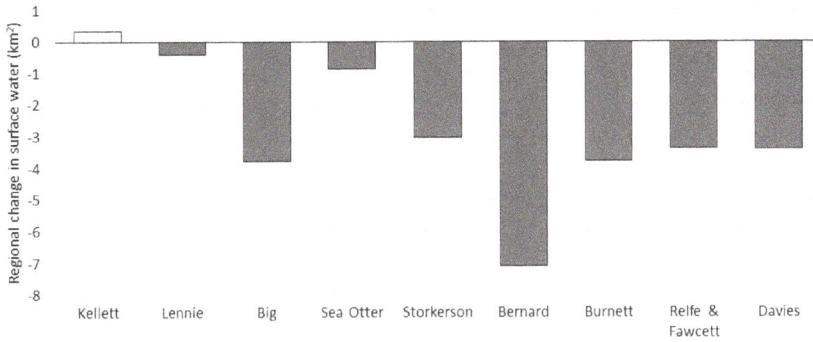

Figure 2. Net change in surface water from 1985–2015 in the 9 major river valleys of western Banks Island. River valleys are ordered by latitude, with the Kellett (at left) being the most southern and the Davies (at right) being the most Northern. SWF trends ($p < 0.05$) were determined using 94 Landsat images between 1985–2015.

3.1.2. SWF Accuracy

Estimates of SWF made using the histogram-breakpoint method were strongly correlated with waterbody areas delineated manually. An ordinary least-squares regression model of the sum of SWF pixel values and the sum of manually-delineated waterbody areas within the 60 accuracy assessment plots produced an r^2 value of 0.942 and a residual standard error of 0.0124 (Figure 3). In our study, the histogram-breakpoint SWF method overestimated surface water in 83.3% of the accuracy assessment plots (Figure 3). On average, SWF calculations overestimated waterbody area by 0.00863 km^2 (14%) compared to manually-delineated waterbody areas. The consistency of overestimation across a range of surface water proportions suggests that it is not likely to have impacted the slopes of SWF trends across the time-series (1985–2015). Overestimation is likely linked to the sensitivity of TCW to plant moisture [43,45] and the occurrence of wet sedge meadows on the landscape, which would not have been delineated as waterbodies using the aerial imagery.

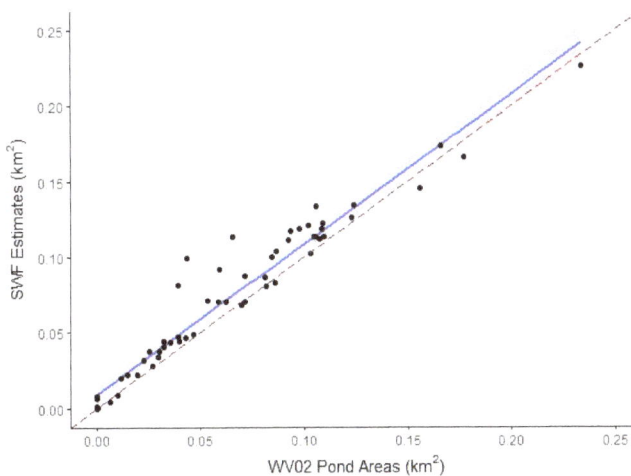

Figure 3. The sum of sub-pixel water fraction pixel values plotted against the sum of the area of manually-delineated waterbodies within 500 m^2 plots. The blue line represents the model predictions (SWF estimates ~ WV02 Pond Areas), the grey bar represents the 95% confidence interval, and the dotted red line shows a 1:1 relationship.

3.2. Fine-Scale Surface Water Change Detection

3.2.1. Waterbody Size Distributions

There were large reductions in the number of waterbodies between 1958 and 2014, in both severe drying and stable plots (Figure 4). Severe drying plots exhibited a complete loss of 732 (48.1%) waterbodies, while stable plots lost 286 (38.1%) waterbodies. Lost waterbodies ranged in their original size, from 58.5 m^2 to 17,708.5 m^2. Only 12 new waterbodies were recorded in the 2014 imagery, and only 19.9% of all waterbodies either increased in surface area or remained stable (-10% to $+10\%$ change).

Severe drying and stable plots exhibited differences in waterbody density and average size. In 1958, severe drying plots had a mean density of 253 waterbodies per km^2 and a mean waterbody size of 1228.2 m^2. Stable plots had a mean density of 149 waterbodies per km^2 and a much larger mean waterbody size of 3319.2 m^2. The chi-square test ($p < 0.001$) confirmed that areas exhibiting severe drying had more small waterbodies (50–900 m^2) and fewer large waterbodies (>900 m^2) than expected. Conversely, stable areas had fewer small waterbodies (50–900 m^2) and more large waterbodies (>900 m^2) than expected (Table 2).

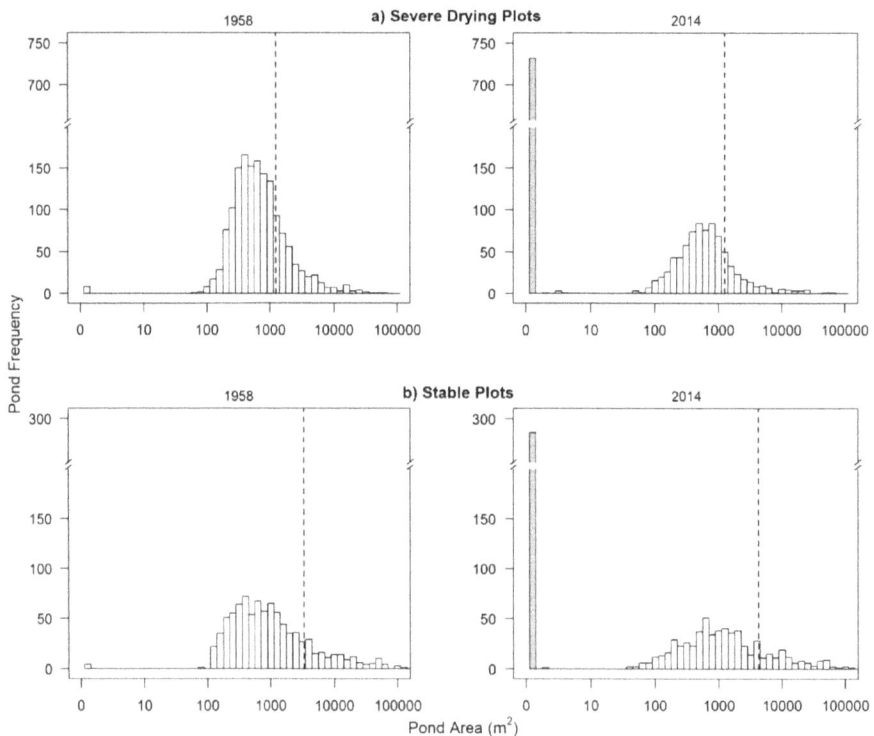

Figure 4. Size distributions of waterbodies mapped using aerial imagery, split by plot type and year. The dashed black lines show the average waterbody size within that year and plot, excluding waterbodies with a size of 0 m^2. The grey bars in 2014 show the number of waterbodies that experienced complete drainage.

Table 2. Observed and expected size class distributions of waterbodies in severe drying and stable plots, based on the chi-square analysis. Bold numbers indicate that the number of observed waterbodies exceeds the number of expected waterbodies. Numbers with an asterisk indicate a significant difference from the expected value, using a Bonferroni correction ($p < 0.0063$).

	Severe Drying		Stable	
	Observed	Expected	Observed	Expected
50–450 m^2	**625**	597	324	352
450–900 m^2	**455 ***	402	184 *	237
900–1800 m^2	249	255	**156**	150
1800–3600 m^2	100 *	121	**93 ***	72
3600–7200 m^2	53	64	**49**	38
7200–14,400 m^2	15 *	35	**41 ***	21
14,400–28,800 m^2	14 *	23	**22 ***	13
28,800–120,000 m^2	4 *	18	**25 ***	11
Column totals	1515	1515	894	894

Across all plots, 47.8% of the total surface water loss occurred in waterbodies smaller than 1800 m^2, despite the fact that these waterbodies only accounted for 20.5% of the total surface area in 1958 (Table 3). Smaller waterbodies lost a higher proportion of their total area, compared to larger waterbodies (Figure 5). The smallest size class (50–450 m^2) lost 90.6% of the total area, compared to the 5% lost by the largest size class (28,800–120,000 m^2).

Table 3. Summary statistics including the total number, area, change in area, and proportion of change for waterbodies in different size classes.

Waterbody Size in 1958 (m^2)	50–450	450–900	900–1800	1800–3600	3600–7200	7200–14,400	14,400–28,800	28,800–120,000
Count	949	639	405	193	102	56	36	29
Total area (m^2)	213,525	431,325	546,750	521,100	550,800	604,800	777,600	2,157,600
Change in total area (m^2)	−193,437	−212,536	−204,329	−153,752	−151,374	−119,682	−142,308	−107,254
Proportion of total area lost	90.59%	49.28%	37.37%	29.51%	27.48%	19.79%	18.30%	4.97%

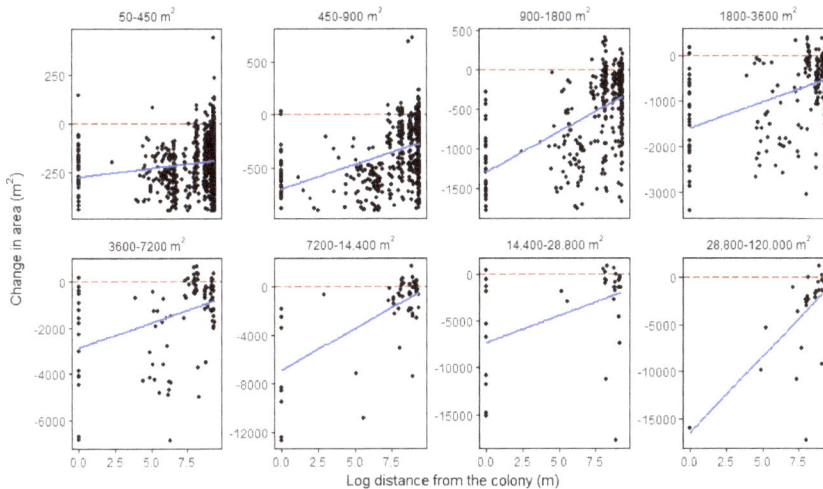

Figure 5. Visualization of the interaction effect (distance from the colony and waterbody size) in the best model (Table 4). Data were divided based on the waterbody size classes indicated above each panel. Each point represents the change in area of a single waterbody. The blue lines show model predictions for waterbody area change within that size class. The dotted red reference lines show no change in waterbody area.

Table 4. Candidate models for change in waterbody area, with goodness-of-fit metrics. The table is ordered by the model fit, and the best model is shown in bold.

Model Number	Explanatory Variables	R^2	AICc	ΔAICc	AICc Weight	Rank
7	**log(Colony Distance) + Pond Size + log(Colony Distance) × Pond Size**	**0.429**	**39960.4**	**0**	**1.0**	**1**
6	log(Colony Distance) + Pond Size	0.291	40482.6	522.2	4.0×10^{-114}	2
4	2015 TCG + Pond Size + 2015 TCG × Pond Size	0.276	40532.4	572.0	6.2×10^{-125}	3
3	2015 TCG + Pond Size	0.220	40710.3	749.9	1.4×10^{-163}	4
1	Pond Size	0.211	40739.5	779.1	6.5×10^{-170}	5
9	Flow Accumulation + Pond Size	0.210	40741.0	780.6	3.1×10^{-170}	6
10	Flow Accumulation + Pond Size + Flow Accumulation × Pond Size	0.210	40742.5	782.1	1.4×10^{-170}	7
5	log(Colony Distance)	0.0995	41056.4	1096.0	9.8×10^{-239}	8
2	2015 TCG	0.00926	41286.4	1326.0	1.1×10^{-288}	9
8	Flow Accumulation	0.00264	41302.5	1342.1	3.6×10^{-292}	10

In stable and severe drying plots, there was an increase in the mean size of waterbodies present in 2014 (Figure 4), which is indicative of a disproportionate loss of small waterbodies. The change in mean waterbody size between 1958–2014 was considerably smaller in severe drying plots (Δ 9.4 m^2), compared to stable plots (Δ 876.9 m^2), which showed a more balanced loss of waterbodies of varying sizes.

3.2.2. Model Selection

The model selection procedure showed that large reductions in waterbody area were most strongly associated with waterbody size and proximity to the nesting colony. The best model included an interaction between the distance from the colony and pond size measurements (Table 4). Waterbody area losses were larger near the colony regardless of waterbody size, but the impact of the colony was most obvious in larger waterbodies. This was evidenced by an increase in the slope of the relationship between the distance from the colony and area loss in larger waterbodies (Figure 5).

3.2.3. Categorical Proportional Area Loss

All waterbody size classes, except the largest (>28,800 m^2), had significantly greater proportional area loss ($p < 0.05$) within 1 km of the nesting colony, compared to waterbodies between 1–5 km and further than 5 km from the nesting colony (Figure 6). At distances greater than 1 km from the nesting colony, waterbodies showed similar proportional changes in all size classes. Regardless of distance from the colony, the smallest waterbody size class had the largest proportional area loss at 95.3% $+/-$ 9.2 for 0–1 km from the nesting colony, 76.3% $+/-$ 11.1 for 1–5 km, and 74.4% $+/-$ 7.8 for greater than 5 km.

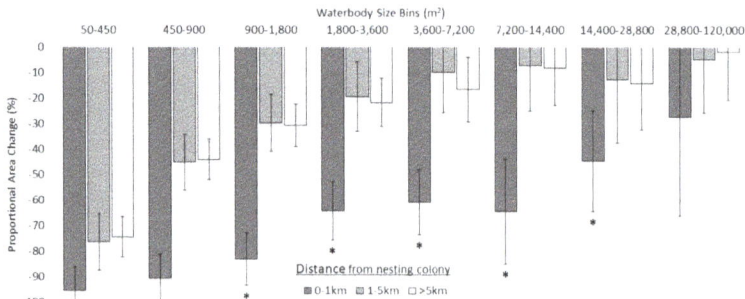

Figure 6. Bar plots showing the proportional area change by distance from the nesting colony and waterbody size intervals. The error bars represent 95% confidence intervals. Asterisks are present when the 0–1 km distance group is significantly different from the other distance groups based on the least-squares means estimates.

3.3. Field Surveys

3.3.1. Pond Basin Transect Intersections

Field surveys showed evidence of widespread drying, as all site types intersected former pond basins to some extent. Colony sites had the highest proportion of former pond basins (75.76%), then drying sites (31.82%), and then control sites (2.05%) (Figure 7). Regardless of site type, former pond basins intersected transect lines exclusively within pixels that experienced negative SWF trends.

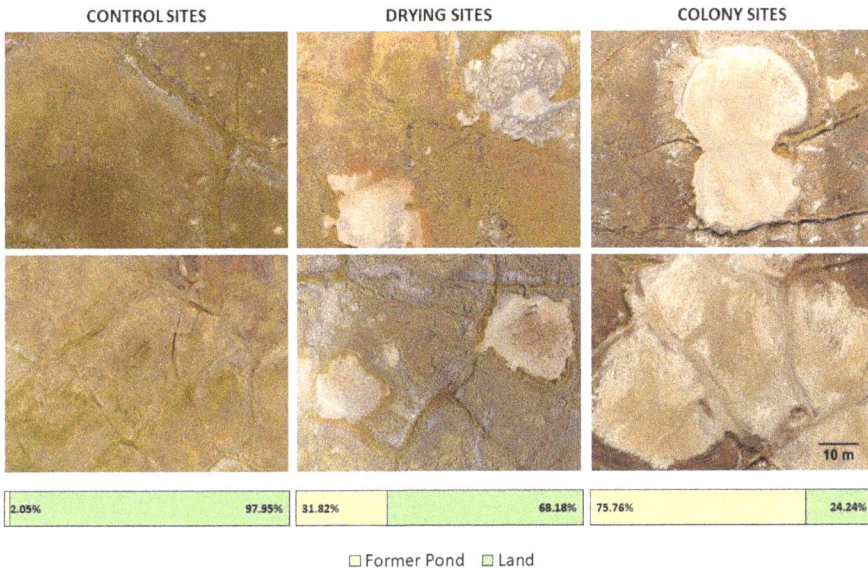

Figure 7. Aerial photographs captured using an unmanned aerial vehicle (UAV) during July 2017 field surveys. The bars below the images show the proportion of field transects classified as former pond basins and regular land.

3.3.2. Biotic and Abiotic Site Differences

Soils were significantly drier at colony sites ($p < 0.05$) compared to the control and drying site types, which were not significantly different from each other (Figure 8). Mean soil volumetric water content was three times lower at colony sites (24.6% $+/-$ 30.4%), compared to drying (78.4% $+/-$ 24.6%) and control sites (72.2% $+/-$ 15.8%). Thaw depth was not significantly different among site types (Figure S2). Vegetation cover was also significantly lower at the colony sites ($p < 0.01$) (Figure 8), where groundcover was dominated by exposed peat. Mean vascular plant cover was 10–15 times lower at colony sites (4.1% $+/-$ 23.6%), compared to drying (43.8% $+/-$ 18.2%) and control sites (59.5% $+/-$ 10.9%). Goose grubbing was not significantly different among site types (Figure S2).

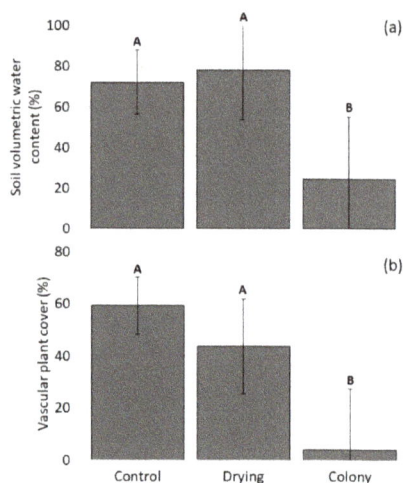

Figure 8. Least-squares means estimates of (**a**) soil volumetric water content and (**b**) vegetation cover from the linear mixed effects models. Error bars represent 95% confidence intervals and bars with different letters are significantly different.

4. Discussion

The results of our analyses confirm our hypothesis that the number and size of waterbodies in the pond complexes of western Banks Island are decreasing. This observation differs from most surface water studies in the low Arctic [17,19], which have reported increases in surface water. In the pond complexes of western Banks Island, only a small portion of existing waterbodies expanded and few new waterbodies emerged. This indicates that surface water responses to climate change in high Arctic regions are distinct and require additional research, as the high Arctic is one of Canada's largest ecozones, covering about 15% of the country. Estimates of change in surface water based on the SWF analysis and manual aerial imagery digitization span different time periods, but results from both sources of information showed similar patterns of substantial surface water loss.

4.1. Smaller Waterbodies Are More Vulnerable

Our observation that small waterbodies are being disproportionately affected is aligned with our second hypothesis and suggests that climate is the main driver of drying in the study area. Almost half of the loss in surface water occurred in small waterbodies (50–1800 m^2), despite only making up 20.5% of the total surface water in 1958. There are several climate-driven processes that may be contributing to the loss of surface water on Banks Island. Warming summer temperatures and extended ice-free seasons increase annual evaporative losses and can lead to the complete desiccation of waterbodies [15,62]. The effects of this would be most evident in small and shallow waterbodies because of their high surface area to volume ratios [15,20,21]. Declines in terrestrial water storage in large river basins across the Arctic have also been linked to increases in evapotranspiration [64]. Warmer and longer summers would also reduce summer snowpack, which might be sustaining wet areas with meltwater throughout the ice-free season [26,28,65]. Most of the summer snowpack observed in the 1958 aerial photographs was not visible in the 2014 aerial imagery, despite having acquisition dates only five days apart. Warming air and ground temperatures can also increase thaw depth and groundwater storage capacities, which might lower the water table [28]. This would reduce hydrological connectivity [28] and could potentially desiccate shallow waterbodies perched above the lowered water table. Degradation of low-centered polygonal terrain, from increasing ground temperatures, would also result in a large loss of small waterbodies [66]. As low-centered polygons degrade into high-centered polygons, their capacity to hold water decreases. Our data suggests this is not occurring because it would also result in an increase in waterbodies in the trough areas of the polygons [66], which we did not observe. Furthermore, areas visited in the field, which experienced concentrated losses of small waterbodies, have remained as low-centered polygonal terrain. Further investigation is needed to evaluate the potential contributions of these processes and to understand how similar terrain types in other regions of the Arctic are being impacted. The knowledge of hunters and land-users can also provide significant insight into these processes [67–69].

Our finding that small ponds are the most vulnerable to change highlights the importance of using fine-scale data or sub-pixel/spectral un-mixing techniques for surface water change detection in areas with small waterbody size distributions. It is important to use data at appropriate scales or remote sensing techniques that are matched with the biophysical variation in the area of interest. In our study area, most waterbodies were smaller than 900 m^2, the size of the Landsat pixel footprint. Previous broad-scale analyses, not considering sub-pixel information [70], were only sensitive to changes impacting the majority of a pixel and therefore did not detect the magnitude of change we observed.

4.2. Intensified Drying in the Nesting Colony

Our results also confirm our hypothesis that surface water changes are following different trajectories in the nesting colony area and indicate that the intensive occupation of lesser snow geese may be reducing the resilience of waterbodies to climate warming by facilitating drying or draining processes. Reductions in waterbody area were larger and more consistent closer to the nesting colony, regardless of original waterbody size. Colony sites were shown to have reduced vegetation cover and soil moisture, which are common impacts of overgrazing in expanding snow goose nesting colonies [34,35]. Vegetation provides a strong insulating layer that stabilizes near-surface ground temperatures and increases soil moisture retention [34,37]. In other regions, reductions in the vegetation layer associated with intense goose activity have been found to increase evaporation in soils [34], which may reduce subsurface hydrological connectivity. Waterbodies isolated from subsurface inputs are also more vulnerable to desiccation over the ice-free season [26,27,62].

Trampling by lesser snow geese, which can create depressions and terraces [36], likely contributes to reduced vegetation cover to increase the risk of lateral drainage through accelerated degradation of ice-wedge polygons. Ice-wedge polygons are one of the most common forms of ground ice in the Arctic [71,72] and were ubiquitous at our field sites. These features have been suggested to be more vulnerable to degradation in higher latitude areas because wedge ice is located closer to the ground surface and is less insulated from changing air temperatures [29,30].

Since lesser snow goose breeding areas are synonymous with wet Arctic habitats, a better understanding of the impacts of these animals on permafrost and surface water dynamics is important. Similar patterns of surface water loss can be seen in the lowland areas of Southampton Island, Nunavut [70], which hosts the third largest lesser snow goose population in the Canadian Arctic [35]. As populations continue to expand, impacts on freshwater systems in the Canadian Arctic are likely to intensify. Further research on snow goose habitat impacts to permafrost and surface water dynamics are necessary, as this topic is largely unstudied to date.

4.3. Implications

Drying of high Arctic wetlands will impact lesser snow geese and other herbivore populations, as these areas provide important breeding habitat in largely arid polar deserts [35]. A recent study projected 30–80% reductions in lake extent within the areas of five Alaskan National Wildlife Refuges over the next 50 years, and anticipated that these areas would not persist as important waterfowl production areas if rates of change continue [23]. Changes in the extent of surface water can also affect permafrost thaw [12–14] and increase methane and carbon dioxide emissions with shoreline expansion [12,73]. Understanding surface water changes in high Arctic environments is therefore critical for making accurate evaluations of greenhouse gas emissions across the Arctic.

5. Conclusions

Based on the data analyzed here, we draw the following conclusions:

- The pond complexes on western Banks Island are drying.
- Wetland drying is being caused by warming climate, but is exacerbated in areas with intensive snow goose habitat use.
- Future studies should explore the mechanisms causing pond desiccation, the impacts of snow geese on these processes, and the impacts of drying on vegetation and permafrost conditions.
- Remote sensing studies must use data and methods that consider the biophysical variation in the area of interest to adequately assess environmental changes.

Supplementary Materials: The following are available online at http://www.mdpi.com/2072-4292/10/12/1892/s1, Figure S1: Study methods flow diagram, Figure S2: Linear mixed effects model outputs for thaw depth and goose grubbing.

Author Contributions: Conceptualization—T.K.F.C. and T.C.L.; methodology—T.K.F.C., T.C.L., and R.H.F.; software—T.K.F.C. and R.H.F.; validation—T.K.F.C., T.C.L., and R.H.F.; formal analysis—T.K.F.C.; investigation—T.K.F.C., T.C.L., and R.H.F.; resources—R.H.F. and T.C.L.; data curation—T.C.L. and R.H.F.; writing—original draft preparation, T.K.F.C.; writing—review and editing, T.C.L. and R.H.F.; visualization—T.K.F.C.; supervision—T.C.L.; project administration—T.C.L.; funding acquisition—T.C.L.

Funding: This research was funded by: The Polar Continental Shelf Program; the Natural Sciences and Engineering Research Council of Canada; ArcticNet; the Northern Scientific Training Program; the Canadian Space Agency Government Related Initiatives Program (GRIP); and the University of Victoria.

Acknowledgments: This work was made possible by the Aurora Research Institute–Western Arctic Research Centre, the Canadian Wildlife Service—Yellowknife, and the Sachs Harbour Hunters and Trappers Committee. We thank Marie Fast, Megan Ross, Danica Hogan, Eric Reed, and Cindy Wood from the Canadian Wildlife Service, as well as Trevor Lucas, from the Sachs Harbour Hunters and Trappers Committee, for their critical in-field and logistical support. We also thank Ian Olthof and Hana Travers-Smith for their assistance with the collection and processing of remote sensing data.

Conflicts of Interest: The authors declare no conflict of interest.

Appendix A

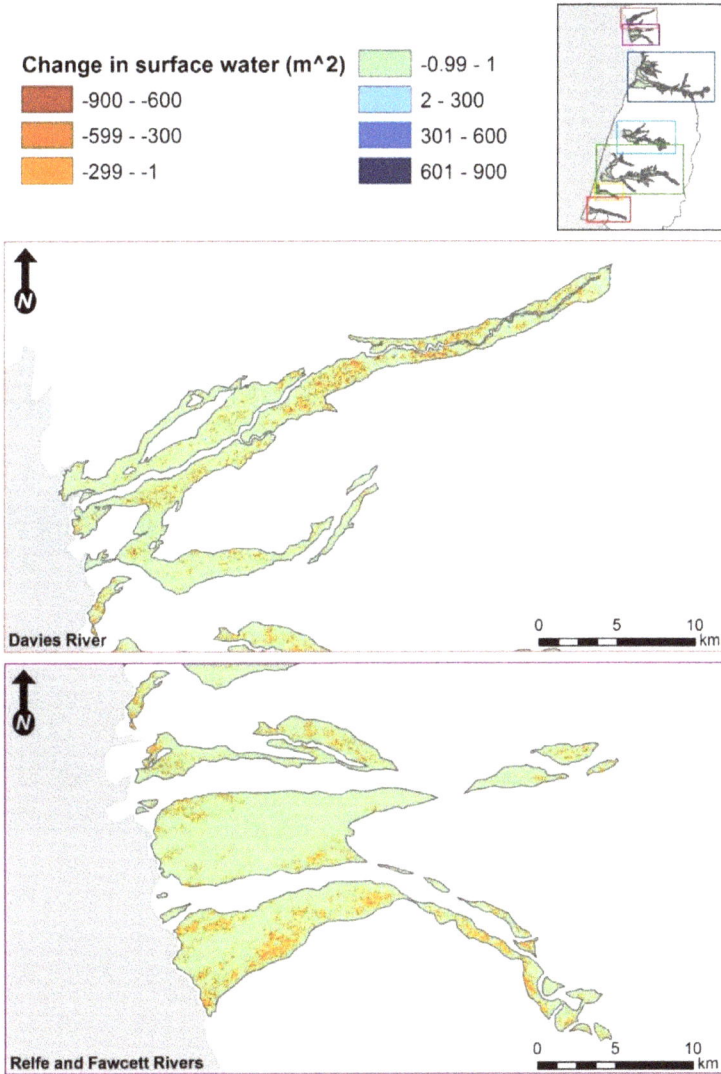

Figure A1. Sub-pixel water fraction trend surfaces of the Davies, Relfe, and Fawcett river valleys of western Banks Island, between 1985–2015.

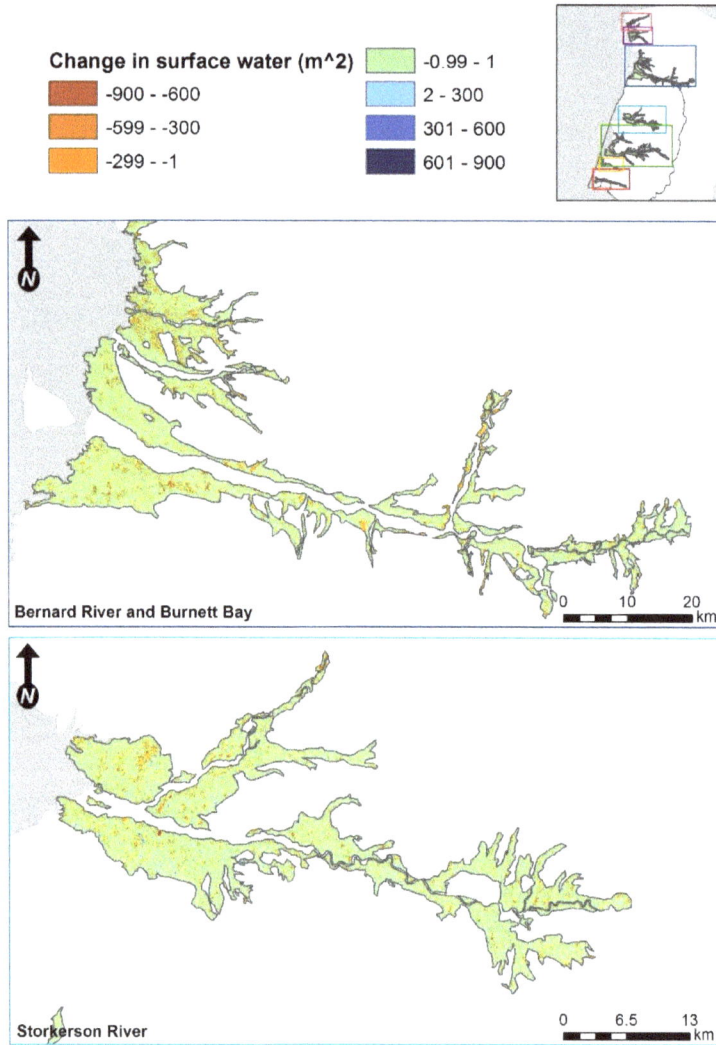

Figure A2. Sub-pixel water fraction trend surfaces of the Burnett Bay area, and the Bernard and Storkerson river valleys of western Banks Island, between 1985–2015.

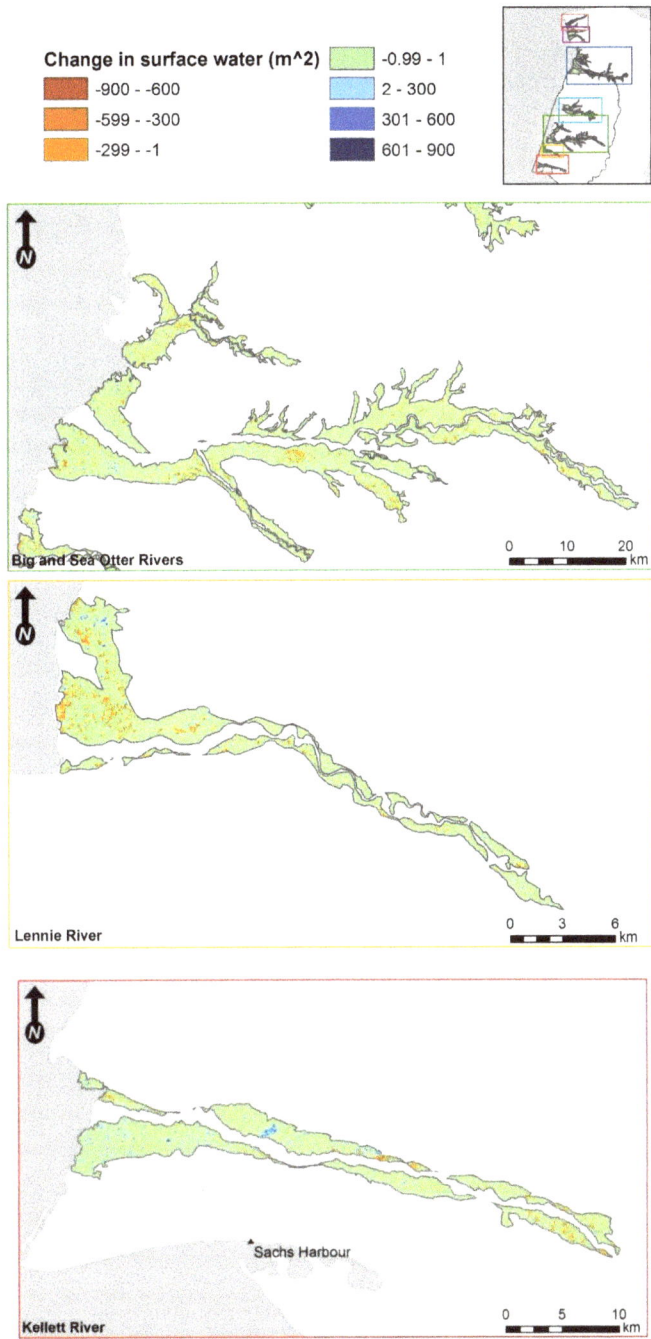

Figure A3. Sub-pixel water fraction trend surfaces of the Sea Otter, Big, Lennie, and Kellett river valleys of western Banks Island, between 1985–2015.

References

1. AMAP (Arctic Monitoring and Assessment Programme). *Arctic Climate Issues 2011: Changes in Arctic Snow, WATER, ice and Permafrost*; SWIPA 2011 Overview Report; AMAP: Oslo, Norway, 2012; p. 97.
2. Pithan, F.; Mauritsen, T. Arctic amplification dominated by temperature feedbacks in contemporary climate models. *Nat. Geosci.* **2014**, *7*, 181–184. [CrossRef]
3. Lantz, T.C.; Turner, K.W. Changes in lake area in response to thermokarst processes and climate in Old Crow Flats, Yukon. *J. Geophys. Res. Biogeosci.* **2015**, *120*, 513–524. [CrossRef]
4. Kaplan, J.O.; New, M. Arctic climate change with a 2 °C global warming: Timing, climate patterns and vegetation change. *Clim. Chang.* **2006**, *79*, 213–241. [CrossRef]
5. Bintanja, R.; Andry, O. Towards a rain-dominated Arctic. *Nat. Clim. Chang.* **2017**, *7*, 263–268. [CrossRef]
6. Yoshikawa, K.; Hinzman, L.D. Shrinking thermokarst ponds and groundwater dynamics in discontinuous permafrost near Council Alaska. *Permafr. Periglac. Process.* **2003**, *14*, 151–160. [CrossRef]
7. Nitze, I.; Grosse, G.; Jones, B.M.; Arp, C.D.; Ulrich, M.; Fedorov, A.; Veremeeva, A. Landsat-based trend analysis of lake dynamics across northern permafrost regions. *Remote Sens.* **2017**, *9*, 640. [CrossRef]
8. Wolfe, B.B.; Light, E.M.; Macrae, M.L.; Hall, R.I.; Eichel, K.; Jasechko, S.; White, J.; Fishback, L.; Edwards, T.W.D. Divergent hydrological responses to 20th century climate change in shallow tundra ponds, western Hudson Bay Lowlands. *Geophys. Res. Lett.* **2011**, *38*, 1–6. [CrossRef]
9. Negandhi, K.; Laurion, I.; Whiticar, M.J.; Galand, P.E.; Xu, X.; Lovejoy, C. Small thaw ponds: An unaccounted source of methane in the Canadian high Arctic. *PLoS ONE* **2013**, *8*, e78204. [CrossRef] [PubMed]
10. Becker, M.S.; Davies, T.J.; Pollard, W.H. Ground ice melt in the high Arctic leads to greater ecological heterogeneity. *J. Ecol.* **2016**, *104*, 114–124. [CrossRef]
11. Slattery, S.M.; Alisauskas, R.T. Distribution and habitat use of Ross' and lesser snow geese during late brood rearing. *J. Wildl. Manag.* **2007**, *71*, 2230–2237. [CrossRef]
12. Anthony, K.W.; Daanen, R.; Anthony, P.; Schneider von Deimling, T.; Ping, C.-L.; Chanton, J.P.; Grosse, G. Methane emissions proportional to permafrost carbon thawed in Arctic lakes since the 1950s. *Nat. Geosci.* **2016**, *9*, 679–685. [CrossRef]
13. White, D.; Hinzman, L.; Alessa, L.; Cassano, J.; Chambers, M.; Falkner, K.; Francis, J.; Gutowski, W.J., Jr.; Holland, M.; Holmes, R.M.; et al. The arctic freshwater system: Changes and impacts. *J. Geophys. Res.* **2007**, *112*, G04S54. [CrossRef]
14. Raymond, P.A.; Hartmann, J.; Lauerwald, R.; Sobek, S.; McDonald, C.; Hoover, M.; Butman, D.; Striegl, R.; Mayorga, E.; Humborg, C.; et al. Global carbon dioxide emissions from inland waters. *Nature* **2013**, *503*, 355–387. [CrossRef] [PubMed]
15. Smol, J.P.; Douglas, S.V. Crossing the final ecological threshold in high Arctic ponds. *PNAS* **2007**, *104*, 12395–12397. [CrossRef] [PubMed]
16. Plug, L.J.; Walls, C.; Scott, B.M. Tundra lake changes from 1978 to 2001 on the Tuktoyaktuk Peninsula, western Canadian Arctic. *Geophys. Res. Lett.* **2008**, *35*, L03502. [CrossRef]
17. Smith, L.C.; Sheng, Y.; MacDonald, G.M.; Hinzman, L.D. Disappearing Arctic lakes. *Science* **2005**, *308*, 1429. [CrossRef] [PubMed]
18. Jones, B.M.; Grosse, G.; Arp, C.D.; Jones, M.C.; Anthony, K.M.W.; Romanovsky, V.E. Modern thermokarst lake dynamics in the continuous permafrost zone, northern Seward Peninsula, Alaska. *J. Geophys. Res.* **2011**, *116*, G00M03. [CrossRef]
19. Olthof, I.; Fraser, R.H.; Schmitt, C. Landsat-based mapping of thermokarst lake dynamics on the Tuktoyaktuk coastal plain: Northwest Territories, Canada since 1985. *Remote Sens. Environ.* **2015**, *168*, 194–204. [CrossRef]
20. Arp, C.D.; Jones, B.M.; Urban, F.E.; Grosse, G. Hydrogeomorphic processes of thermokarst lakes with grounded-ice and floating-ice regimes on the Arctic coastal plain, Alaska. *Hydrol. Process.* **2011**, *25*, 2422–2438. [CrossRef]
21. Marsh, P.; Bigras, S.C. Evaporation from Mackenzie Delta lakes, N.W.T., Canada. *Arct. Alp. Res.* **1988**, *20*, 220–229. [CrossRef]
22. Turner, K.W.; Wolfe, B.B.; Edwards, T.E.; Lantz, T.C.; Hall, R.I.; La Rocque, G. Controls on water balance of shallow thermokarst lakes and their relations with catchment characteristics: A multi-year, landscape-scale assessment based on water isotope tracers and remote sensing in Old Crow Flats, Yukon (Canada). *Glob. Chang. Boil.* **2014**, *20*, 1585–1603. [CrossRef]

23. Roach, J.K.; Griffith, B.; Verbyla, D. Landscape influences on climate-related lake shrinkage at high latitudes. *Glob. Chang. Boil.* **2013**, *19*, 2276–2284. [CrossRef] [PubMed]

24. Carroll, M.L.; Townshend, J.R.G.; DiMiceli, C.M.; Loboda, T.; Sohlberg, R.A. Shrinking lakes of the Arctic: Spatial relationships and trajectory of change. *Geophys. Res. Lett.* **2011**, *38*, 1–5. [CrossRef]

25. Riordan, B.; Verbyla, D.; McGuire, A.D. Shrinking ponds in subarctic Alaska based on 1950–2002 remotely sensed images. *J. Geophys. Res.* **2006**, *111*, G04002. [CrossRef]

26. Woo, M.-K.; Young, K.L. High Arctic wetlands: Their occurrence, hydrological characteristics and sustainability. *J. Hydrol.* **2006**, *320*, 432–450. [CrossRef]

27. Abnizova, A.; Young, K.L. Sustainability of high Arctic ponds in a polar desert environment. *Arctic* **2010**, *63*, 67–84. [CrossRef]

28. Woo, M.-K.; Young, K.L.; Brown, L. High Arctic patchy wetlands: Hydrological variability and their sustainability. *Phys. Geogr.* **2006**, *27*, 297–307. [CrossRef]

29. Steedman, A.E.; Lantz, T.C.; Kokelj, S.V. Spatio-temporal variation in high-centre polygons and ice-wedge melt ponds, Tuktoyaktuk Coastlands, Northwest Territories. *Permafr. Periglac. Process.* **2017**, *28*, 66–78. [CrossRef]

30. Fraser, R.H.; Kokelj, S.V.; Lantz, T.C.; McFarlane-Winchester, M.; Olthof, I.; Lacelle, D. Climate sensitivity of high Arctic permafrost terrain demonstrated by widespread ice-wedge thermokarst on Banks Island. *Remote Sens.* **2018**, *10*, 954. [CrossRef]

31. Hines, J.E.; Latour, P.B.; Squires-Taylor, C.; Moore, S. *The Effects on Lowland Habitat in the Banks Island Bird Sanctuary Number 1, Northwest Territories, by the Growing Colony of Lesser Snow Geese (Chen caerulescens caerulescens)*; Environment Canada Occasional Paper 118; Environment Canada, Canadian Wildlife Service: Edmonton, AB, Canada, 2010; pp. 8–26.

32. Batt, B.D.J. *Arctic Ecosystems in Peril: Report of the Arctic Goose Habitat Working Group*; Arctic Goose Joint Venture Special Publication; U.S. Fish and Wildlife Service: Washington, DC, USA; Canadian Wildlife Service: Ottawa, ON, Canada, 1997; pp. 1–12. ISBN 0961727934.

33. Kotanen, P.M.; Jefferies, R.L. Long-term destruction of sub-arctic wetland vegetation by lesser snow geese. *Ecoscience* **1997**, *4*, 179–182. [CrossRef]

34. Srivastava, D.S.; Jefferies, R.L. A positive feedback: Herbivory, plant growth, salinity, and the desertification of an Arctic salt-marsh. *J. Ecol.* **1996**, *84*, 31–42. [CrossRef]

35. Calvert, A.M. *Interactions between Light Geese and Northern Flora and Fauna: Synthesis and Assessment of Potential Impacts*; Unpublished Report; Environment Canada: Ottawa, ON, Canada, 2015; pp. 1–37.

36. Jefferies, R.L.; Jensen, A.; Abraham, K.F. Vegetational development and the effect of geese on vegetation at La Perouse Bay, Manitoba. *Can. J. B* **1979**, *57*, 1439–1450. [CrossRef]

37. Iacobelli, A.; Jefferies, R.L. Inverse salinity gradients in coastal marshes and the death of stands of *Salix*: The effects of grubbing by geese. *J. Ecol.* **1991**, *79*, 61–73. [CrossRef]

38. Park, J.S. A race against time: Habitat alteration by snow geese prunes the seasonal sequence of mosquito emergence in a subarctic brackish landscape. *Polar Biol.* **2017**, *40*, 553–561. [CrossRef]

39. Mudryk, L.R.; Derksen, C.; Howell, S.; Laliberte, F.; Thackeray, C.; Sospedra-Alfonso, R.; Vionnet, V.; Kushner, P.J.; Brown, R. Canadian snow and sea ice: Historical trends and projections. *Cryosphere* **2018**, *12*, 1157–1176. [CrossRef]

40. Lakeman, T.R.; England, J.H. Late Wisconsinan glaciation and postglacial relative sea-level change on western Banks Island, Canadian Arctic Archipelago. *Quat. Res.* **2013**, *80*, 99–112. [CrossRef]

41. Ecosystem Classification Group. *Ecological Regions of the Northwest Territories—Northern Arctic*; Department of Environment and Natural Resources, Government of the Northwest Territories: Yellowknife, NT, Canada, 2013; pp. 1–157, ISBN 978-0-7708-0205-9.

42. Chander, G.; Markham, B.L.; Helder, D.L. Summary of current radiometric calibration coefficients for Landsat MSS, TM, ETM+, and EO-1 ALI sensors. *Remote Sens. Environ.* **2009**, *113*, 893–903. [CrossRef]

43. Crist, E.P.; Cicone, R.C. A physically-based transformation of thematic mapper data—The TM tasseled cap. *IEEE Trans. Geosci. Remote Sens.* **1984**, *GE-22*, 256–263. [CrossRef]

44. Huang, C.; Wylie, B.; Yang, L.; Homer, C.; Zylstra, G. Derivation of a tasseled cap transformation based on Landsat 7 at-satellite reflectance. *Int. J. Remote Sens.* **2002**, *23*, 1741–1748. [CrossRef]

45. Kauth, R.J.; Thomas, G.S. The tasseled cap—A graphic description of the spectral-temporal development of agricultural crops as seen by Landsat. *LARS Symp.* **1976**, *159*, 41–51.

46. Zeileis, A.; Leisch, F.; Hornik, K.; Kleiber, C. Strucchange: An R package for testing for structural change in linear regression models. *J. Stat. Softw.* **2002**, *7*, 1–38. [CrossRef]

47. R Core Team. R: A Language and Environment for Statistical Computing. R Foundation for Statistical Computing, Vienna, Austria, 2016. Available online: https://www.R-project.org/ (accessed on 18 March 2018).

48. Bai, J.; Perron, P. Computation and analysis of multiple structural change models. *J. Appl. Econ.* **2003**, *18*, 1–22. [CrossRef]

49. Fraser, R.H.; Olthof, I.; Kokelj, S.V.; Lantz, T.C.; Lacelle, D.; Brooker, A.; Wolfe, S.; Schwarz, S. Detecting landscape changes in high latitude environments using Landsat trend analysis: 1. Visualization. *Remote Sens.* **2014**, *6*, 11533–11557. [CrossRef]

50. Kendall, M.G.; Stuart, A.S. *Advanced Theory of Statistics*; Charles Griffin and Company: London, UK, 1967; Volume 2.

51. Noh, M.-J.; Howat, I.M. Automated stereo-photogrammetric DEM generation at high latitudes: Surface extraction with TIN-based search-space minimization (SETSM) validation and demonstration over glaciated regions. *GISci. Remote Sens.* **2015**, *52*, 198–217. [CrossRef]

52. Anselin, L. Local indicators of spatial association—LISA. *Geogr. Anal.* **1995**, *27*, 93–115. [CrossRef]

53. Anselin, L. GeoDa: An introduction to spatial data analysis. *Geogr. Anal.* **2005**, *38*, 5–22. [CrossRef]

54. Burnham, K.P.; Anderson, D.R. *Model Selection and Multimodel Inference: A Practical Information-Theoretic Approach*, 2nd ed.; Springer: New York, NY, USA, 2002; ISBN 0-387-95364-7.

55. Zuur, A.F.; Ieno, E.N.; Elphick, C.S. A protocol for data exploration to avoid common statistical problems. *Methods Ecol. Evol.* **2010**, *1*, 3–14. [CrossRef]

56. Graham, M.H. Confronting multicollinearity in ecological multiple regression. *Ecology* **2003**, *84*, 2809–2815. [CrossRef]

57. Littell, R.C.; Milliken, G.A.; Stroup, W.W.; Wolfinger, R.D.; Schabenberger, O. *SAS for Mixed Models*, 2nd ed.; SAS Institute Inc.: Carry, NC, USA, 2006; ISBN 978-1-59047-500-3.

58. Samelius, G.; Alisauskas, R.T.; Hines, J.E. *Productivity of Lesser Snow Geese on Banks Island, Northwest Territories, Canada, in 1995–1998*; Environment Canada Occasional Paper 115; Environment Canada, Canadian Wildlife Service: Edmonton, AB, Canada, 2008; pp. 3–33.

59. Price, L.W. Vegetation, microtopography, and depth of active layer on different exposures in subarctic alpine tundra. *Ecology* **1971**, *52*, 638–647. [CrossRef] [PubMed]

60. Fisher, J.P.; Estop-Aragones, C.; Thierry, A.; Charman, D.J.; Wolfe, S.A.; Hartley, I.P.; Murton, J.B.; Williams, M.; Phoenix, G.K. The influence of vegetation and soil characteristics on active-layer thickness of permafrost soils in boreal forest. *Glob. Chang. Boil.* **2016**, *22*, 3127–3140. [CrossRef] [PubMed]

61. Gornall, J.L.; Jonsdottir, I.S.; Woodin, S.J.; Van der Wal, R. Arctic mosses govern below-ground environment and ecosystem processes. *Oecologia* **2007**, *153*, 931–941. [CrossRef] [PubMed]

62. Woo, M.-K.; Guan, X.J. Hydrological connectivity and seasonal storage change of tundra ponds in a polar oasis environment, Canadian high Arctic. *Permafr. Periglac. Process.* **2006**, *17*, 309–323. [CrossRef]

63. Kokelj, S.V.; Jorgenson, M.T. Advances in thermokarst research. *Permafr. Periglac. Process.* **2013**, *24*, 108–119. [CrossRef]

64. Suzuki, K.; Matsuo, K.; Yamakazi, D.; Ichii, K.; Iijima, Y.; Papa, F.; Yanagi, Y.; Hiyama, T. Hydrological variability and changes in the Arctic circumpolar tundra and the three largest pan-Arctic river basins from 2002 to 2016. *Remote Sens.* **2018**, *10*, 402. [CrossRef]

65. Brown, L.; Young, K.L. Assessment of three mapping techniques to delineate lakes and ponds in a Canadian high Arctic wetland complex. *Arctic* **2006**, *59*, 283–293. [CrossRef]

66. Liljedahl, A.K.; Boike, J.; Daanen, R.P.; Fedorov, A.N.; Frost, G.V.; Grosse, G.; Hinzman, L.D.; Iijima, Y.; Jorgenson, J.C.; Matveyeva, N.; et al. Pan-Arctic ice-wedge degradation in warming permafrost and its influence on tundra hydrology. *Nat. Geosci.* **2016**, *9*, 312–319. [CrossRef]

67. Riedlinger, D.; Berkes, F. Contributions of traditional knowledge to understanding climate change in the Canadian Arctic. *Polar Rec.* **2001**, *37*, 315–328. [CrossRef]

68. Ashford, G.; Catleden, J. *Inuit Observations on Climate Change: Final Report*; International Institute for Sustainable Development: Winnipeg, MB, Canada, 2001; pp. 1–31.

69. Ferguson, M.A.D.; Messier, F. Collection and analysis of traditional ecological knowledge about a population of Arctic tundra caribou. *Arctic* **1997**, *50*, 17–28. [CrossRef]

70. Pekel, J.-F.; Cottam, A.; Gorelick, N.; Belward, A.S. High-resolution mapping of global surface water and its long-term changes. *Nature* **2016**, *540*, 418–422. [CrossRef] [PubMed]

71. French, H.M. *The Periglacial Environment*, 3rd ed.; John Wiley & Sons: Chichester, UK, 2007; ISBN 978-0-47086-588-0.

72. Lachenbruch, A.H. *Mechanics of Thermal Contraction Cracks and Ice-Wedge Polygons in Permafrost*; Geological Society of America Special Papers; Geological Society of America: Boulder, CO, USA, 1962; Volume 70, pp. 1–66. [CrossRef]

73. Martin, A.F.; Lantz, T.C.; Humphreys, E.R. Ice wedge degradation and CO_2 and CH_4 emissions in the Tuktoyaktuk coastlands, Northwest Territories. *Arct. Sci.* **2018**, *4*, 130–145. [CrossRef]

remote sensing

MDPI

Article

Recovery Rates of Wetland Vegetation Greenness in Severely Burned Ecosystems of Alaska Derived from Satellite Image Analysis

Christopher Potter

NASA Ames Research Center, Moffett Field, CA 94035, USA; chris.potter@nasa.gov; Tel.: +1-650-604-6164

Received: 27 July 2018; Accepted: 8 September 2018; Published: 12 September 2018

Abstract: The analysis of wildfire impacts at the scale of less than a square kilometer can reveal important patterns of vegetation recovery and regrowth in freshwater Arctic and boreal regions. For this study, NASA Landsat burned area products since the year 2000, and a near 20-year record of vegetation green cover from the MODIS (Moderate Resolution Imaging Spectroradiometer) satellite sensor were combined to reconstruct the recovery rates and seasonal profiles of burned wetland ecosystems in Alaska. Region-wide breakpoint analysis results showed that significant structural change could be detected in the 250-m normalized difference vegetation index (NDVI) time series for the vast majority of wetland locations in the major Yukon river drainages of interior Alaska that had burned at high severity since the year 2001. Additional comparisons showed that wetland cover locations across Alaska that have burned at high severity subsequently recovered their green cover seasonal profiles to relatively stable pre-fire levels in less than 10 years. Negative changes in the MODIS NDVI, namely lower greenness in 2017 than pre-fire and incomplete greenness recovery, were more commonly detected in burned wetland areas after 2013. In the years prior to 2013, the NDVI change tended to be positive (higher greenness in 2017 than pre-fire) at burned wetland elevations lower than 400 m, whereas burned wetland locations at higher elevation showed relatively few positive greenness recovery changes by 2017.

Keywords: wildfire; wetlands; elevation; MODIS; Landsat; Alaska

---

1. Introduction

High-severity wildfires have been shown to have long-term impacts on freshwater ecosystems; as nutrients are mobilized, runoff and erosion can increase, and soil properties may be modified [1]. While there is a growing literature for the effects of fire on upland vegetation types [2], the existing information on vegetation removal by burning remains limited for most freshwater plant communities globally.

To extend this knowledge base, satellite remote sensing can be used to effectively monitor changes in high-latitude (boreal and tundra) wetland vegetation cover and productivity, especially following disturbance events such as wildfires [3–7]. Most of these remote sensing studies have been carried out for non-wetland (interior boreal forest and upland tundra) vegetation cover types. Nonetheless, Potter et al. [6] reported that the wetland tundra areas of Alaska that burned since the year 1980 had a 3:2 ratio coverage of significant positive versus negative vegetation greening trends between 2000–2010, whereas non-wetland tundra areas that burned since 1980 had a 2:5 coverage ratio of significant positive versus negative vegetation greening trends between 2000–2010. This result suggested that the wetland areas of Alaska can recover more completely and rapidly in greenness cover from recent wildfires than non-wetland land cover types; however, this supposition remains to be tested region-wide over longer time periods.

Over the past several decades, there has been an increase in the frequency and severity of boreal region wildfires in Alaska [8]. During the 2000s, an average of 767,000 ha per year were burned statewide, which is 50% higher than in any previous decade since the 1940s. In the extreme wildfire year of 2015, nearly 60% of Alaska's burned area was consumed at moderate-to-high severity levels [9].

Most of the wildfires in the spruce forest ecosystems of Alaska are either crown or ground fires with a high enough severity to kill over-story trees [10–12]. Usually, some of the organic layer of the forest floor remains, but fires in late summer following exceptionally dry or windy conditions may consume all of the organic layer, exposing mineral soil [13]. Jiang et al. [14] and Brown et al. [15] reported that the post-fire thickness of the soil organic layer and its impact on soil thermal conductivity was the most important factor determining post-fire soil temperatures and thaw depth. In moderately burned sites, the presence of permafrost can mitigate the loss of the insulating soil organic layer, decrease soil drying, and increase surface water pooling.

The objective of this study was to analyze the vegetation recovery patterns of all of the Alaska wetlands that have burned at high severity since the year 2000 using a combination of the Landsat and MODIS (Moderate Resolution Imaging Spectroradiometer) satellite datasets. A statistical analysis of the changes in the MODIS vegetation index time series was conducted using the "Breaks for Additive Seasonal and Trend" method (BFAST, Verbesselt et al. [16,17]). de Jong et al. [18] analyzed trends in the normalized difference vegetation index (NDVI) satellite time series using the BFAST procedure, and detected both abrupt and gradual changes in large parts of the world, especially in shrubland and grassland biomes where abrupt greening was often followed by gradual browning.

This study was undertaken as a contribution to the NASA Arctic Boreal Vulnerability Experiment (ABoVE) field campaign, chiefly to better understand changes in related hydrologic and biogeochemical mechanisms in the years following high-latitude wildfires. One of the major questions being addressed by ABoVE, and in this type of Landsat/MODIS study, is "What processes are controlling changes in boreal–Arctic land cover properties, and what are the impacts of these changes?"

2. Materials and Methods

2.1. Landsat Burn Severity Classes

Digital maps of burn severity classes at 30-m spatial resolution were obtained from the Monitoring Trends in Burn Severity (MTBS; www.mtbs.gov) project, which has consistently mapped fires greater than 1000 acres (405 ha) across the United States from 1984 to the present [19]. The MTBS project is conducted through a partnership between the United States (US) Geological Survey (USGS) National Center for Earth Resources Observation and Science (EROS) and the Unites States Department of Agriculture (USDA) Forest Service. Landsat data have been analyzed through a standardized and consistent methodology by the MTBS project.

The normalized burn ratio (NBR) index was calculated by MTBS using approximately one-year pre-fire and post-fire images from the near infrared (NIR) and shortwave infrared (SWIR) bands of the Landsat sensors, with reflectance values scaled to between 0–10,000 NBR units.

$$NBR = (NIR - SWIR)/(NIR + SWIR)$$

Pre-fire and post-fire NBR images were next differenced for each Landsat scene pair to generate the Relative dNBR (RdNBR) [20].

$$RdNBR = [(NBRpre\text{-}fire - NBRpost\text{-}fire)]/\sqrt{ABS}\,(NBRpre\text{-}fire)$$

The RdNBR severity classes of low, moderate, and high (LBS, MBS, HBS) have been defined previously by Miller and Thode [20] and cover a range of −500 to +1200 over burned land surfaces. Positive RdNBR values represent a decrease in vegetation cover and a higher burn severity, while negative values would represent an increase in live vegetation cover following the fire event.

2.2. MODIS Vegetation Index Time Series

NASA's MODIS (Moderate Resolution Imaging Spectroradiometer) satellite sensors Terra and Aqua have been used to generate a 250-m resolution NDVI (MOD13) global product on 16-day intervals since the year 2000 [21,22]. The MODIS Collection 6 NDVI data set provides consistent spatial and temporal profiles of vegetation canopy greenness according to the equation:

$$NDVI = (NIR - Red)/(NIR + Red)$$

where NIR is the reflectance of wavelengths from 0.7 μm to 1.0 μm, and Red is the reflectance from 0.6 μm to 0.7 μm, with values scaled to between 0 and 10,000 NDVI units to preserve decimal places in integer file storage. Low values of NDVI (near 0) indicate barren land cover, whereas high values of NDVI (above 8000) indicate dense canopy greenness cover.

The MOD13 250-m vegetation indices (VIs) have been retrieved from daily, atmosphere-corrected, bidirectional surface reflectance. The VIs were computed from MODIS-specific compositing methods based on product quality assurance metrics to remove all of the low-quality pixels from the final NDVI value reported. Cloud and water pixels were identified and excluded using other MODIS atmospheric data masks. From the remaining good-quality NDVI values, a constrained view-angle approach (closest to nadir) then selected the optimal pixel value to represent each 16-day compositing period. These MOD13 data sets were downloaded from the files that were available at modis.gsfc.nasa.gov/data/dataprod/mod13.php for time series analysis across Alaska wetland locations.

2.3. Elevation and Land Cover Map Layers

Digital elevation (in vertical meters) for Alaska was derived from USGS [23] mapping at 300-m ground resolution. Wetland cover was mapped for the state at 30-m ground resolution from the 2011 National Land Cover Dataset (NLCD) of Alaska ([24]; available at www.mrlc.gov/nlcd11_leg.php). The overall thematic accuracy for the previous Alaska NLCD was 76% at Level II (12 classes evaluated). For contextual comparison purposes, the open water (class 11), barren land (class 31), and evergreen forest (class 42) classes of this NLCD were mapped with high user's accuracy, while the herbaceous wetland (class 95) was mapped with moderate user's accuracy.

For this study, the NLCD woody wetland (class 90) together with all of the herbaceous wetland pixels were combined into one class, and were all overlaid 200 × 200-m resolution areas with a majority of the wetland surface coverage identified and mapped for the entire state (Figure 1). This combined wetland coverage was overlaid with statewide MTBS high burn severity (HBS) class pixels from the years 2001 to 2015, and with MODIS 250-m summer season NDVI (from the composite Julian day 177; 26 June) images for each of these years to carry out a time trend analysis of the burned wetland area NDVI changes statewide. "Pre-fire" MODIS NDVI values were all derived from the Julian day 177 NDVI from the year before the fire date for change detection.

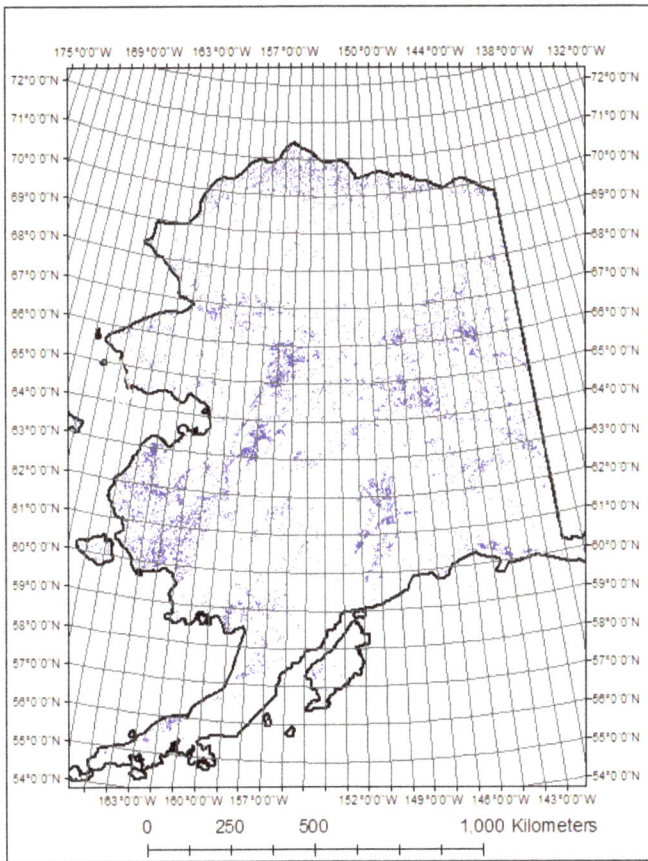

Figure 1. Alaska wetland cover (blue pixels) at 200-m resolution derived from the 2011 National Land Cover Dataset (NLCD) map.

The section of the Julian date 177 for NDVI change detection over time was not an arbitrary choice, but rather was determined to be a seasonally consistent metric of green cover change, since 26 June is nearly always near the seasonal maximum in interior Alaska for green cover, which was verified by examining thousands of pixels in time-stacked NDVI maps of Alaska wetland locations. Wetland areas that covered less than a majority of 200 × 200-m resolution areas in the statewide grid were too small to be matched consistently with MODIS 250-m summer season NDVI, and were therefore not included in the results.

2.4. Statistical Analysis Methods

The BFAST (Breaks for Additive Seasonal and Trend) methodology was applied to a MODIS NDVI monthly time series for selected wetland locations that covered the majority of a 250 × 250 m pixel area within severely burned locations. BFAST was developed by Verbesselt et al. [16,17] for detecting and characterizing abrupt changes within a time series, while also adjusting for regular seasonal cycles. A harmonic seasonal model is first applied in BFAST to account for regular seasonal phenological variations. BFAST next computes the Ordinary Least Squares Moving Sum (OLS-MOSUM) by considering that the moving sums of the residuals after the harmonic seasonal model have been removed from the time series data values. MOSUM tests for structural change using the null hypothesis that all regression coefficients are equal i.e., every observed value can be expressed as a linear function with the same slope [25]. If the null hypothesis is true, the values can be modeled by one line with that slope, and the sum of residuals will have a zero mean. MOSUM compares moving sums of residuals to test the likelihood of the regression coefficient for a certain time period based on a user's input stating the minimum time between potential "breakpoints". A rejection of the null hypothesis indicates that the regression coefficient changes at that point in time.

The MOSUM uses a default *p*-value of 0.05, meaning that the probability of it detecting a structural change when none has occurred is less than 5%. If MOSUM does not detect some structural change with a confidence level of 95%, it returns a "no breakpoints" result. If MOSUM detects some structural change with a confidence level of 95%, it then processes the time series through a second test, which is used to determine where the breakpoints are located in time. The output of this function is a 95% confidence interval for each breakpoint (expressed as two date numbers that define a range, before and after a breakpoint.).

For BFAST timer-series analysis, MOD13 NDVI data values (2000 to 2017) from Alaska wetland locations were subsampled to include only the growing season values during the low snow cover period of 1 May to 1 October, leaving about 10 observations per year. If a "no data" value was present in the growing season MOD13 record, then the NDVI from the previous 16-day period was substituted. Change metrics generated by BFAST from the time series analysis results included the number of breakpoints, date of each breakpoint, and the slope of the NDVI between breakpoint dates.

3. Results

3.1. Wetland NDVI Changes within Large Wildfires 1999–2009

Based on yearly MTBS RdNBR map collections, the names and locations of the largest contiguous wetland area that burned at high severity in Alaska were determined for each year since 2000. Four of the largest wildfires were selected (Figure 2) to generate examples of post-fire NDVI time series analysis results using BFAST.

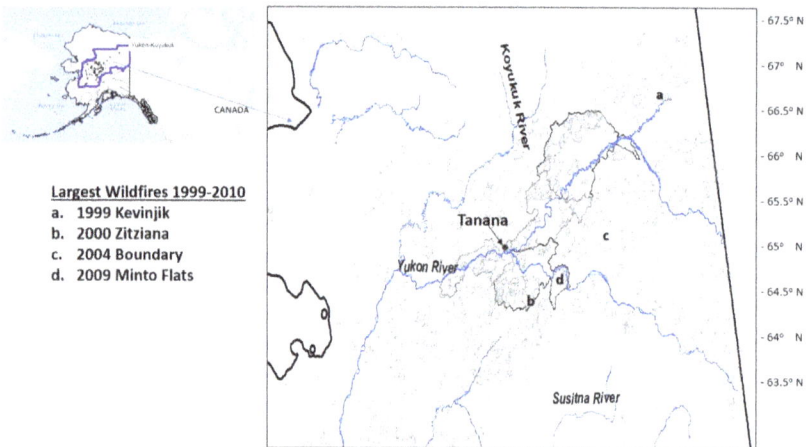

Figure 2. Wildfire boundaries (as gray outlines) mapped by the Monitoring Trends in Burn Severity (MTBS) project for interior Alaska in from 2000 to 2015, along with the locations of the largest wildfires recorded from the years 1999, 2000, 2004, and 2009. Yukon River drainage basins comprised of the Yukon Flats, Ramparts, Lower Tanana River, and the Klatsuta River sub-basins (USGS Level 8 Hydrologic Units, [26]), are delineated in shaded boundaries.

The complete MODIS 16-day NDVI record was plotted for these wetland-dominated pixels from 2000 to 2017, starting with the 102,385 ha 1999 Kevinjik Fire (Figure 3a). Two HBS wetland cover locations were plotted for each large wildfire, and the time series showed that the recovery of NDVI seasonal profiles following this 1999 fire was gradual and relatively stable by about the year 2012. Comparing the HBS wetland vegetation profiles following the 67,987 ha Zitziana Fire from the year 2000, NDVI recovery appeared to be even more rapid (than for the Kevinjik Fire) and became relatively stable by about the year 2007.

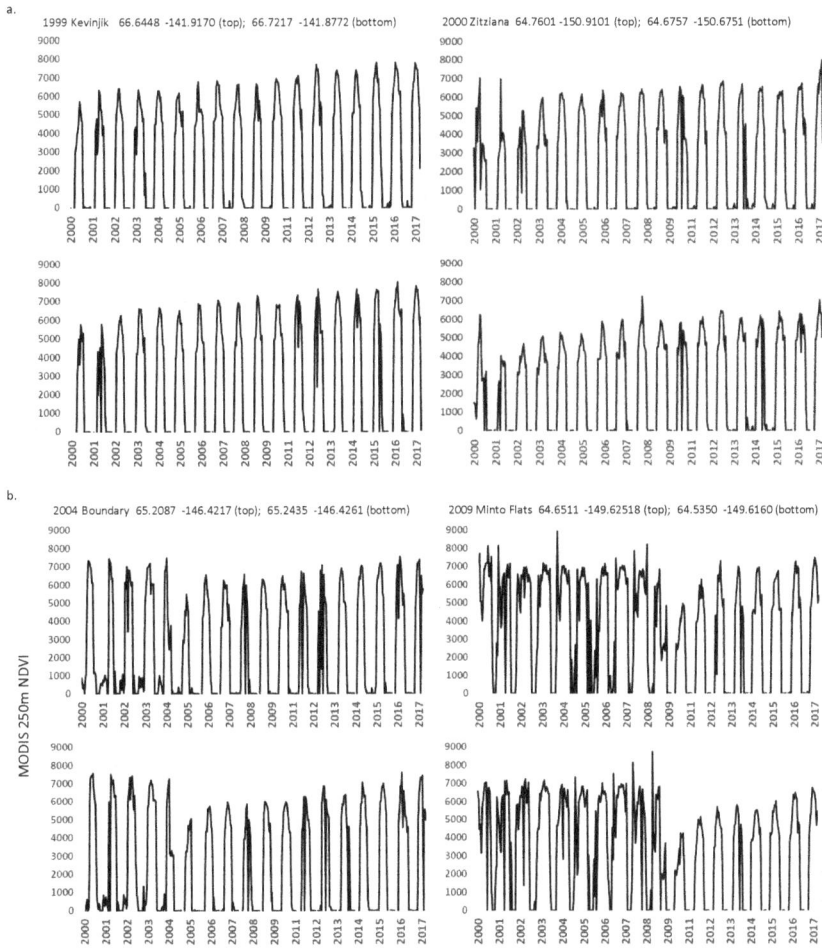

Figure 3. Yearly time series plots (16-days) of 250-m resolution normalized difference vegetation index (NDVI) from Moderate Resolution Imaging Spectroradiometer (MODIS) for all of the wetland cover pixels within the four largest Alaska wildfires from the years 1999, 2000, 2004, and 2009. The latitude and longitude of each pixel center is shown.

The largest contiguous wetland areas that had burned at high severity within the 217,720 ha Boundary Fire of 2004 recovered to near pre-fire NDVI profile levels by 2014 (Figure 3b visual assessment), whereas the largest contiguous wetland areas that had burned at high severity within the 212,050 ha Minto Flats Fire in 2009 had yet to fully recover to pre-fire NDVI profile levels by 2017.

It was noteworthy that the change that can be visually detected in seasonal NDVI profiles before and after the Minto Flats Fire may be indicative of a shift from the longer (broader) green-season evergreen (conifer forest) profile to a mixed deciduous (birch, alder, willow) shrub and herbaceous cover profile.

The BFAST results for these four large fires for wetland-dominated MODIS pixels showed significant ($p < 0.01$) downward breakpoint shifts in NDVI (Table 1 and Figure 4) during the MTBS-document years of the most severe burning (2001, 2004, and 2009) detected in the Landsat NBR records. As expected, the BFAST results for the selected pixel location of the 1999 Kevinjik fire showed no breakpoints, but instead a showed strong upward slope in the de-seasonalized NDVI values for the years following 2000. The significant wildfire-related breakpoints detected for the Zitziana, Boundary, and Minto Flats wetland fire areas were commonly followed by one to two years of relatively rapid recovery of NDVI, and then by about five years of relatively slower NDVI recovery (Figure 4).

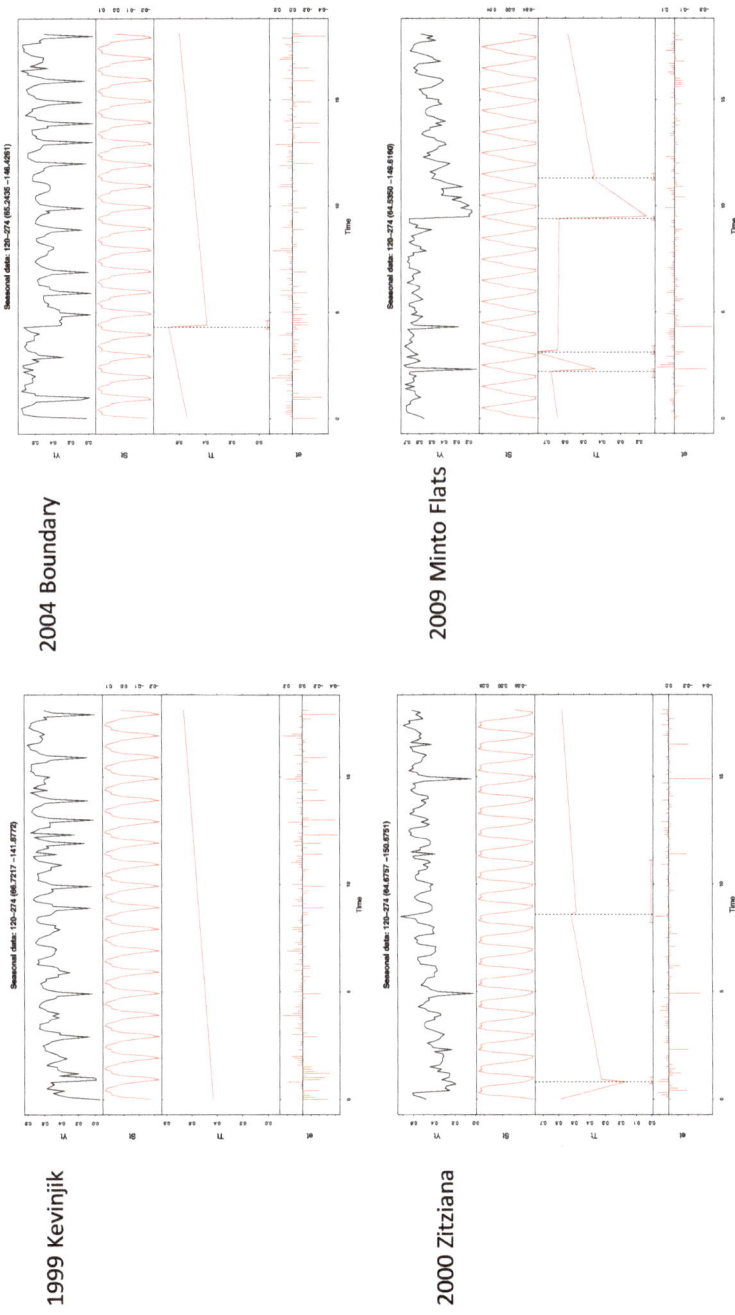

Figure 4. Breaks for Additive Seasonal and Trend (BFAST) plot outputs for large fire locations (labelled in Figure 3) between 1999–2009 covering wetland–dominated Moderate Resolution Imaging Spectroradiometer (MODIS) pixels in Alaska. Y_t is the time-series MODIS NDVI value; S_t is the fitted seasonal component; T_t is the fitted trend component; e_t is the noise component [16]; statistical breakpoints ($p < 0.01$) are identified by vertical dashed lines. Year numbers on the horizontal axis start at 1 = 2000 and end in early 2018.

Table 1. BFAST results for large wildfire locations between 1999–2009 covering wetland–dominated MODIS pixels in Alaska. Breakpoint dates that corresponded to Monitoring Trends in Burn Severity (MTBS) severe burn events are shown in boldface, followed by the NDVI slope after each date. For locations with no breakpoint detected, the overall slope of the entire NDVI time series, 2000 to 2017, was listed in the second column.

		Breakpoint 1		Breakpoint 2		Breakpoint 3		Breakpoint 4	
	Overall Slope (da⁻¹)	Date 1	Slope 1	Date 2	Slope 2	Date 3	Slope 3	Date 4	Slope 4
Kevinjik 1999	0.0012 0.0013	none none							
Zitziana 2000		**10 June 2001** **14 September 2000**	−0.021 0.003	12 July 2002 13 August 2008	0.001 0.001				
Boundary 2004		26 June 2001 **26 June 2004**	0.003 0.002	**26 June 2004**	0.015	30 September 2005	0.001		
Minto Flats 2009	0.0005	none 10 June 2002	0.039	25 May 2003	0.000	**12 July 2009**	0.011	26 June 2011	0.002

To expand the BFAST results to a regional level for interior Alaska, breakpoint analysis was applied to a total of 3200 wetland locations (mapped at 200-m resolution) that were recorded as burned at high severity by large wildfires [19] between 2001–2015 within the following Yukon River drainage sub-basins: Yukon Flats, Ramparts, Lower Tanana River, and the Klatsuta River (USGS Level 8 Hydrologic Units, [26], as shown in Figure 2). Results showed that at least one breakpoint was detected by BFAST analysis in the MODIS NDVI time series at 85% of these HBS wetland locations in interior Alaska. The distribution of dates (binned by year) of burning from Landsat HBS mapping [19] at these wetland locations closely matched the distribution of dates of breakpoints from the MODIS NDVI time series BFAST analysis, with the exception of 2015, for which BFAST did not detect the highest number of breakpoints (Figure 5).

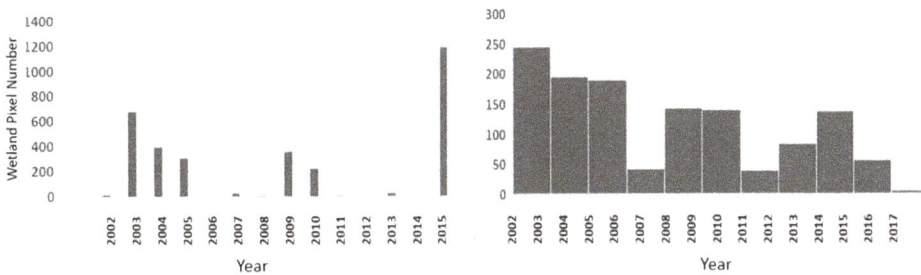

Figure 5. Histograms of the years of fire dates for wetland areas that burned at high severity in interior Alaska from the MTBS Landsat [19] (**left**) and the BFAST analysis of the MODIS NDVI time series (**right**).

The histogram of the BFAST 18-year trend results of MODIS NDVI for 3200 wetland locations that were recorded as having burned at high severity between 2001–2015 within the Yukon River drainage basins (detected with at least one breakpoint) showed a strong skewness (2.8) and the predominance of positive greening trends that reflect regrowth from disturbance (Figure 6). For comparison, the BFAST results for 1000 wetland locations in the Upper Yukon Flats sub-drainage basin that had not been burned by large wildfires between the years 2001–2015 [19] showed the opposite frequency distribution to Figure 6, with a strong negative skewness (−3.0) in slope values and almost no positive NDVI slope values greater than 100 units per year.

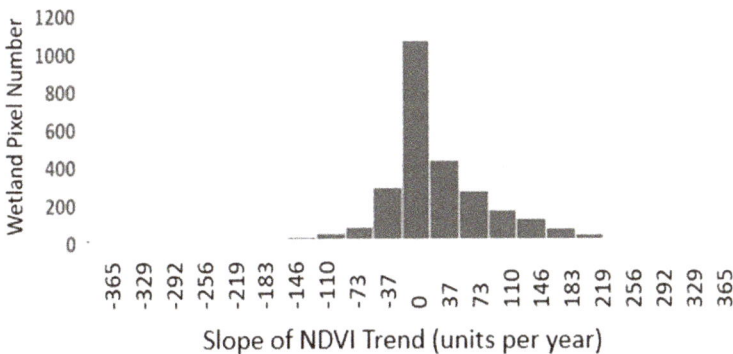

Figure 6. Histogram of the slopes of MODIS NDVI trends for wetland areas that burned at high severity in interior Alaska between 2001–2015 from BFAST breakpoint analysis.

3.2. Statewide Wetland NDVI Change for HBS Areas 2000–2015

For standardized comparisons, "pre-fire" MODIS NDVI values were all derived from the Julian day 177 NDVI from the year before the fire date for change detection. The resulting plots of the NDVI change (pre-fire to post-fire summer of 2017, both on Julian day 177) for years when the total HBS wetland areas were greater than 2000 ha, statewide, showed that the fraction of negative change in the NDVI (lower greenness in 2017 than pre-fire and incomplete greenness recovery) increased markedly after the 2013 wildfires (Figure 7). In years prior to 2013, the NDVI change tended to be positive (higher greenness in 2017 than pre-fire) at HBS wetland elevations lower than 400 m, whereas higher elevation HBS wetland locations showed relatively few positive greenness recovery changes by 2017.

The selection of the NDVI from the summer of 2017 as the post-fire comparison year made the change in greenness comparisons consistent among all of the previous years of recorded wetland wildfires. According the long-term weather station records from Fairbanks (available at w1.weather.gov/obhistory/PAFA.html), 2017 had a total of 40.6 cm of precipitation, compared to the average annual total of 29.7 cm since 1999. The mean annual temperature in 2017 was recorded in Fairbanks as 29.5 °C versus the annual average of 28.1 °C since the year 1999. Therefore, 2017 would not have been a particularly dry year with poor growing conditions for these post-fire NDVI computations.

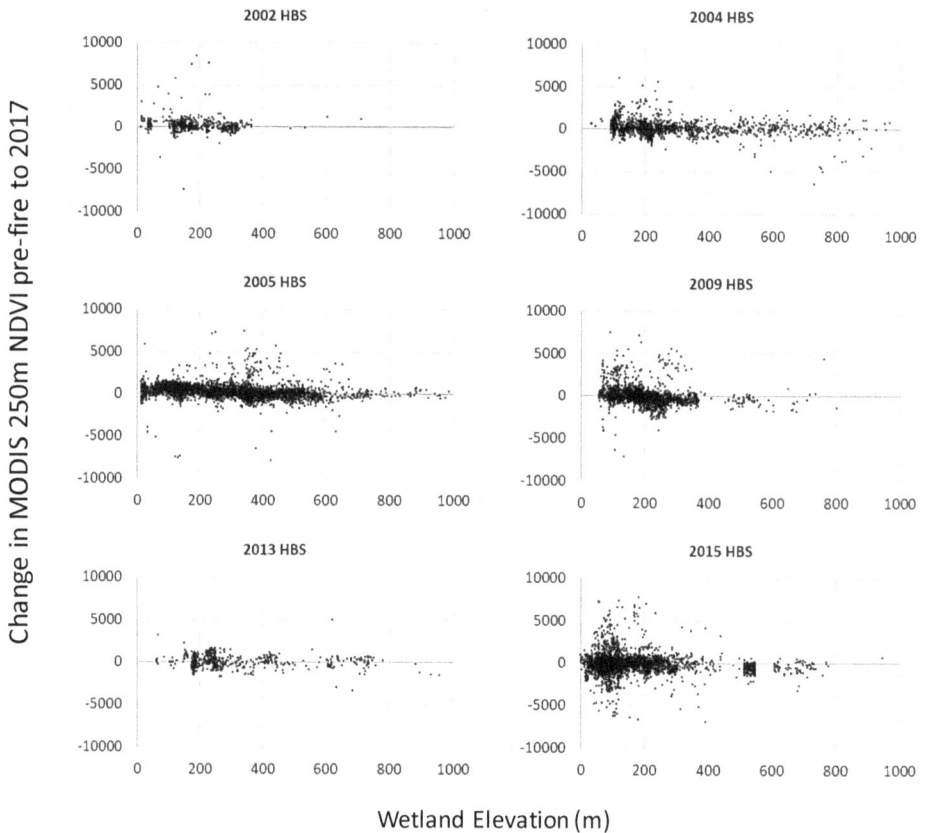

Figure 7. Yearly plots of change in MODIS 250-m resolution wetland NDVI between pre-fire years and 2017 versus wetland elevation for years with the largest areas that burned at high severity.

The year (since 2000) that was calculated with the highest area of HBS wetland vegetation loss was 2005 with 20,700 ha, followed by 2015 with 16,540 ha, 2009 with 9280 ha, and 2004 with 6460 ha of wetland vegetation consumed by wildfire across the state of Alaska. The average change in NDVI from pre-fire levels to post-fire (2017) levels, along with the within-year variability, that was estimated across these acreages showed that severely burned Alaska wetlands from the years 2002 and 2005 have had a significant (t-test, $p < 0.05$) positive recovery of green vegetation cover since wildfire. For all of the other years before 2014, the average wetland change in NDVI since wildfire was not different from zero, indicating a full recovery of green vegetation cover by 2017. On the other hand, severely burned wetlands from the year 2015 still had a significantly ($p < 0.05$) lower average level of green vegetation cover than was estimated from the years before the wildfires (Figure 8).

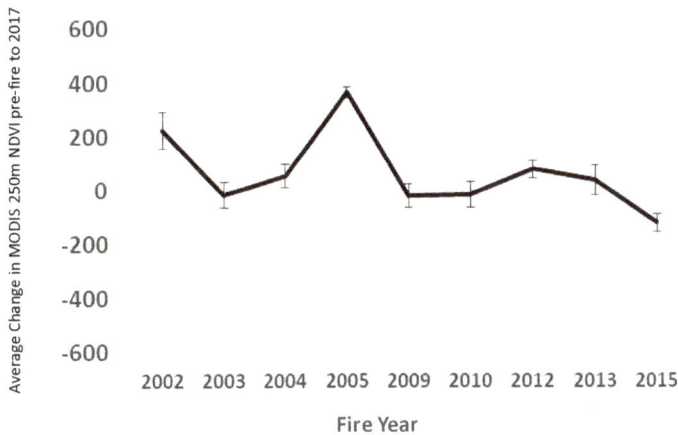

Figure 8. Average change in MODIS 250-m resolution wetland NDVI between pre-fire years and 2017, for all of the MTBS wildfires in Alaska greater than 2000 ha in the National Land Cover Dataset (NLCD) wetland area burned at high severity. Error bars of post-fire NDVI minus pre-fire NDVI represent two standard errors of the average change in NDVI.

4. Discussion

The principal findings of this study were that wetland cover locations across Alaska that burned at high severity subsequently recovered their green cover seasonal profiles to relatively stable pre-fire levels in less than a decade. The large wetland fires in Alaska from 2013 to 2016 showed an incomplete greenness recovery compared to earlier fires. In the years prior to 2013, the NDVI change tended to be positive at HBS wetland elevations lower than 400 m, whereas higher elevation HBS wetland locations showed much weaker greenness recovery changes by 2017. This elevation threshold of 400 m for positive post-fire NDVI recovery is not obviously related to any known topoecological changes for high-latitude wetlands, which is a new finding that merits more field research to understand this remote sensing observation. By all accounts, this is the first statewide or regional study of wetland burning and greenness recovery for Alaska that has been published, making comparisons to previous published results of a similar nature unattainable.

Nonetheless, the outcomes of recovery and regrowth pathways after high severity burning over the next few decades will be of significant consequence to the local community members in Alaska who have depended on wetland ecosystems for subsistence hunting and trapping. In and around mesic soil locations, the deep surface organic material of low bulk density in evergreen tree stands generally precludes deciduous boreal species from establishing seedlings [10].

However, relatively thin post-fire organic layer depths, such as those measured by Potter [9] in field surveys in the Tanana Area fires of 2015, may cause notable alterations in the successional outcomes of severely burned ecosystems, including a shift from the conifer-dominated thick organic layer to an increase in the dominance of deciduous or shrub species [10,26]. Barrett et al. [27] reported that the areas with less than 3 cm of surface organic layer depth after boreal forest fires will be susceptible to deciduous-dominated regeneration, whereas areas with 3–10 cm of organic layer depth will be susceptible to co-dominant regeneration by both coniferous and deciduous trees.

The aboveground biomass levels in the tundra wetlands of Alaska have been positively correlated with NDVI, and with elevated ecosystem carbon (CO_2) fluxes, including net ecosystem production and ecosystem respiration [3]. Anywhere that most of the live vegetation is consumed by intense wildfire, nutrients can be mobilized, surface water temperature can become elevated, and soil erosion may increase [1]. The recovery of wetland vegetation production and live green foliage cover after wildfires can help stabilize hydrological and thermal regimes, promote biodiversity, and reduce the seepage of dissolved nutrients to adjacent fluvial systems [28].

In previously published studies of satellite greenness (NDVI) in Alaska using the BFAST method, Forkel et al. [29] found the region to be of special interest for the analysis of trend change detection, because of greening NDVI trends in the tundra ecosystems of the North Slope as well as browning trends in the interior boreal forests. These authors reported that most of the breakpoints in NDVI time series coincided with large wildfire events. As in the present study of wetland NDVI trends and wildfire, BFAST methods detected stronger greening and browning trends if snow-affected values were excluded from the analysis or when only peak seasonal NDVI values were used. Breakpoints with abrupt changes, i.e., higher magnitudes, were detected more frequently than were breakpoints with gradual changes, i.e., low magnitudes.

Forkel et al. [29] further reported that downward (browning) trends in the NDVI between 20–30 years long occurred in some of the boreal regions of central Alaska and in southwestern Alaska, usually with uncertainties of up to four years. The detection of breakpoints in the NDVI time series in 2004 agreed with the spatial distribution of Landsat-mapped large wildfires and other field-based observations. Seasonal NDVI patterns suggested that the conifer forests that burned in 2004 tended to be replaced by broad-leaved shrubs (dwarf birch and aspen) and grasses during post-fire recovery years, which resulted in a structural change in the NDVI time series.

The region-wide BFAST analysis results from the present study similarly indicated that significant structural change (in the form of breakpoints) could be detected in the 250-m NDVI time series for the vast majority of wetland locations in the major Yukon river drainages of interior Alaska that had burned at high severity since the year 2001. The lower-than-expected number of breakpoints in the MODIS NDVI time series detected by BFAST in 2015 may be explained by (1) there being fewer dates in 2016 and 2017 to compare to pre-2015 NDVI levels than for other fire years, and (2) the presence of fires prior to 2015 at the same wetland location, which would have depressed NDVI values in 2015 more than unburned locations. The predominance of positive overall slopes in the NDVI time series of wetlands with breakpoints indicated rates of vegetation recovery that are typically in the range of 40 to 200 NDVI units per year (scaled from 0–1 by a factor or 10,000).

5. Conclusions

Breakpoint analysis was able to detect significant structural change and increasing yearly slopes (since fire) in the 250-m NDVI time series for most of the wetland locations in interior Alaska that had burned at high severity since the year 2001. The results from this study of greenness recovery in Alaska wetlands following high severity burns support the supposition that vegetation cover density in these ecosystems, albeit possibly with a higher fraction of annual herbaceous cover than was present before the fire, will make a nearly complete recovery within 10 years, and often within five years post-fire. A corollary hypothesis worth testing in future field and remote sensing studies is that relatively low-elevation wetlands are not as susceptible to drying and warming trends in the Arctic and boreal regions as are upland forest and tundra shrublands, and therefore wetlands on the whole will recover aboveground biomass more rapidly than non-wetlands over the same period of climate change.

Remote Sens. **2018**, *10*, 1456

Funding: This research received no external funding.

Acknowledgments: This work was supported by NASA Ames Research Center and the NASA ABoVE Logistics Office in Fairbanks, Alaska.

Conflicts of Interest: The author declares no conflict of interest.

References

1. Bixby, R.J.; Cooper, S.D.; Gresswell, R.E.; Brown, L.E.; Clifford, D.N.; Dwire, K.A. Fire effects on aquatic ecosystems: An assessment of the current state of the science. *Fresh. Sci.* **2015**, *34*, 1340–1350. [CrossRef]
2. Richardson, C.J. The Everglades: North America's subtropical wetland. *Wetl. Ecol. Manag.* **2010**, *18*, 517–542. [CrossRef]
3. Boelman, N.T.; Stieglitz, M.; Rueth, H.M.; Sommerkorn, M.; Griffin, K.L.; Shaver, G.; Gamon, J. Response of NDVI, biomass, and ecosystem gas exchange to long-term warming and fertilization in wet sedge tundra. *Oecologia* **2003**, *135*, 414–421. [CrossRef] [PubMed]
4. Epting, J.; Verbyla, D.L. Landscape level interactions of pre-fire vegetation, burn severity, and post-fire vegetation over a 16-year period in interior Alaska. *Can. J. For. Res.* **2005**, *35*, 1367–1377. [CrossRef]
5. Goetz, S.J.; Bunn, A.G.; Fiske, G.J.; Houghton, R.A. Satellite observed photosynthetic trends across boreal North America associated with climate and fire disturbance. *Proc. Natl. Acad. Sci. USA* **2005**, *103*, 13521–13525. [CrossRef] [PubMed]
6. Potter, C. Regional analysis of MODIS satellite greenness trends for ecosystems of interior Alaska. *GISci. Remote Sens.* **2014**, *51*, 390–402. [CrossRef]
7. Potter, C.; Li, S.; Crabtree, R. Changes in Alaskan tundra ecosystems estimated from MODIS greenness trends, 2000 to 2010. *J. Geophys. Remote Sens.* **2013**, *2*, 107. [CrossRef]
8. Kasischke, E.S.; Turetsky, M.R. Recent changes in the fire regime across the North American boreal region: Spatial and temporal patterns of burning across Canada and Alaska. *Geophys. Res. Lett.* **2006**, *33*, L09703. [CrossRef]
9. Potter, C.S. Ecosystem carbon emissions from 2015 forest fires in interior Alaska. *Carbon Balance Manag.* **2018**, *13*, 2. [CrossRef] [PubMed]
10. Johnstone, J.F.; Chapin, F.S.; Hollingsworth, T.N.; Mack, M.C.; Romanovsky, V.; Turetsky, M. Fire, climate change, and forest resilience in interior Alaska. *Can. J. For. Res.* **2010**, *40*, 1302–1312. [CrossRef]
11. Kasischke, E.S.; Johnstone, J.F. Variation in postfire organic layer thickness in a black spruce forest complex in interior Alaska and its effects on soil temperature and moisture. *Can. J. For. Res.* **2005**, *35*, 2164–2177. [CrossRef]
12. Kasischke, E.S.; Verbyla, D.L.; Rupp, T.S.; McGuire, A.D.; Murphy, K.A.; Jandt, R.; Barnes, J.L.; Hoy, E.E.; Duffy, P.A.; Calef, M.; et al. Alaska's changing fire regime -implications for the vulnerability of its boreal forests. *Can. J. For. Res.* **2010**, *40*, 1313–1324. [CrossRef]
13. Viereck, L. The Effects of Fire in Black Spruce Ecosystems of Alaska and Northern Canada. In *The Role of Fire in Northern Circumpolar Ecosystems*; Wein, R.W., MacLean, D.A., Eds.; Wiley: Chichester, UK, 1983; pp. 201–220.
14. Jiang, Y.; Rocha, A.V.; O'Donnell, J.A.; Drysdale, J.A.; Rastetter, E.B.; Shaver, G.R.; Zhuang, Q. Contrasting soil thermal responses to fire in Alaskan tundra and boreal forest. *J. Geophys. Res. Earth Surf.* **2015**, *120*, 363–378. [CrossRef]
15. Brown, D.R.N.; Jorgenson, M.T.; Kielland, K.; Verbyla, D.L.; Praka, A.; Koch, J.C. Landscape effects of wildfire on permafrost distribution in Interior Alaska derived from remote sensing. *Remote Sens.* **2016**, *8*, 654. [CrossRef]
16. Verbesselt, J.; Hyndman, R.; Newnham, G.; Culvenor, D. Detecting Trend and Seasonal Changes in Satellite Image Time Series. *Remote Sens. Environ.* **2010**, *114*, 106–115. [CrossRef]
17. Verbesselt, J.; Hyndman, R.; Zeileis, A.; Culvenor, D. Phenological change detection while accounting for abrupt and gradual trends in satellite image time series. *Remote Sens. Environ.* **2010**, *114*, 2970–2980. [CrossRef]
18. De Jong, R.; Verbesselt, J.; Schaepman, M.E.; de Bruin, S. Trend changes in global greening and browning: Contribution of short-term trends to longer-term change. *Glob. Chang. Biol.* **2012**, *18*, 642–655. [CrossRef]

19. Eidenshink, J.; Schwind, B.; Brewer, K.; Zhu, Z.; Quayle, B.; Howard, S. A project for monitoring trends in burn severity. *Fire Ecol. Spec. Issue* **2007**, *3*, 3–21. [CrossRef]

20. Miller, J.D.; Thode, A.E. Quantifying burn severity in a heterogeneous landscape with a relative version of the delta Normalized Burn Ratio (dNBR). *Remote Sens. Environ.* **2007**, *109*, 66–80. [CrossRef]

21. Didan, K.; Munoz, A.B.; Solano, R.; Huete, A. *MODIS Vegetation Index User's Guide (MOD13 Series) 2016*; Version 3.00, June 2015 (Collection 6); University of Arizona, Vegetation Index and Phenology Lab: Tucson, AZ, USA, 2016.

22. Shao, Y.; Lunetta, R.; Wheeler, B.; Iiames, J.; Campbell, J. An evaluation of time-series smoothing algorithms for land cover classifications using MODIS-NDVI multi-temporal data. *Remote Sens. Environ.* **2016**, *174*, 258–265. [CrossRef]

23. United States Geological Survey (USGS). *Digital Elevation Model 300M Grid for Alaska, from Defense Mapping Agency 3-Arc Second 1x1 Degree 1, 250,000 Scale, Digitial Elevation Models. Source Data for Digital Shaded-Relief Image of Alaska, USGS Map I-2585*; U.S. Geological Survey EROS Alaska Field Office: Anchorage, AK, USA, 2016.

24. Selkowitz, D.J.; Stehman, S.V. Thematic accuracy of the National Land Cover Database (NLCD) 2001 land cover for Alaska. *Remote Sens. Environ.* **2011**, *115*, 1401–1407. [CrossRef]

25. Kleiber, C.; Hornik, K.; Leisch, F.; Zeileis, A. Strucchange: An R package for testing for structural change in linear regression models. *J. Stat. Softw.* **2002**, *7*, 1–38.

26. Seaber, P.R.; Kapinos, F.P.; Knapp, G.L. *Hydrologic Unit Maps: U.S. Geological Survey Water-Supply Paper 2294*; US Geological Survey: Denver, CO, USA, 1987; p. 63.

27. Barrett, K.; McGuire, A.D.; Hoy, E.E.; Kasischke, E.S. Potential shifts in dominant forest cover in interior Alaska driven by variations in fire severity. *Ecol. Appl.* **2011**, *21*, 2380–2396. [CrossRef] [PubMed]

28. Hood, E.; Fellman, J.; Spencer, R.; Hermes, P.; Edwards, R.; D'Amore, D.; Scott, S. Glaciers as a source of ancient and labile organic matter to the marine environment. *Nature* **2009**, *462*, 1044. [CrossRef] [PubMed]

29. Forkel, M.; Carvalhais, N.; Verbesselt, J.; Mahecha, M.D.; Neigh, C.S.; Reichstein, M. Trend change detection in NDVI time series: Effects of inter-annual variability and methodology. *Remote Sens.* **2013**, *5*, 2113–2144. [CrossRef]

remote sensing

MDPI

Review

Remote Sensing of Environmental Changes in Cold Regions: Methods, Achievements and Challenges

Jinyang Du [1], Jennifer D. Watts [2], Lingmei Jiang [3,*], Hui Lu [4], Xiao Cheng [5], Claude Duguay [6], Mary Farina [2], Yubao Qiu [7,8], Youngwook Kim [1], John S. Kimball [1] and Paolo Tarolli [9]

[1]   Numerical Terradynamic Simulation Group, W.A. Franke College of Forestry and Conservation, The University of Montana, Missoula, MT 59812, USA
[2]   Woods Hole Research Center, Falmouth, MA 02540, USA
[3]   State Key Laboratory of Remote Sensing Science, Jointly Sponsored by Beijing Normal University and Institute of Remote Sensing and Digital Earth of Chinese Academy of Sciences, Faculty of Geographical Science, Beijing Normal University, Beijing 100875, China
[4]   Department of Earth System Science, Tsinghua University, Beijing100084, China
[5]   School of Geospatial Engineering and Science, Sun Yat-Sen University, Guangzhou 510275, China
[6]   Department of Geography & Environmental Management, University of Waterloo, Waterloo, ON N2L 3G1, Canada
[7]   Aerospace Information Research Institute, Chinese Academy of Sciences, Beijing 100101, China
[8]   Group on Earth Observations Cold Regions Initiative (GEO CRI), 7 bis, avenue de la Paix, Case postale 2300, CH-1211 Geneva, Switzerland
[9]   Department of Land, Environment, Agriculture and Forestry, University of Padova, viale dell'Università 16, 35020 Legnaro (PD), Italy
*   Correspondence: jiang@bnu.edu.cn; Tel.: +86-10-5880-5042

Received: 18 July 2019; Accepted: 17 August 2019; Published: 20 August 2019

Abstract: Cold regions, including high-latitude and high-altitude landscapes, are experiencing profound environmental changes driven by global warming. With the advance of earth observation technology, remote sensing has become increasingly important for detecting, monitoring, and understanding environmental changes over vast and remote regions. This paper provides an overview of recent achievements, challenges, and opportunities for land remote sensing of cold regions by (a) summarizing the physical principles and methods in remote sensing of selected key variables related to ice, snow, permafrost, water bodies, and vegetation; (b) highlighting recent environmental nonstationarity occurring in the Arctic, Tibetan Plateau, and Antarctica as detected from satellite observations; (c) discussing the limits of available remote sensing data and approaches for regional monitoring; and (d) exploring new opportunities from next-generation satellite missions and emerging methods for accurate, timely, and multi-scale mapping of cold regions.

Keywords: remote sensing; cryosphere; climate change; northern high latitudes; Antarctica; Tibetan Plateau

---

1. Introduction

Cold regions, including high-latitude and high-altitude landscapes, are experiencing climate warming with amplification at roughly twice the global rate for the Arctic region (>60°N) [1,2] and Tibetan Plateau (TP) [3,4]. Cold regions are typically characterized by the presence of permafrost, extensive snow and ice cover, and rich reserves of stored soil organic carbon in the northern regions and TP [5,6]. Ecosystems within these regions are highly vulnerable to changes resulting from the rapid destabilization and melting of ice above the 0 °C isotherm [7,8], lengthening the annual non-frozen season [9,10], and thawing of carbon-rich permafrost soils [11].

Recent region-wide warming trends have altered vegetation, and interactions and feedbacks between the water, energy, and carbon cycles [12–14]; these changes have also resulted in a myriad of impacts to landscape function and ecosystem services [15]. Warmer temperatures have reduced the duration of seasonal snow and ice cover over land, ocean, and inland water bodies [16–19]. Longer snow and ice-free seasons have led to lower surface albedos and greater net energy loading, reinforcing regional warming trends [20–22]. The alteration of seasonal snow and ice cover has also altered wildlife habitats and human mobility, including degrading the stability of snow and ice cover for winter travel [23].

Permafrost soils are estimated to store up to 1,600 billion tonnes of soil carbon, representing roughly twice the amount of carbon stored in the atmosphere [11]. Warming soils have promoted permafrost degradation and active layer deepening, enhancing the mobilization and potential transfer of soil carbon to the atmosphere [24,25]. Ground surface deformation from degrading permafrost has also increased the risk of damage to human infrastructure, including roads, pipelines, and buildings [26]. Changes in permafrost properties greatly impact the surface water budget because the soil ice layers form a relatively impermeable barrier to soil drainage [27]. Surface subsidence into the water table driven by the thawing of ice rich soils has increased surface water inundation and lake expansion in continuous permafrost areas (where more than 80% of the ground is underlain by permafrost). In contrast, extensive draining of lakes and wetlands has been observed in more degraded permafrost areas [28–30].

Satellite observations have indicated vegetation greening over northern latitude tundra, attributed to enhanced vegetation growth from a longer frost-free season, contrasting with vegetation browning in boreal forest and some tundra regions, that may result from greater drought stress due to warmer temperatures and a longer frost-free season [31]. Boreal forests have also been affected by an increase in the frequency and severity of wildfires exacerbated by warmer and drier conditions, and insect-related disturbances [32,33]. The net effect of these changes is a complex snow/ice, vegetation, soil, and wetland mosaic where the terrestrial water, carbon, and energy cycles are strongly coupled and interactive with the climate.

Remote sensing provides an unprecedented approach for characterizing the timing, magnitude, and patterns of environmental changes. This is especially advantageous for geographically remote cold regions, where site observations are often spatially sparse and temporally limited. The multi-scale nature of remote sensing also provides insight into the often emergent spatial and temporal patterns and properties of ecosystems that may not be fully identified nor understood when approached from the perspective of a local region. Earth parameter data records derived from optical-infrared (optical-IR) and microwave satellite observations spanning multiple decades are particularly valuable in distinguishing large characteristic natural climate variability from more subtle environmental trends in cold regions [14,19,34,35] and for the detection of local to regional disturbances [36,37]. Finer spatial-resolution optical-IR and active microwave sensors are essential for distinguishing the complex spatial heterogeneity in permafrost landscapes [38–40] and for near-real time applications such as monitoring river ice jams [41] and glacial lake outburst flooding [42].

This paper provides an overview of recent progress and prospects in remote sensing of cold regions by first reviewing general principles and methods in measuring a selection of key environmental variables, including glacier ice; snow; surface water bodies; permafrost and surface deformation; vegetation and terrestrial carbon process. We then summarize recent environmental changes documented by the remote sensing data record. Finally, we explore opportunities for leveraging available and future satellite missions, and integrating emerging remote sensing and big data techniques to establish a next-generation monitoring system for cold regions.

2. Principles and Methods

Satellite optical-IR and microwave sensors (Supplementary Table S1) have provided complementary observations of cold regions since the 1970s. In general, optical-IR sensors are well suited for mapping environmental variables over heterogeneous landscapes due to their relatively high-resolution (sub-meter to 1 km) imaging capability, though the signal-to-noise ratio of the observations may be degraded by cloud–atmosphere aerosol contamination, low solar elevation, and long periods of seasonal darkness at higher latitudes. Microwave remote sensing is less affected by atmospheric conditions and provides earth observations day-or-night under nearly all-weather conditions [43]. The microwave penetration ability is generally superior to optical-IR wavelengths and depends on sensor frequency and landscape conditions. Lower-frequency (e.g. 1–2GHz or L-band, 0.3–1GHz or P-band) observations provide better measurements of forest biomass and enhanced soil sensitivity under low to moderate vegetation and snow cover, while higher frequency (e.g., 12–18GHz or Ku-, 26.5–40 GHz or Ka-band) signals are more suitable to detect sparsely vegetated soil, snow properties (e.g., snow depth, surface roughness, stratification, and microstructures) [44,45] and ephemeral surface freeze–thaw (FT) conditions [38,46–48]. Among microwave sensors, satellite synthetic aperture radar (SAR) measures backscatter signals at relatively high spatial resolution (1–100s m), though the utility of operational SARs has been constrained by limited data access, incomplete global coverage and low temporal sampling. Alternatively, satellite microwave scatterometers and radiometers provide global coverage and frequent sampling (i.e., every 1–3 days) valuable for monitoring environmental dynamics over large regions, but at relatively coarse spatial resolution (~5–36 km).

A variety of sensor configurations and remote sensing techniques have been applied for monitoring cold regions, based on radiative transfer theory and the unique spectral signatures of various target variables. These approaches are summarized in the following subsections for selected variables where remote sensing has been used to document significant environmental changes attributed to global warming.

2.1. Remote Sensing of Ice

2.1.1. Glacier Mass and Movement

Glaciers are slow moving masses of ice formed over time by the accumulation and compaction of snow, holding 75% of Earth's freshwater [49] and 10% of the global land area, including most of Greenland and Antarctica [50]. Glacier mass balance is highly sensitive to climate change and controls a glacier's long-term behavior and evolution. A glacier flows under its own weight due to the pull of gravity, and thus transports ice mass to lower altitudes. Remote sensing is the only practical approach for inferring glacier mass and movement over large regions.

Glacier mass is commonly estimated through independent or combined gravimetry and altimetry measurements [51]. Satellite and aircraft-based gravimetry measurements have been widely used in glacier mass change assessments [52]. Glacier mass can also be indirectly estimated through glacier area and thickness measurements. Glacier area change events, such as ice calving, can be precisely detected using satellite images acquired over different times [53]. Glacier thickness change caused by ice melting or accumulation can be measured via geodetic approaches, including point measurements from altimetry and digital elevation models (DEMs) derived from photogrammetry or interferometric SAR (InSAR) techniques (Supplementary Table S2). Satellite laser altimeter measurements can achieve decimeter to centimeter accuracy levels; for example, the Geoscience Laser Altimeter System (GLAS) onboard the Ice, Cloud, and land Elevation Satellite (ICESat) provided decimeter-accuracy elevation data with a 70-m ground footprint over the global ice sheets [54]. The ICESat-2 satellite was launched in late 2018 and has a significantly improved laser system providing observations with enhanced spatial resolution, temporal sampling, and measurement accuracy [55]. Alternatively, satellite photogrammetry using optical-IR (e.g., ASTER, IKONOS) and SAR/InSAR (e.g., ERS-1/2 and ENVISAT) measurements have been successful in providing glacier raster DEMs [56,57]. In particular, the SAR Interferometer Radar Altimeter (SIRAL) onboard Cryosat-2 is capable of measuring changes in the thickness of both sea ice and land ice under three different measurement modes (low resolution, SAR and InSAR) [58].

Glacier movement can be detected from repeat-pass satellite images using feature tracking and Differential InSAR (DInSAR) techniques. The feature tracking method identifies and matches ice surface features from satellite images and calculates the moving distance of the features over different acquisition times. The methods have been applied to both optical-IR and SAR image series, including Landsat Operational Land Imager (OLI) [59], Moderate Resolution Imaging Spectroradiometer (MODIS) [60], and ERS-1/2 SAR [61] sensors. DInSAR uses repeat-pass SAR imagery to calculate glacier motion velocity after removing the topography signals from the sensor interferograms using an external DEM or a combination of interferograms [62,63]. The accuracy of the feature-tracking method can be within the sub-pixel level, while that of DInSAR is up to half of the radar wavelength.

2.1.2. Lake Ice Cover

Ice cover plays an important role in lake-atmosphere interactions at high latitudes. The presence (or absence) and extent/concentration of ice cover on large lakes has a significant impact on regional weather and climate (e.g., lake-effect snowfall, thermal moderation effect) [64–68]. Ice cover (extent) and ice thickness have recently been identified as Essential Climate Variables by the Global Climate Observing System (GCOS) of the World Meteorological Organization [69]. Both ice extent, from which ice phenology (i.e., ice dates during freeze-up and break-up, and ice cover duration) can be determined, and ice thickness are sensitive indicators of climate change [70,71]. Not identified by GCOS is the bedfast ice regime of shallow Arctic/sub-Arctic lakes (less than about 3-m). Such lakes are widespread across permafrost regions of Alaska, Northern Canada, and Siberia. Determining if and when entire lakes or lake sections become bedfast (i.e., ice cover is thick enough to reach lake bottoms) in winter has been shown to also be relevant for climate monitoring [72,73]. Winters with a larger (smaller) fraction of bedfast ice are generally indicative of colder (warmer) air temperature and/or lower (higher) on-ice snow depth conditions which can lead to thicker (thinner) ice. Considering the sparse distribution of weather stations in northern high latitudes, whose temperature measurements are not representative for large areas, satellite remote sensing provides an alternative to measure regional ice cover extent (phenology), ice thickness, and bedfast ice as summarized below.

Ice cover extent and phenology—Satellite remote sensing has assumed a greater role in lake ice observations in recent years due to the dramatic reduction in ground-based observational recordings and the availability of increasingly longer satellite time series, particularly from the 2000s onward [74]. Ice cover extent products are either generated manually, largely from visual interpretation of multi-source/frequency satellite imagery such as the National Oceanic and Atmospheric Administration (NOAA) National Ice Center Interactive Multisensor Snow and Ice Mapping System (IMS). The IMS products are produced manually through assimilation of various sources of data, including polar-orbit and geostationary satellite imagery and in situ data. In some cases, automated algorithms are applied to these data to facilitate analysis. The IMS products are available at various resolutions (1 km, 4 km, and 24 km) [75]. Ice phenology dates (freeze-up/ice-on and break-up/ice-off dates) and ice cover duration can also be derived from the IMS products [74,76]. MODIS 500-m snow (MOD10A1/MYD10A1) and 250-m surface reflectance (MOD09GQ) products have been used in a few recent studies, alone or in combination with each other and 1-km MODIS (MOD11A1/MYD11A1) Land Surface Temperature (LST), to derive ice dates (start and end of break-up and freeze-up dates) and their associated trends (2001-2017 or shorter) [77–82]. Approaches that use top-of-atmosphere or surface reflectance (e.g., MODIS near-infrared and red bands) are based on threshold values where ice is determined to be present/absent above/below a certain value. High solar zenith angle, which is important during freeze-up for high-latitude lakes, and cloud cover are two factors that affect the quality of lake ice products. Hence, research has also focused on developing ice phenology retrieval algorithms from passive and active microwave observations.

At passive microwave frequencies (e.g., 18–37 GHz) used for satellite remote sensing of lake ice cover, nadir emissivity from open water is low (ε = 0.443– 0.504 at 24 GHz) compared to that of ice (ε = 0.858–0.908 at 24 GHz) [83]. This makes the determination of the timing of ice formation and decay on lakes feasible from brightness temperature (Tb) measurements. The emissivity of ice, and therefore Tb, further increases during ice formation, as the influence of radiometrically cold water under the ice cover decreases with ice thickening [84]. Kang et al. (2012) found AMSR-E (Advanced Microwave Scanning Radiometer for Earth Observing System) 18.7 GHz (H- pol) Tb data (interpolated onto a 10-km grid) to be the most suitable for estimating ice dates (freeze-onset, ice-on dates, melt-onset, and ice-off dates), as well as the duration of ice cover and ice- free seasons using a thresholding approach. Du et al. [17] further demonstrated that ice dates could be derived from AMSR-E and AMSR2 (Advanced Microwave Scanning Radiometer 2) 36.5 GHz (H-pol) Tb (re-gridded at 5-km) data using a moving *t*-test algorithm. Derived ice dates compared favorably with those obtained from ground-based observations and other satellite products such as IMS.

Threshold-based and semi-automated (region-based segmentation followed by manual labelling of ice/open water) approaches have also been developed to generate ice cover extent and phenology products using SAR data. Wang et al. [85] evaluated the semi-automated segmentation algorithm "glocal" Iterative Region Growing with Semantics (IRGS) [86] for lake ice classification using dual polarized (HH and HV) RADARSAT-2 imagery acquired over Lake Erie. Their analysis showed that the algorithm could provide reliable discrimination between ice and open water with high overall classification accuracy (90.4%) when compared to Great Lakes image analysis charts from the Canadian Ice Service. Murfitt et al. [87] developed a threshold-based approach for estimating ice phenology events for mid-latitude lakes in Central Ontario by tracking the temporal evolution in backscatter from HH-polarization RADARSAT-2 imagery (2008–2017). The authors reported mean absolute errors of 2.5–10 days for freeze events and 1.5–7.1 days for water clear-of-ice when compared to MODIS imagery. The method was also successful in detecting multiple freeze (high backscatter) and melt (low backscatter) events throughout the ice season. By combining acquisitions from ENVISAT Advanced SAR (ASAR) wide swath and RADARSAT-2 ScanSAR data, Surdu et al. [88] showed the advantage of more frequent sampling (i.e., every 2–5 days over the 2005–2011 study period), but also the need for sensor incidence angle correction for more precise ice phenology detection from backscatter thresholds over Alaskan North Slope lakes.

Ice thickness—Field measurements of ice thickness are spatially and temporally sparse in cold regions. Recent investigations have developed approaches to estimate ice thickness from passive microwave, active microwave (altimetry and SAR) and thermal remote sensing data. Kang et al. [84] showed that the temporal evolution of Tb measurements from AMSR-E at 10.7 GHz and 18.7 GHz frequency (V polarization) during the ice growth season on Great Bear Lake (GBL) and Great Slave Lake (GSL), Canada, is strongly related to ice thickness. The authors proposed simple linear regression equations to estimate ice thickness for the lakes using 18.7 GHz V-pol data (2002–2009), while the estimated ice thicknesses compared favorably with in situ measurements (Mean Bias Error, MBE, 6 cm; Root Mean Square Error, RMSE, 19 cm). Beckers et al. [89] explored waveforms from CryoSat-2 Ku-band radar altimetry to estimate ice thickness also on GBL and GSL. Their study obtained ice thickness estimates with RMSE of 33 cm when compared to in-situ measurements obtained at GSL. Murfitt et al. [90] evaluated RADARSAT-2 data for estimating lake ice thickness in Central Ontario, Canada. They reported RMSE values of 11.7 cm and attributed the uncertainty to unexplored questions about scattering mechanisms and the interaction of the radar signal with lake ice having complex structure within the ice layer and at the ice–water interface. In addition to the radar-based investigations, lake surface (ice/snow) temperature observations from MODIS have also been evaluated for estimating lake ice thickness [91]. Using heat balance terms derived from the Canadian Lake Ice Model [92], the authors retrieved ice thicknesses up to 1.7 m from MODIS with RMSE of 17 cm and MBE of 7 cm when compared to field measurements acquired on GSL and Baker Lake, Canada. Work on the estimation of ice thickness from satellite remote sensing is still in its infancy. Biases in retrievals are relatively large from some spaceborne instruments, while the remote sensing time series are generally too short to analyze climate trends in satellite-derived ice thickness.

Bedfast ice—Radar remote sensing allows for distinguishing lakes with bedfast (grounded) ice due to the difference in backscatter intensities between floating ice (generally higher backscatter return) and bedfast ice (lower return) [93]. Recent analyses of polarimetric SAR (X-, C-, and L-band) satellite and ground-based scatterometer (Ku- and X-band) measurements, supported by radiative transfer modeling experiments, have revealed that the high backscatter of floating ice on shallow Arctic lakes is from the ice–water interface (due to appreciable surface roughness or preferentially oriented ice facets), dominated by single-bounce scattering [94,95]. Areas of bedfast and floating lake ice are monitored/mapped from SAR using image thresholding [73] or region-based segmentation approaches [96]. Analyses of C-band SAR time series (ERS-1/2, RADARSAT-1/2, ENVISAT ASAR, and Sentinel-1) have been used to document trends and variability in bedfast ice across Alaska, over the last 20–25 years [73,88]. Antonova et al. [97] have also shown the potential of a unique time series of three-year repeat-pass TerraSAR-X imagery with higher temporal (11 days) and spatial (10 m) resolutions than available in past studies for monitoring both bedfast ice and lake ice phenology in the Lena River Delta, Siberia. The authors also analyzed an 11-day sequential interferometric coherence time series from TerraSAR-X as a supplementary approach for bedfast ice monitoring. Coherence time series have been found to detect most areas of bedfast ice as well as spring snow/ice melt onset.

2.2. Remote Sensing of Snow

Snow and glaciers provide one-sixth of the world's population with fresh drinking water, and seasonal snow is the main fresh-water source at mid-latitudes [98]. Snow is also a crucial factor controlling the seasonal radiation balance of the land surface, and a sensitive indicator of global climate change. Snow measurement is essential to snowmelt driven runoff predictions, water resources management, flood control, and climate change studies [99]. Key snow properties derived from remote sensing include snow cover area or extent, structure (e.g., depth, density), and water equivalent.

2.2.1. Snow Cover Area

Snow cover area has been estimated using satellite optical sensors such as Landsat TM, Aqua/Terra MODIS, and NOAA AVHRR (advanced very-high-resolution radiometer). Snow cover can be identified under clear-sky conditions using the Normalized Difference Snow Index (NDSI), which exploits the contrasting reflectance of snow in the visible and short-wave infrared bands [100,101]. Utilizing ancillary spectral indices such as Normalized Difference Vegetation Index (NDVI) and Normalized Difference Forest Snow Index (NDFSI) helps incorporate vegetation information in snow detection and improves performance in mapping forest snow cover [102,103]. Radiation transfer models can also be used for improved snow mapping over forest areas [104]. Recent snow mapping efforts have focused on generating long-term snow cover products using observations from multiple satellite sensors [105], high-spatial resolution snow mapping using Sentinel-2 and Landsat optical imagery [106,107] and machine-learning techniques [108,109]. In addition, satellite sub-pixel snow cover is valuable for improved estimation of snowmelt runoff and understanding energy exchanges between the land surface and atmosphere [110]. Two main approaches to derive sub-pixel snow fraction include empirical linear regression of NDSI [111] and spectral mixture analysis [112–116].

Cloud contamination can significantly limit the signal quality of snow property detections made by satellite optical-IR remote sensing. Daily composites of half-hourly to hourly observations from geostationary satellites enable optimized snow detection while suppressing cloud contamination effects. Automated snow mapping has been achieved for a variety of geostationary satellites including GOES (Geostationary Operational Environmental Satellite) over North America [117]; Meteosat Second Generation (MSG) satellites over Europe [118,119]; and Multifunctional Transport Satellites (MTSAT)-2, Himawari-8, and Feng Yun (FY)-2 over Asia [120,121]. A linear interpolation method [117] has been used to derive fractional snow cover from FY-2 satellite observations [122]. Cloud-free snow cover products have also been derived using combined observations from polar-orbiting and geostationary satellites [121], additional observations from satellite microwave sensors [123–126];

and through processing of satellite data using advanced spatiotemporal filters and interpolation techniques [127,128]. Optical-IR sensors have limitations in distinguishing between dry and wet snow, which can be effectively addressed by incorporating SAR observations.

2.2.2. Snow Water Equivalent

Snow water equivalent (SWE) describes the amount of water contained in the snowpack when completely melted. To estimate SWE from satellite microwave observations, the scattering and emission contributions from intervening atmosphere, snow surface, snowpack, and underlying soil and snow–soil interactions need to be distinguished and accounted for, which can be done by exploiting the frequency-dependent sensitivity of microwaves to land surface components [129,130] (Figure 1).

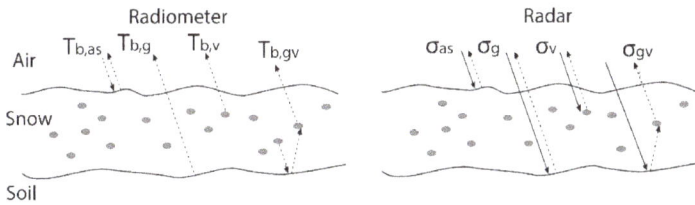

Figure 1. Components of snow emissions and scattering observed by space-borne microwave radiometer and radar observations. By neglecting atmosphere scattering and upward emission, the satellite signals mainly consist of contributions from the air–snow surface (as), snow pack (v), underlying soil (g), and snow–soil interactions (gv).

Satellite passive microwave SWE estimation relies on multi-frequency observations from SMMR (Scanning Multichannel Microwave Radiometer), SSM/I (Special Sensor Microwave/Imager), SSMIS (Special Sensor Microwave Imager Sounder), AMSR-E/2, and FY-3 MWRI (Microwave Radiation Imager) [125] sensors. The associated SWE algorithms include (a) static [130] and dynamic semi-empirical algorithms [131,132]; (b) iterative algorithms [133]; (c) physically based statistical algorithms [134]; (d) probabilistic approaches [135]; (e) machine learning methods [136,137]; and (f) data assimilation methods [138,139].

The satellite SAR SWE retrieval algorithms can be grouped into two categories: (a) physical inversion algorithms and (b) interferometry methods. By utilizing the frequency-dependent sensitivity to snow and underlying soil properties, combined multi-frequency SAR observations (e.g., X- and Ku-band) are capable of SWE retrievals as demonstrated by model simulations and field experiments [140–143]. The uncertainties related to snow density, ice microstructure, snow layer stratification, vegetation, and terrain effects are the main issues affecting the performance of both passive and active microwave snow retrieval algorithms [144–146]. Alternatively, SAR interferometry techniques show promise in overcoming many of the above difficulties by utilizing interferometry phase difference information for SWE estimation [147,148].

2.3. Remote Sensing of Frozen Soil

2.3.1. Landscape Freeze/Thaw States

The landscape FT status is closely linked to ecosystem carbon, water and energy exchanges, snow melt dynamics, and permafrost extent and stability [21,34,149,150]. Global FT observational data records spanning the modern satellite era have been used to document environmental trends from global warming, including earlier and longer non-frozen seasons as a driver of northern vegetation greening, increased trends in damaging frost events in early spring [9], degrading permafrost, and an earlier spring flood pulse across the pan-Arctic [151]. By availing of the high sensitivity of microwaves to significant dielectric shifts between solid and liquid water phase transitions, the FT signals can be captured by classifying Tb or radar backscatter coefficient (σ^0) changes relative to frozen or non-frozen reference conditions [35].

FT algorithms applied to both active and passive microwave observations include threshold-based methods [35,38,152], change-detection approaches [48,153], and multi-channel data fusion or machine-learning methods [154–158]. The threshold-based methods determine FT conditions by comparing the satellite observations with reference Tb or σ^0 values representative of seasonal frozen and non-frozen conditions. Such approaches are robust and relatively easy to implement for operational FT retrievals. More sophisticated multi-channel combinations, decision tree, and probabilistic model methods can effectively distinguish FT conditions from precipitation events in sparsely vegetated and drier climate zones. As a recent development, data fusion, and machine learning methods show promise in providing potentially enhanced FT retrievals by exploiting massive archives of satellite observations and ancillary data [157–159].

Multiple global FT data records have been developed using observations from satellite microwave radiometers and scatterometers, including SMMR and SSM/I[S] [160], AMSR-E/2 [161], Aquarius [162], SMOS (Soil Moisture and Ocean Salinity) [46], SMAP [163,164], and ASCAT [165]. Long-term (>39-year) global daily FT data records have been developed using similar overlapping 37 GHz Tb retrievals from SMMR and SSM/I[S] sensors with moderate (~25km) spatial resolution [160]. Finer (~12km) resolution FT data have also been developed using calibrated 36.5 GHz orbital swath Tb records from the AMSR-E/2 sensors [161,166]. The SMAP mission provides an operational FT data record derived from L-band (1.4 GHz) Tb retrievals with global coverage, 1–3-day temporal fidelity, and 9-km and 36-km resolution gridding [163,164,167]. Example maps of the observed non-frozen days in 2017 are shown (Figure 2) from three different FT data products and operational satellite sensors, including SMAP [167], SSM/I [160], and AMSR2 [161]. The SSM/I and AMSR2 FT records are derived from vertically polarized Tb retrievals and a modified single channel algorithm. The SMAP FT record is derived using a dual algorithm approach including a normalized polarization ratio (NPR) of vertically and horizontally polarized Tb differences from NPR reference states, and a single channel algorithm where conditions are unfavorable for the NPR. All of these records show similar FT regional patterns, including generally fewer frost days at lower latitudes and altitudes, and in coastal areas relative to higher latitude, alpine, and inland areas. However, the SMAP and AMSR2 products show generally enhanced spatial delineation of FT patterns due to the respective finer 9-km and 6-km resolution gridding of these products, relative to the 25-km resolution SSM/I global grid product. The SMAP L-band products also have greater soil FT sensitivity than the K-band retrievals from AMSR2 or SSM/I [152].

In addition, the SMAP radar produced operational FT retrievals over northern (≥45°N) land areas with ~3-km resolution and 1–3-day fidelity until the radar transmitter failed in July, 2015; however, these data are overlapping with SMAP radiometer based FT records which share the same satellite antenna receiver [152]. FT retrievals have also been acquired from L-band Global Navigation Satellite System (GNSS) signals captured from the SMAP radar receiver, which has provided kilometer scale observations of FT seasonal transitions.

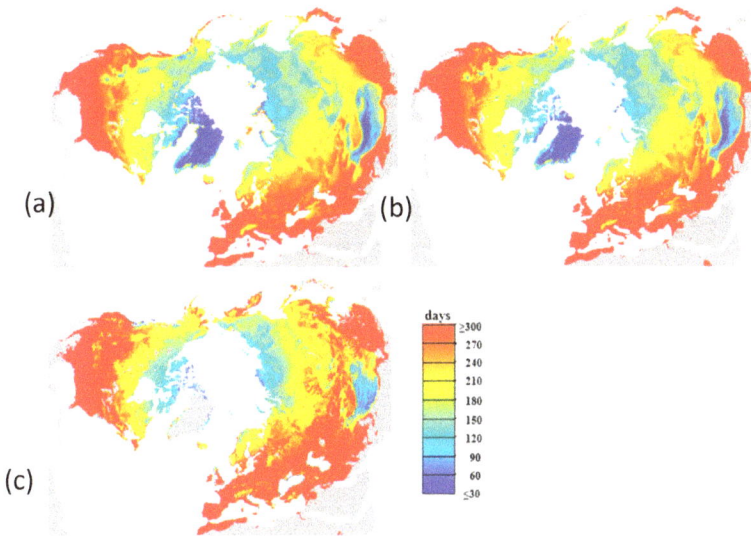

Figure 2. Estimated annual non-frozen season in 2017 derived from three operational satellite FT data products, including: (**a**) SSM/I (37 GHz; 25-km), (**b**) AMSR2 (36.5 GHz; 6-km), and (**c**) SMAP (1.4 GHz; 9-km). Areas outside of the FT global domain for each product are shown in grey and white.

2.3.2. Surface Deformation

In cold regions, climate change could significantly affect surface morphology (e.g., permafrost melting due to global warming), which may pose a threat to 70% of current infrastructure in Arctic permafrost regions by 2050 (Figure 3) [26]. Remote sensing technologies could offer a useful platform to better understand these geomorphic changes, especially those that guarantee high-resolution topography analysis [168].

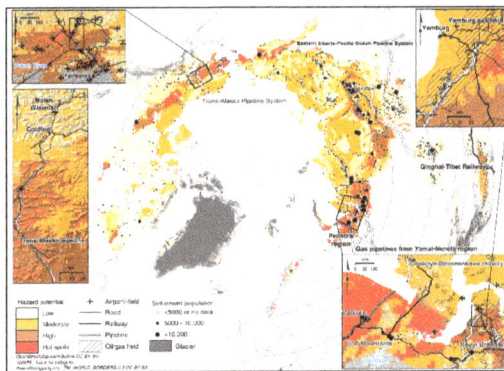

Figure 3. Projected infrastructure hazard areas from degrading permafrost over the Northern Hemisphere by year 2050 [26]. The figure was reproduced with permission from [26], which is licensed under a Creative Commons Attribution 4.0 International License.

Liu et al. [169] used spaceborne InSAR data to map surface subsidence trends at a thermokarst landform located near Deadhorse on the North Slope of Alaska. The motivation of this study was the fact that the intrinsic dynamic thermokarst process of surface subsidence remains a challenge to quantify and is seldom examined using remote sensing methods. Subsequent InSAR analysis using

Phased Array type L-band SAR (PALSAR) images revealed localized thermokarst subsidence of a few cm yr^{-1} between 2006 and 2010. Luo et al. [170] used terrestrial laser scanning for quantifying surface deformation pertaining to underlying hydrological–thermal processes affecting active layer FT conditions in the Tibetan Autonomous Region (China). The Terrestrial Laser Scanner and six Trimble 5700 GNSS systems were deployed in the region between May 2014 and October 2015, where the site was monitored four times. The results indicated that as air temperature and precipitation increase with climate warming the active layer will become more unstable, exacerbating slope instability as phase changes (thawing and freezing) occur. Jorgenson and Grosse [171] summarized recent developments (2010–2015) in remote sensing applications to detect and monitor landscape changes in permafrost regions, analyzing surface temperatures, snow cover, topography, surface water, vegetation cover, and disturbances from fire and human activities. According to this review, repeated light detection and ranging (LIDAR), InSAR, and airborne geophysics will be key tools for monitoring permafrost changes (topographic and subsurface) in Arctic and boreal regions. Arenson et al. [172] stressed the fact that in situ monitoring of periglacial dynamics is essential for the study of periglacial morphology and the design of mitigation and adaptation measures for infrastructure in permafrost zones. The application of structure-from-motion photogrammetric techniques is relatively low-cost and easy to use (e.g., see [173] for details), and provide capabilities for multi-temporal surveys of surface deformation. Meng et al. [174] used X-band SAR Interferometry for the detection of surface deformation in the Sichuan–Tibet Grid Connection Project Area (China). In this area, landslides, and debris flows triggered by climate change are becoming a major threat. Surface deformation time series observations were obtained through sequential TerraSAR X-band images. The analysis suggested that the deformation rate tends to increase dramatically during the late spring and late autumn, but with reduced deformation during colder winter months. Stettner et al. [175] discussed the capability of high spatiotemporal resolution X-band microwave satellite data (obtained from the TerraSAR-X satellite) to quantify cliff-top erosion of an ice-rich permafrost riverbank in the central Lena Delta. The results indicated continuous erosion from June to September in 2014 and 2015 with no significant seasonality across the thawing season. The authors identified X-band backscatter time series as a useful tool complementing optical remote sensing and in situ monitoring for rapid analysis of tundra permafrost erosion along riverbanks and coastal areas. Chen et al. [176] used Persistent Scatterer Interferometry (PSI) to map and quantify permafrost thaw subsidence in the Qinghai–Tibet Plateau. According to the authors, the PSI approach is less affected by temporal or geometric decorrelation, while their results indicated that permafrost areas near gullies are more vulnerable to gradual thawing and degradation.

2.4. Remote Sensing of Water Bodies

Surface water (SW) over inland areas is a key component of the water and energy cycles, covering about three percent (4.46 million km^2) of Earth's land [177]. The dynamics of SW in cold land regions are closely linked to terrestrial water storage changes [178], flood and drought events [179], seasonal thawing and the spring flood pulse [180], microtopography, underlying geology and permafrost conditions [181]. SW changes are also occurring in Arctic-boreal wetlands, lakes, rivers, and streams as permafrost degrades with regional climate warming [29,180,182]; surface subsidence during the initial stages of permafrost degradation leads to increased inundation, while later stages of permafrost thaw lead to surface drying and reduced wetland extent as drainage pathways increase [27]. The emerging glacier and thermokarst lakes formed as ice melts have a strong climate feedback and may increase regional hazards from outburst flooding [183,184].

Clear and calm water appears dark and is readily distinguished from surrounding land features using optical-IR, microwave radiometer, and mono-static radar sensors. Optical-IR satellite sensors such as PlanetScope multispectral cameras, MODIS, AVHRR, and Landsat provide potential daily to 16-day repeat global observations of SW cover at moderate to high-resolution (3–1000 m), while screening and temporal compositing of the data to reduce the influence of cloud and atmosphere contamination may degrade temporal fidelity [177,181,185].

Potential drawbacks from optical-IR sensors can be partially overcome by active [186] and passive [178,187–189] microwave remote sensing. Daily satellite passive microwave observations were used to monitor spatial variability and multi-year trends in surface inundation in permafrost affected regions [29,188]. Lower frequency (e.g., L-band) microwave retrievals have shown greater sensitivity and detection of surface water even under low to moderate vegetation cover [190]. The synergistic use of optical-IR and microwave observations, and ancillary hydrologic information has shown promise in producing optimum SW mapping results in terms of accuracy, and spatial and temporal resolution [190–195].

2.5. Remote Sensing of Terrestrial Ecosystems

Vegetation growth in cold regions is limited by multiple environmental constraints including low temperatures, frozen sub-surface soils in permafrost affected terrain, low light levels in winter and shoulder seasons, water stress in summer, and nutrient limitations due to the very slow release of plant available nitrogen and phosphorus from seasonally frozen or inundated soils [196,197]. Although ground-based measurements are needed to inform local-scale investigations, remote sensing is especially important for the cold regions because it provides spatially and temporally resolved data over broader scales, allowing for synoptic investigations of vegetation ranging from individual plots to regional and global extents.

2.5.1. Vegetation Mapping

Land cover type is a general term encompassing a range of important information about ecosystems, including biotic and abiotic properties related to vegetation, energy balance, and carbon exchanges. Satellite remote sensing has been used throughout the cold regions to provide various maps of land cover that group vegetation according to geobotanical themes, including physiology. The spatial patterns in the vegetation maps can provide useful insight into how microclimates, soil type, and hydrology, disturbance and plant succession contribute to variations in plant community characteristics at landscape to regional levels. These maps are also used to parameterize ecosystem process models by means of parameter look-up tables aggregated according to generalized plant functional types [198,199].

Classification algorithms trained on satellite optical-IR spectral information have been used to map land cover and vegetation type over large areas. For example, the United States Geological Survey (USGS) provides 30-m land cover data for the state of Alaska for years 2001 and 2011, using the C5 decision-tree classifier applied to Landsat TM and ETM+ (Enhanced Thematic Mapper) imagery (http://www.mrlc.gov). For the Anderson Level II classification (19 classes, defining different types of forest, shrub, herbaceous, and wetland), the overall accuracy ranged from 59% to 76%, depending on the definition of agreement with reference data [200]. The Earth Observation for Sustainable Development of Forests (EOSD) project used an unsupervised *k*-means clustering approach with Landsat ETM+ data to map land cover across Canada at 25-m resolution for the year 2000, using a detailed 23-class system [201]. In another study, phenological data derived from Landsat 8 NDVI timeseries, along with other inputs, were used to classify land cover across ice-free portions of Greenland at 30-m resolution [202]. Mapped classes included fen, dry heath and grasslands, wet heath, and copse and tall shrubs, with an overall classification accuracy of 89%. Other studies have used higher resolution imagery to map vegetation communities in heterogenous tundra landscapes (e.g., using IKONOS imagery [203]). For sites across the North Slope of Alaska, tundra vegetation communities were shown to be separable based on visible wavelengths collected through field spectroscopy, and vegetation type was mapped at 2-m resolution with ~70% accuracy using linear discriminant analysis applied to WorldView-2 data [40].

In addition to optical-IR data, airborne and satellite SAR data can be effective for separating land cover classes due to its sensitivity to vegetation structure and water content [204–206]. A supervised classification approach was applied to airborne, multifrequency polarimetric SAR imagery to map functional vegetation types at 30-m resolution across a boreal forest area [204]. Classes such as jack pine, black spruce, and trembling aspen were mapped with high accuracy (> 90%). Radar data can also be used to distinguish different types of wetland vegetation, as higher frequency microwave data can detect flooded short vegetation (e.g., fens and bogs), while lower frequency data are better at detecting flooded tall vegetation (e.g., forests and swamps) [206]. The fusion of microwave and optical data through traditional supervised classification (e.g., maximum-likelihood classifier) [205] and machine-learning approaches [206,207] enables improved land cover classification over the high-latitudes. Deep-learning approaches also show promise in utilizing semantic information of satellite imagery for classifying complex ecosystems [208].

2.5.2. Vegetation Growth and Photosynthetic Carbon Assimilation

A number of remote sensing vegetation indices such as NDVI, leaf area index (LAI), and enhanced vegetation index (EVI) have been developed to characterize vegetation properties related to photosynthesis on a per-pixel basis [209]. In Arctic-boreal regions, multispectral satellite data have been used to quantify variables related to vegetation growth [210], biomass [211,212], and carbon fluxes [213]. Satellite microwave systems can also provide information about vegetation growth and carbon assimilation. For example, vegetation optical depth (VOD) derived from satellite microwave Tb observations is sensitive to vegetation water content and provides information on both canopy biomass (photosynthetic and non-photosynthetic) phenology and drought stress [214–218]. VOD has also been used to monitor vegetation growth and recovery after fires in boreal forests [219].

Much of the remote sensing-based work for plant growth and carbon cycling has focused on modeling gross primary productivity (GPP), which represents carbon biomass created by plants through the process of photosynthesis over a given length of time and space (e.g., m^2 day^{-1}). GPP also represents the amount of atmospheric CO_2 sequestered by plants within biomass. GPP models are often based on a light-use efficiency (LUE) framework, which estimates GPP as a function of APAR, the fraction of absorbed photosynthetically active radiation (PAR), and a photosynthetic efficiency parameter which describes the rate at which absorbed radiation is used for carbon fixation [220]. To model APAR, inputs of LAI and fPAR (the fraction of canopy absorbed PAR) are needed, which can be modeled using spectral reflectance data combined with radiative transfer algorithms [221]. Alternatively, NDVI can be used as a proxy for fPAR [222]. The efficiency parameter is more difficult to model and can vary based on vegetation type, water limitations, and light conditions [220]. However, GPP has also been shown to be directly related to the EVI, which tends to be less saturated than NDVI over higher canopy densities and less sensitive to soil background noise [220]. Relationships between EVI and GPP were shown to vary among sites, with correlations generally stronger for deciduous sites than for evergreen sites [220]. Another important and newer proxy for GPP is solar-induced fluorescence (SIF) [223]. SIF quantifies the amount of light reemitted by chlorophyll molecules as a byproduct of photosynthesis, and has been shown to be directly proportional to GPP [224]. Satellite-based SIF (e.g., from the Global Ozone Monitoring Experiment 2 [225] and Orbiting Carbon Observatory-2 [226]), allows for global monitoring of terrestrial GPP and the carbon cycle. However, the SIF signal has relatively small magnitude and generally requires a large sensor footprint and coarse temporal compositing of the data to obtain an adequate signal-to-noise ratio relative to NDVI and EVI observations [227].

Aboveground biomass (AGB) is an important component of global carbon accounting. Remote sensing techniques are used to estimate AGB by capturing both vertical structure and spatial variability of canopies. For example, AGB has been estimated from LIDAR-derived canopy height and vertical structure metrics [228,229]. Biomass was mapped for the circumboreal zone using multi-step modeling that combined field-based biomass data, airborne and satellite LIDAR data [230,231], and similarly at the regional scale for boreal forest in Québec [232]. Other studies demonstrated the potential of low-frequency (e.g., L- and P-band) polarimetric SAR/InSAR data for estimating AGB in boreal forests and over complex terrain [233–237].

3. Changes and Trends

Climate warming in cold regions has been altering the phenology of snow, lake, and river ice, and vegetation, causing accelerated ice melting in polar regions, and leading to complex wetting and drying trends, and greening and browning patterns in northern high latitudes and TP. Multi-decade remote sensing observations are essential in documenting the environmental changes and revealing the long-term trends

3.1. Northern High Latitudes

The long-term (1979–2017) anomalies of the annual non-frozen season derived from the satellite microwave FT observational record show a lengthening non-frozen season trend over vegetated lands (excluding large water bodies and permanent ice/snow covered areas) in the high-northern latitudes (3.30 day decade^{-1}; *p*-value <0.001), with similar trends over the Northern Hemisphere (3.98 day decade^{-1}), Southern Hemisphere (3.61 day decade^{-1}), and global domains (3.93 day decade^{-1}) as shown in Figure 4.

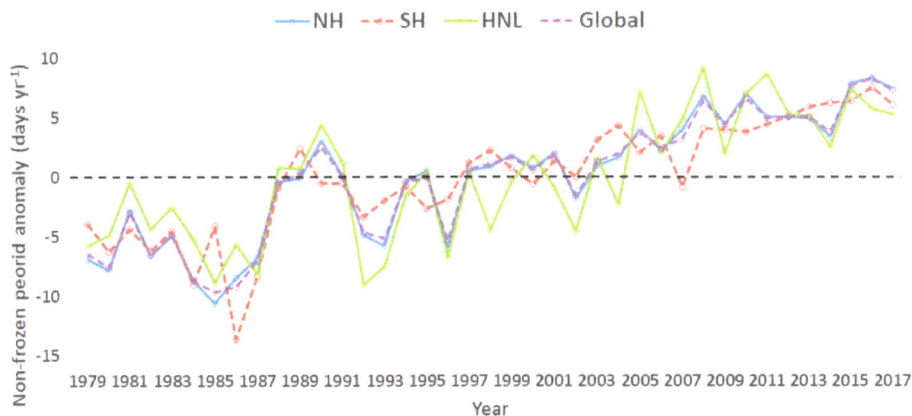

Figure 4. Non-frozen season trend (1979–2017) for the Northern Hemisphere (NH), Southern Hemisphere (SH), high-northern latitudes (HNL), and global domain. The anomalies were calculated as annual differences from the long-term mean. Grid cells dominated by permanent snow/ice cover and large water bodies (open water fraction ≥ 20%) are excluded from the analysis.

The changes in the annual non-frozen season length reflect the overall warming trend in the climate system. The timing and duration of the non-frozen season is an important factor affecting water, carbon, and energy budgets in cold land areas. Recent trends toward earlier and longer non-frozen seasons coincide with global warming and have been shown to be a major driver of northern vegetation greening, active layer deepening and permafrost degradation, enhanced evapotranspiration, earlier snowmelt onset, and associated changes in terrestrial water and energy budgets [21,34,149]. For example, satellite observations indicate that the snow end date in spring advanced by 5.11 days from 2001 to 2014 in the high northern latitudes (52–75°N) [238], along with shorter lake ice cover duration at higher latitudes from 2002 to 2015 [17]. The melting of permafrost ice provides the water supply to thermokarst lakes [239] and leads to surface water expansion detected within continuous and discontinuous permafrost zones [29].

Satellite remote sensing observations have been used to detect changes in warming Arctic–boreal ecosystems. Increasing temperatures and atmospheric CO_2 concentrations can impact the production, dynamics, and composition of vegetation, as well as soil moisture and other soil properties [240]. One of the most pronounced ongoing changes is tundra shrub expansion [203]. Changes to vegetation are likely to occur around vegetation ecotones [203,240]. Much effort has been given to identifying regions of vegetation production increase (greening) or decrease (browning). A 10-year time series of monthly average AVHRR NDVI indicated an increase in photosynthetic activity in the northern high latitudes from 1981 to 1991 driven by earlier growing season onset [241]. A subsequent study using a longer AVHRR time series indicated an increase in tundra vegetation growth in boreal North America, largely driven by longer growing seasons, but decreasing growth in interior forests [242]. Recent work used time series of Landsat vegetation indices in Arctic sites, with results indicating more areas with increasing growth than decreasing growth [243]. Analysis of a 28-year Landsat NDVI record indicated that greening and browning of Canadian boreal forests were largely driven by disturbances from wild fire and insect damages; and, in forests not affected by disturbance, climate changes were associated with both areas of greening and browning [240]. Regarding phenological changes, a 30-year Landsat record was used to detect an earlier/heterogeneous leaf emergence trend in temperate/boreal deciduous forests [197], while a 33-year AVHRR NDVI time series revealed regional trends toward earlier growing season onset, later growing season end, and longer growing season duration over the high northern latitudes [31]. Based on recent satellite optical and microwave observations for years 2003 through 2017 over the high northern latitudes, the mean summer (JJA) MODIS NDVI record (MYD13A1.006; [244]) showed similar spatial patterns with the AMSR-E/2 VOD record [216], despite the different spatial resolutions and underlying physics of the observations (Figure 5a,b). Major greening and biomass growth trends are found in northern taiga and tundra regions, where both NDVI and VOD show significant increases (*p*-value < 0.05) (Figure 5c,d). However, declining NDVI and VOD trends are also widespread, indicating decreasing productivity. The declining trend areas largely occur in boreal forest but are generally less significant than positive trend areas.

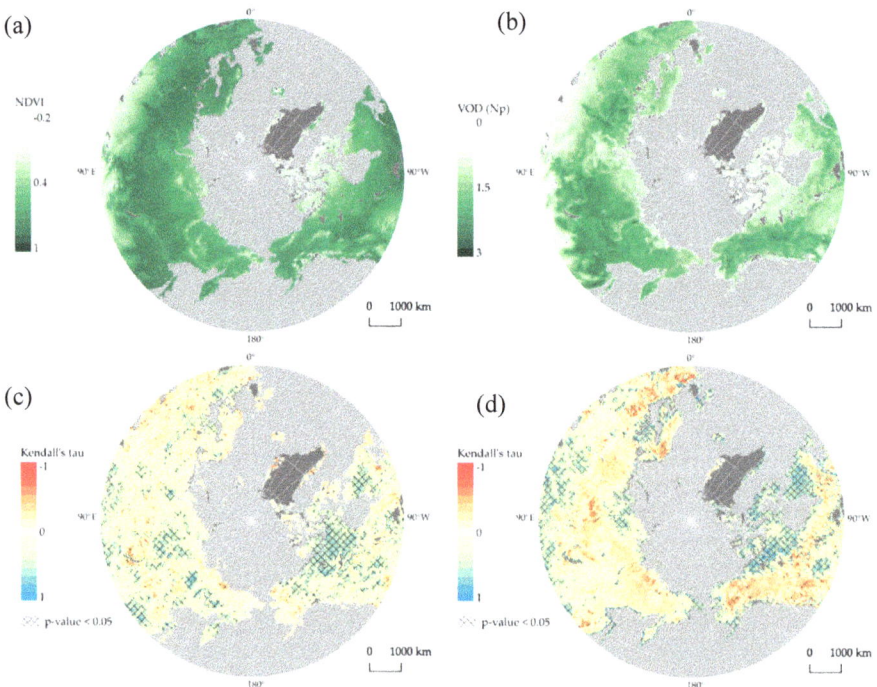

Figure 5. Vegetation conditions and growth trend over the high northern latitudes from 2003 through 2017. (**a**) Mean summer MODIS NDVI; (**b**) Mean summer AMSR-E/2 VOD; (**c**) Trends in mean summer NDVI; (**d**) Trends in mean summer VOD. In c and d, positive values indicate an increasing trend and negative values indicate a decreasing trend; pixels with significant trend (*p*-value < 0.05) are shown with crosshatch.

3.2. Antarctic and Greenland Ice

The Antarctic and Greenland ice sheets are the largest ice bodies on Earth and are significantly affected by changing air temperatures and solar radiation. If completely melted, the polar ice sheets would raise global sea level by 70 m [245]. Recent studies show that the magnitude of recent melting of the Greenland ice sheet is exceptional over at least the last 350 years [246]. Greenland's ice is melting so fast that it could become a major contributor to sea-level rise within two decades. Greenland ice loss mainly occurred in the southeast and northwest margins of the ice sheet in the 2000–2010 period; while the largest sustained acceleration (~10 years) in ice loss was detected in southwest Greenland from GRACE (Gravity Recovery and Climate Experiment) observations [247]. The overall transformation of ice into liquid water appears to be accelerating and Greenland loses an average of 270 billion tons of ice each year [248].

The recent loss of continental ice includes both the northern and southern hemispheres. An estimate of the mass balance of the entire Antarctic ice sheet over a 25-year record (1992 to 2017) shows that the Antarctic Peninsula, the smallest ice sheet in Antarctica, has lost an average of 20 Gigatonnes (Gt) of ice per year. The loss rate increased during the study period especially after the year 2000. The West Antarctic Ice Sheet lost 53 ± 29 Gt yr^{-1} from 1992-1997, and the loss rate accelerated to 159 ± 26 Gt yr^{-1} from 2012-2017. The East Antarctic Ice Sheet is relatively stable, with small gains over the study period [249]. The changes of polar mass balance are associated with snow and ice surface darkening [250], warmer atmosphere and ice surface conditions [251–253], and increased surface melt duration and extent [253–255] as observed by satellite optical and microwave sensors.

The velocity of ice flow in the Antarctic has been closely monitored using optical and radar remote sensing due to its importance in determining ice discharge and sea level rise [60,256]. The velocity map derived from the MODIS-based Mosaic of Antarctica data showed the general flow patterns of glaciers and ice sheets moving from interior Antarctica toward the ocean for the periods from 2003–2004 and 2008–2009 (Figure 6). Continuous monitoring of the widespread ice flow over the entire continent is highly needed for improving our understanding of ice sheet dynamics and evolution in a warming climate [60,256].

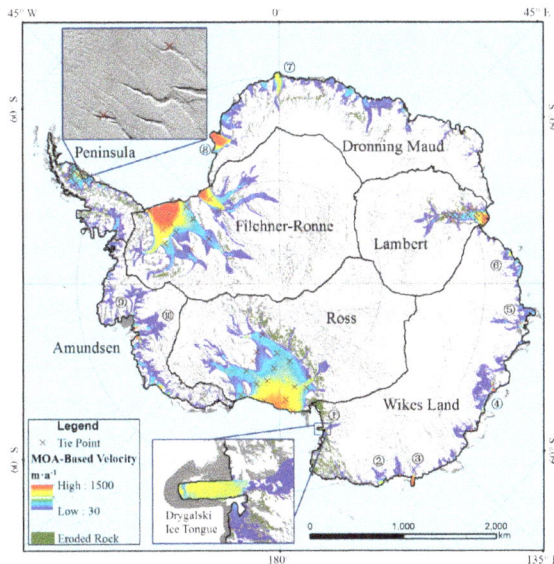

Figure 6. Antarctic surface ice velocity map. The floating velocity field of the Drygalski Ice Tongue is enlarged in the close-up below. The figure was reproduced with permission from [60].

3.3. Tibetan Plateau

The Tibetan Plateau (TP), the most extensive highland in the world, has an area of approximately 2.5×10^6 km^2 and an average elevation of over 4000 m. The TP also has the Earth's largest storage of ice outside of the north and south polar regions. The climate of the TP is changing rapidly with temperatures warming at a rate of around 0.36 °C/decade [3], which is twice the mean global trend [4]. Consequently, the impacts of climate change on the TP environment are most pronounced, leading to earlier onset of seasonal thawing, accelerating glacier melting, permafrost degradation, and complex changes in snow, lakes, and vegetation.

Glacier melting has been observed over the TP and larger High Mountain Asia region. Kaab et al. [257] used ICESat to analyze glacier mass change in the Hindu Kush–Karakoram–Himalaya region during 2003–2008 and found a mass loss rate of -12.8 ± 3.5 Gt yr^{-1} in this region, which is faster than the rate previously estimated using GRACE [258]. For the whole TP region, Neckel et al. [259] estimated an overall mass loss rate of -15.6 ± 10.1 Gt yr^{-1} using ICESat observations from 2003 to 2009. For the same period, Gardner et al. [260] estimated a total mass change of -29 ± 13 Gt yr^{-1} over High Mountain Asia by integrating GRACE and ICESat observations.

Global and localized satellite snow products have been used for studying environmental changes in the TP, including snow impacts on the regional water cycle, ecosystems, and atmospheric circulation [261].

Based on the MODIS snow cover product (MOD10A2) and observations from 37 meteorological stations, a significant trend of earlier onset of snow ablation during the 2001-2015 period was detected over the TP [262]. By analyzing MODIS snow cover data from 2001 to 2011, other research [263] found about 34.14% (5.56%) of the TP area having a declining (significant declining) trend in snow duration, while 24.75% (3.9%) of the region showed increasing snow duration. To further enhance the accuracy of TP snow products, Chen et al. [264] integrated snow cover data from multiple sources to generate a gap-filled daily 5-km Tibetan Plateau snow cover extent record (TPSCE) from 1981–2016. As revealed by the TPSCE, the snow cover fraction increased in the northern interior TP river basins and upper reaches of the Yangtze, Mekong, and Brahmaputra River basins (Figure 7).

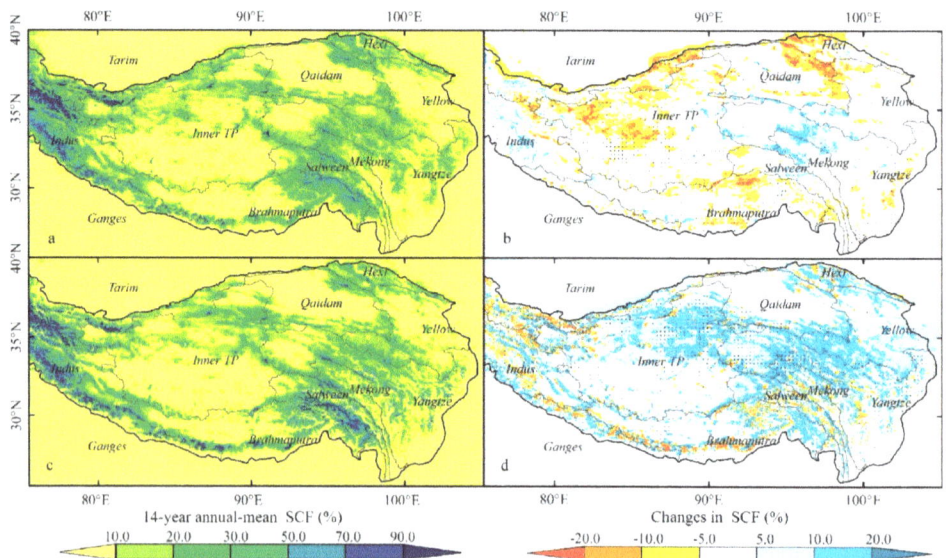

Figure 7. Climatology and changes of the TP snow cover fraction (%) calculated using MCD10A1-TP (**a**, **b**) and TPSCE (**c**, **d**) from 2001–2014. Black dots in (**b**) and (**d**) indicate the changes are statistically significant at the 95% level. The figure was reproduced with permission from [264].

Beside the global FT products available from satellite microwave observations [160], Kou et al. [265] developed an enhanced resolution (0.05 degree) FT product over the TP by merging MODIS LST with AMSR-E Tb records. Li et al. [266] analyzed changes in the soil FT cycle over the TP using SSM/I data from 1988-2007. They identified a trend toward earlier onset of soil thaw by approximately 14 days/decade and later onset of soil freeze by approximately 10 days/decade. The observed changes in FT patterns over the TP are also closely related to regional climate warming [265].

There are more than 1000 lakes with an area greater than 1 km² on the TP. Generally, the total TP lake area is expanding, from 41,800 km² in 2005–2006 [267] to 46,600 km² in 2015 [268]. The lake expansion can be directly measured through optical satellite remote sensing. Wang et al. [269] analyzed the trend of lake area changes during 1960-2000 by integrating aerial photos, satellite images (Landsat and CBERS-1 (China-Brazil Earth Resources Satellite)), and topography information. They found that most lakes in the central TP were expanding, while the lakes in the source regions of the Yellow River were shrinking. For lakes in the central TP, Wan et al. [270] analyzed Landsat TM/ETM+ and CBERS images, and found that lakes southeast of the Qiangtang area were expanding from 1975 to 2005. Another study [271] analyzed Landsat images from 1970 to 2010 and found a shrinking lake trend in the southwest TP contrasting with rapid lake expansion in the northeast TP. Yang et al. [272] investigated lake extent fluctuations in the Hindu Kush–Himalaya–Tibetan (HKHT) regions over the past 40 years using Landsat images obtained from the 1970s to 2014. They showed that the TP lake trends are distinct from region to region, with the most intensive lake shrinking observed in northeastern HKHT (HKHT Interior, Tarim, Yellow, Yangtze), while the most extensive expansion was observed in the western and southwestern HKHT (Amu Darya, Ganges Indus, and Brahmaputra), largely caused by the proliferation of small lakes in high-altitude regions from the 1970s to 1995.

Lake water levels measured from LIDAR/Radar altimetry can also be used to quantify lake changes. Based on 10-year TOPEX/Poseidon altimetry data, the water level of La'nga lake in the western TP was found to decrease steadily from 1993 to 2001, while that of Ngangzi lake in the eastern TP decreased from 1993 to 1997 and then increased monotonically afterwards [273]. Another study [274] used ICESat altimetry data to analyze water level changes from 2003–2009 for 74 TP lakes and identified an increasing lake level trend (~0.23 m yr^{-1}) over 84% of the lakes represented. Consistent with the findings of [275], an average water level increase of 0.20 m yr^{-1} was detected from GLAS data over 154 TP lakes for the same period [276]. By integrating multi-altimeter data from Envisat/RA-2, Cryosat-2/Siral, Jason-1/Poseidon-2, and Jason-2/Poseidon-3, Gao et al. [277] found that water levels increased by about 0.275 m yr^{-1} for over 82% of 51 TP lakes sampled, while major lake expansion and shrinking were identified over the northern and southern TP, respectively.

A general greening trend from the 1980s to 2010s was detected from satellite remote sensing over the TP [278]. Seasonal analysis based on GIMMS (Global Inventory Monitoring and Modeling System) NDVI data further revealed that the largest NDVI increase occurred in autumn over 61% of the TP, while the smallest increase occurred in spring over 41% of the region [279]. Vegetation changes in the northeast, southwest, mid-eastern, and southern TP regions were driven by three different factors, including changes in surface air temperature, water availability, and solar radiation, respectively. Anthropogenic disturbances may offset climate-driven vegetation greening and exacerbate vegetation browning, while ecosystem conservation efforts contributed to vegetation recovery in the TP Three-River Headwaters Region [280].

Based on the NDVI data from 1982 to 2014, an advancing start of growing season, delayed end of season and increasing length of growing season were identified for meadow areas in the eastern TP; while the opposite changes were found for the steppe and sparse herbaceous or sparse shrub areas in the northwest and western edges of the TP. The satellite-observed phenology changes were driven by a number of environmental factors including temperature [281–283], precipitation [284], sunshine duration [285], and snow cover [286]; and may be partially attributed to aerosol contamination in the satellite observations [287].

4. Challenges and Opportunities

4.1. Limitations of Current Approaches

Despite great achievements in cold land remote sensing, comprehensive assessment of both long-term trends and relatively abrupt environmental changes requires quantifying hydrological and ecological variables with greater precision, including clearer delineation of different land components, better accuracy, data consistency, spatial resolution, and temporal sampling.

Ice—The spatial resolution of satellite gravimetry observations of glaciers and ice sheets are generally very coarse (around 100 km), which limits their ability to precisely detect and locate glacier mass change at finer scales. LIDAR-based measurements of glacier elevations are of high accuracy (decimeter level) but with relatively poor spatial coverage, which consequently requires data interpolation and extrapolation for applications over large areas. For monitoring glacier movement, feature tracking fails when low coherence occurs in fast moving glaciers or over prolonged time intervals between observations.

Snow—Snow covered area is mainly derived using optical sensors onboard polar-orbiting or geostationary satellites. However, it is still challenging to distinguish between clouds and snow in satellite optical snow mapping. Long-term (>40-year) SWE data records have been generated for the globe through satellite passive microwave sensors, but the retrieval spatial resolution and accuracy has generally not met the requirements for regional climate, numerical weather prediction, and hydrological research. The retrieval uncertainties mainly come from four sources, including mixed pixel effects, terrain effects, large diversity in snow physical properties, and low microwave sensitivity to shallow snow.

Landscape Freeze/Thaw States—Most of the available global FT products represent aggregate landscape FT conditions that do not distinguish land components of the FT signal within a satellite footprint at several-10s km. Accurate estimates of FT metrics (e.g., spring thaw timing, frost days) pertaining to soil, snow, and vegetation components isolated from the integrated microwave FT signals with improved spatial resolution are required for better understanding of land and atmosphere interactions, including carbon, water, and energy exchanges. It is also needed to understand the scaling effects for heterogeneous terrain to bridge multi-resolution satellite observations [38].

Water Bodies—It is challenging to detect water within mixed pixels or under overlying vegetation for both optical-IR and microwave sensors. In addition, near real-time and fine-scale mapping of regional SW dynamics are urgently needed for monitoring flood hazards and arctic wetting and drying trends.

Vegetation—The short growing season and generally low vegetation productivity of Arctic–boreal regions can create challenges in terms of detecting changes in vegetation growth through use of remote sensing time series data. Other challenges in Arctic–boreal regions include saturated satellite signals over high-biomass vegetation, high occurrence of shallow water bodies, the presence of snow and ice cover, and spatially heterogenous snowmelt resulting in pixels mixed with snow and vegetation. Additionally, long-term and fine-scale delineation (1–200 m) of vegetation (e.g., shrub) patch distributions is lacking, but greatly needed in high-latitude ecosystem studies [288].

4.2. Opportunities

Progress in remote sensing and information technologies such as the Advanced Topographic Laser Altimeter System (ATLAS) on-board ICESat-2 [289,290]; near-nadir SAR [291,292]; C-, L- and P-band SAR [293–295]; small satellites [181,296]; and cloud computation and artificial intelligence [297] will likely provide opportunities to overcome current challenges.

It is anticipated that SAR will play a greater role in cold land studies. The European Space Agency's Sentinel-1A/B satellites and the Canadian Space Agency's RADARSAT Constellation Mission (RCM) launched on 12 June 2019 [293] can provide multi-resolution (1–3 m to 100-m) C-band SAR and InSAR measurements every 1–6 days globally and at low cost for monitoring glacier movement, river ice conditions, water body dynamics, snow, and vegetation parameters [298]. The next generation NISAR (National Aeronautics and Space Administration - Indian Space Research Organisation SAR) mission is expected to improve capabilities for dynamic mapping of global surface water at resolutions from 3–10 m [294] and provide new opportunities to derive SWE in complex terrain from single- and repeat-pass radar (L-band) interferometry [16]. The surface water ocean topography (SWOT) mission has a projected launch in 2021 and will enable estimation of the changing volumes of global water bodies whose surface area exceeds 250 m by 250 m at sub-monthly time scales [178,291]. The upcoming BIOMASS mission (launch 2021) offers opportunities for mapping vegetation biomass with enhanced P-band penetration capability [299], though National restrictions on P-band transmissions will eliminate BIOMASS coverage over North America [300]. Another future direction for analyzing environmental changes in remote areas under harsh climatic conditions could also involve relatively low-cost multi-temporal remote sensing surveys. Unmanned aerial vehicle (UAV)-based remote sensing (e.g., thermal imaging, structure from motion photogrammetric techniques), given relatively low application barriers (e.g., low cost for repeat seasonal or yearly surveys), is expected to become a strategic tool for better monitoring and understanding of the geomorphologic dynamics of cold regions, and related impacts on ecosystems and infrastructures.

Improved retrieval algorithms will better leverage the remote sensing observations. For example, entropy-based multi-scale image matching and optical flow techniques potentially help overcome the limitations of conventional image matching approaches for monitoring fast moving glaciers [301,302]. Backward reconstruction techniques using a temperature-index or energy-balance model provide another promising tool to estimate SWE through the melt season in mountainous regions and elsewhere [303,304]. There is also great potential for improving snow depth and SWE retrievals using LIDAR measurements [305,306], microwave interferometry [147,307], and GNSS techniques [308], which largely avoid issues related to snow microstructure. Considering the complementarity of LIDAR, optical-IR, and microwave remote sensing, the use of data integration techniques such as deep-learning approaches [208] from a collection of multi-sensor observations may enable enhanced delineations of snow, water, soil, and vegetation elements, and accurate mapping of environmental variables at high spatial–temporal resolutions.

5. Conclusions

The rapid environmental changes in cold land regions have profound impacts on ecosystems, geomorphology, animal habitats, and human lives. Remote sensing is the most valuable and indispensable technique for large-scale monitoring of such changes, accurately quantifying both transient anomalies and longer-term trends, while providing observational benchmarks to test earth system model projections. Calibrated long-term observations from these sensors have produced relatively precise satellite data records documenting significant changes in landscape FT states, snow extent and depth, glacier mass and movement, water body dynamics, and lake ice [309] and vegetation phenology. Earlier onset of snow melt [310], soil thaw, and lake ice break-up, longer potential growing seasons, and expanding lakes were identified in both the northern high latitudes and TP; and continuous ice loss has been observed in both Greenland and Antarctica. Despite these achievements, pressing issues such as melting permafrost, disappearing glaciers, shrinking ice cover, and structural and functional changes in ecosystems require more timely and accurate interpretations from remote sensing observations. Multi-sensor data fusion approaches show promise in overcoming the drawbacks of single sensor observations, while strengthening capabilities for monitoring and detecting environmental changes in cold regions. Next generation satellite missions including SWOT, NISAR, and BIOMASS; emerging techniques such as micro-satellites and data mining; and coordinated research activities [311–313] will enable cold land mapping with unprecedented sampling frequency, spatial resolution, and accuracy.

Remote Sens. **2019**, *11*, 1952

Supplementary Materials: The following are available online at http://www.mdpi.com/2072-4292/11/16/1952/s1, Table S1. Overview of satellite missions/sensors for cold land remote sensing and summarized in the review, Table S2. Vertical accuracy of DEMs from different sensors and methods over glacier surface.

Author Contributions: All authors contributed equally to the manuscript.

Funding: This research received no external funding.

Conflicts of Interest: The authors declare no conflict of interest. The funders had no role in the design of the study; in the collection, analyses, or interpretation of data; in the writing of the manuscript, or in the decision to publish the results.

Acronym list

AGB	Aboveground biomass
APAR	Absorbed PAR
AMSR2	Advanced Microwave Scanning Radiometer 2
AMSR-E	Advanced Microwave Scanning Radiometer for Earth Observing System
ASAR	Advanced SAR
ATLAS	Advanced Topographic Laser Altimeter System
AVHRR	Advanced very-high-resolution radiometer
AMSR-E/2	AMSR-E and AMSR2
Tb	Brightness temperature
China–Brazil Earth Resources Satellite	CBERS-1
DEM	Digital Elevation Model
DInSAR	Differential Interferometric Synthetic Aperture Radar
DMSP	Defense Meteorological Satellite Program
ERS	European remote sensing satellite
ETM	Enhanced Thematic Mapper
EVI	Enhanced Vegetation Index
FT	Freeze–thaw
FY	Feng Yun
GLAS	Geoscience Laser Altimeter System
GIMMS	Global Inventory Monitoring and Modeling System
GCOS	Global Climate Observing System
GNSS	Global Navigation Satellite System
GOES	Geostationary Operational Environmental Satellite
GRACE	Gravity Recovery and Climate Experiment
GBL	Great Bear Lake
GSL	Great Slave Lake
GPP	Gross Primary Productivity
ICESat	Ice, Cloud, and land Elevation Satellite
IMS	Interactive Multisensor Snow and Ice Mapping System
InSAR	Interferometric Synthetic Aperture Radar
ISRO	Indian Space Research Organisation
HKHT	Kush-Himalaya-Tibetan
LST	Land Surface Temperature
LAI	Leaf Area Index
LIDAR	Light Detection and Ranging
LUE	Light Use Efficiency
MBE	Mean Bias Error
MSG	Meteosat Second Generation
MWRI	Microwave Radiation Imager
MODIS	Moderate Resolution Imaging Spectroradiometer
MTSAT	Multifunctional Transport Satellites
NASA	National Aeronautics and Space Administration

NISAR	NASA-ISRO Synthetic Aperture Radar
NOAA	National Oceanic and Atmospheric Administration
NDSI	Normalized Difference Snow Index
NDVI	Normalized Difference Vegetation Index
NDFSI	Normalized Difference Forest Snow Index
Optical-IR	Optical and Infrared
OLI	Operational Land Imager
PSI	Persistent Scattered Interferometry
PALSAR	Phased Array type L-band Synthetic Aperture Radar
PAR	Photosynthetically active radiation
RMSE	Root Mean Square Error
SIRAL	SAR Interferometer Radar Altimeter
SMMR	Scanning Multichannel Microwave Radiometer
SWE	Snow water equivalent
SMAP	Soil Moisture Active Passive
SMOS	Soil Moisture and Ocean Salinity
SIF	Solar Induced Fluorescence
SSM/I	Special Sensor Microwave/Imager
SSMIS	Special Sensor Microwave Imager Sounder
SW	Surface water
SWOT	Surface Water Ocean Topography
SAR	Synthetic Aperture Radar
fPAR	The fraction of absorbed PAR
TM	Thematic Mapper
TP	Tibetan Plateau
TPSCE	Tibetan Plateau Snow Cover Extent record
UAV	Unmanned aerial vehicle
USGS	United States Geological Survey
VOD	Vegetation Optical Depth

References

1. Cohen, J.; Screen, J.A.; Furtado, J.C.; Barlow, M.; Whittleston, D.; Coumou, D.; Francis, J.; Dethloff, K.; Entekhabi, D.; Overland, J.; et al. Recent Arctic amplification and extreme mid-latitude weather. *Nat. Geosci.* **2014**, *7*, 627. [CrossRef]
2. Stuecker, M.F.; Bitz, C.M.; Armour, K.C.; Proistosescu, C.; Kang, S.M.; Xie, S.P.; Kim, D.; McGregor, S.; Zhang, W.; Zhao, S.; et al. Polar amplification dominated by local forcing and feedbacks. *Nat. Clim. Chang.* **2018**, *8*, 1076. [CrossRef]
3. Wang, B.; Bao, Q.; Hoskins, B.; Wu, G.X.; Liu, Y.M. Tibetan plateau warming and precipitation changes in East Asia. *Geophys. Res. Lett.* **2008**, *35*. [CrossRef]
4. Xu, B.Q.; Cao, J.J.; Hansen, J.; Yao, T.D.; Joswia, D.R.; Wang, N.L.; Wu, G.J.; Wang, M.; Zhao, H.B.; Yang, W.; et al. Black soot and the survival of Tibetan glaciers. *Proc. Natl. Acad. Sci. USA* **2009**, *106*, 22114–22118. [CrossRef]
5. Hugelius, G.; Routh, J.; Kuhry, P.; Crill, P. Mapping the degree of decomposition and thaw remobilization potential of soil organic matter in discontinuous permafrost terrain. *J. Geophys. Res. Biogeosci.* **2012**, *117*. [CrossRef]
6. Olefeldt, D.; Goswami, S.; Grosse, G.; Hayes, D.; Hugelius, G.; Kuhry, P.; McGuire, A.D.; Romanovsky, V.E.; Sannel, A.B.K.; Schuur, E.A.G.; et al. Circumpolar distribution and carbon storage of thermokarst landscapes. *Nat. Commun.* **2016**, *7*, 13043. [CrossRef]
7. Vaughan, D.G.; Marshall, G.J.; Connolley, W.M.; Parkinson, C.; Mulvaney, R.; Hodgson, D.A.; King, J.C.; Pudsey, C.J.; Turner, J. Recent rapid regional climate warming on the Antarctic Peninsula. *Clim. Chang.* **2003**, *60*, 243–274. [CrossRef]

8. Huang, J.; Zhang, X.; Zhang, Q.; Lin, Y.; Hao, M.; Luo, Y.; Zhao, Z.; Yao, Y.; Chen, X.; Wang, L.; et al. Recently amplified arctic warming has contributed to a continual global warming trend. *Nat. Clim. Chang.* **2017**, *7*, 875. [CrossRef]

9. Kim, Y.; Kimball, J.S.; Zhang, K.; Didan, K.; Velicogna, I.; McDonald, K.C. Attribution of divergent northern vegetation growth responses to lengthening non-frozen seasons using satellite optical-NIR and microwave Remote Sens. *Int. J. Remote Sens.* **2014**, *35*, 3700–3721. [CrossRef]

10. Kim, Y.; Kimball, J.S.; Robinson, D.A.; Derksen, C. New satellite climate data records indicate strong coupling between recent frozen season changes and snow cover over high northern latitudes. *Environ. Res. Lett.* **2015**, *10*, 084004. [CrossRef]

11. Schuur, E.A.; McGuire, A.D.; Schädel, C.; Grosse, G.; Harden, J.W.; Hayes, D.J.; Hugelius, G.; Koven, C.D.; Kuhry, P.; Lawrence, D.M.; et al. Climate change and the permafrost carbon feedback. *Nature* **2015**, *520*, 171. [CrossRef]

12. Van Huissteden, J.; Dolman, A.J. Soil carbon in the Arctic and the permafrost carbon feedback. *Curr. Opin. Environ. Sustain.* **2012**, *4*, 545–551. [CrossRef]

13. Pepin, N.; Bradley, R.S.; Diaz, H.F.; Baraër, M.; Caceres, E.B.; Forsythe, N.; Fowler, H.; Greenwood, G.; Hashmi, M.Z.; Liu, X.D.; et al. Elevation-dependent warming in mountain regions of the world. *Nat. Clim. Chang.* **2015**, *5*, 424.

14. Zhu, Z.; Piao, S.; Myneni, R.B.; Huang, M.; Zeng, Z.; Canadell, J.G.; Ciais, P.; Sitch, S.; Friedlingstein, P.; Arneth, A.; et al. Greening of the Earth and its drivers. *Nat. Clim. Chang.* **2016**, *6*, 791–795. [CrossRef]

15. Schuur, E.A.; Mack, M.C. Ecological response to permafrost thaw and consequences for local and global ecosystem services. *Annu. Rev. Ecol. Evol. Syst.* **2018**, *49*, 279–301. [CrossRef]

16. Bormann, K.J.; Ross, D.B.; Chris, D.; Thomas, H. Painter. Estimating snow-cover trends from space. *Nat. Clim. Chang.* **2018**, *8*, 924–928. [CrossRef]

17. Du, J.; Kimball, J.S.; Duguay, C.R.; Kim, Y.; Watts, J. Satellite microwave assessment of Northern Hemisphere lake ice phenology from 2002 to 2015. *Cryosphere* **2017**, *11*, 47–63. [CrossRef]

18. Serreze, M.C.; Stroeve, J. Arctic sea ice trends, variability and implications for seasonal ice forecasting. *Philos. Trans. R. Soc. A Math. Phys. Eng. Sci.* **2015**, *373*, 20140159. [CrossRef]

19. Rignot, E.; Mouginot, J.; Scheuchl, B.; van den Broeke, M.; van Wessem, M.J.; Morlighem, M. Four decades of Antarctic Ice Sheet mass balance from 1979–2017. *Proc. Natl. Acad. Sci. USA* **2019**, *116*, 1095–1103. [CrossRef]

20. Flanner, M.G.; Shell, K.M.; Barlage, M.; Perovich, D.K.; Tschudi, M.A. Radiative forcing and albedo feedback from the Northern Hemisphere cryosphere between 1979 and 2008. *Nat. Geosci.* **2011**, *4*, 151. [CrossRef]

21. Kim, Y.; Kimball, J.S.; Du, J.; Schaaf, C.L.B.; Kirchner, P.B. Quantifying the effects of freeze-thaw transitions and snowpack melt on land surface albedo and energy exchange over Alaska and Western Canada. *Environ. Res. Lett.* **2018**, *13*, 075009. [CrossRef]

22. Duan, L.; Cao, L.; Caldeira, K. Estimating Contributions of Sea Ice and Land Snow to Climate Feedback. *J. Geophys. Res. Atmos.* **2019**, *124*, 199–208. [CrossRef]

23. Boelman, N.T.; Liston, G.E.; Gurarie, E.; Meddens, A.J.; Mahoney, P.J.; Kirchner, P.B.; Bohrer, G.; Brinkman, T.J.; Cosgrove, C.L.; Eitel, J.U.; et al. Integrating snow science and wildlife ecology in Arctic-boreal North America. *Environ. Res. Lett.* **2019**, *14*, 010401. [CrossRef]

24. Liljedahl, A.K.; Boike, J.; Daanen, R.P.; Fedorov, A.N.; Frost, G.V.; Grosse, G.; Hinzman, L.D.; Iijma, Y.; Jorgenson, J.C.; Matveyeva, N.; et al. Pan-Arctic ice-wedge degradation in warming permafrost and its influence on tundra hydrology. *Nat. Geosci.* **2016**, *9*, 312. [CrossRef]

25. Turetsky, M.R.; Abbott, B.W.; Jones, M.C.; Anthony, K.W.; Olefeldt, D.; Schuur, E.A.; Koven, C.; McGuire, A.D.; Grosse, G.; Kuhry, P.; et al. Permafrost collapse is accelerating carbon release. *Nature* **2019**, *569*, 32–34. [CrossRef] [PubMed]

26. Hjort, J.; Karjalainen, O.; Aalto, J.; Westermann, S.; Romanovsky, V.E.; Nelson, F.E.; Etzelmüller, B.; Luoto, M. Degrading permafrost puts Arctic infrastructure at risk by mid-century. *Nat. Commun.* **2018**, *9*, 5147. [CrossRef] [PubMed]

27. Walvoord, M.A.; Kurylyk, B.L. Hydrologic impacts of thawing permafrost—A review. *Vadose Zone J.* **2016**, *15*. [CrossRef]

28. Smith, L.C.; Sheng, Y.; MacDonald, G.M.; Hinzman, L.D. Disappearing arctic lakes. *Science* **2005**, *308*, 1429. [CrossRef]

29. Watts, J.D.; Kimball, J.S.; Jones, L.A.; Schroeder, R.; McDonald, K.C. Satellite Microwave remote sensing of contrasting surface water inundation changes within the Arctic–Boreal Region. *Remote Sens. Environ.* **2012**, *127*, 223–236. [CrossRef]

30. Andresen, C.G.; Lougheed, V.L. Disappearing Arctic tundra ponds: Fine-scale analysis of surface hydrology in drained thaw lake basins over a 65 year period (1948–2013). *J. Geophys. Res. Biogeosci.* **2015**, *120*, 466–479. [CrossRef]

31. Park, T.; Ganguly, S.; Tømmervik, H.; Euskirchen, E.S.; Høgda, K.-A.; Karlsen, S.R.; Brovkin, V.; Nemani, R.R.; Myneni, R.B. Changes in growing season duration and productivity of northern vegetation inferred from long-term remote sensing data. *Environ. Res. Lett.* **2016**, *11*. [CrossRef]

32. Soja, A.J.; Tchebakova, N.M.; French, N.H.; Flannigan, M.D.; Shugart, H.H.; Stocks, B.J.; Sukhinin, A.I.; Parfenova, E.I.; Chapin, F.S., III; Stackhouse, P.W., Jr. Climate-induced boreal forest change: Predictions versus current observations. *Glob. Planet. Chang.* **2007**, *56*, 274–296. [CrossRef]

33. Loranty, M.M.; Lieberman-Cribbin, W.; Berner, L.T.; Natali, S.M.; Goetz, S.J.; Alexander, H.D.; Kholodov, A.L. Spatial variation in vegetation productivity trends, fire disturbance, and soil carbon across arctic-boreal permafrost ecosystems. *Environ. Res. Lett.* **2016**, *11*, 095008. [CrossRef]

34. Park, H.; Kim, Y.; Kimball, J.S. Widespread permafrost vulnerability and soil active layer increases over the high northern latitudes inferred from satellite remote sensing and process model assessments. *Remote Sens. Environ.* **2016**, *175*, 349–358. [CrossRef]

35. Kim, Y.; Kimball, J.S.; Glassy, J.; Du, J. An extended global earth system data record on daily landscape freeze-thaw status determined from satellite passive microwave Remote Sens. *Earth Syst. Sci. Data* **2017**, *9*, 133–147. [CrossRef]

36. Potter, C. Recovery Rates of Wetland Vegetation Greenness in Severely Burned Ecosystems of Alaska Derived from Satellite Image Analysis. *Remote Sens.* **2018**, *10*, 1456. [CrossRef]

37. Pan, C.G.; Kirchner, P.B.; Kimball, J.S.; Kim, Y.; Du, J. Rain-on-snow events in Alaska, their frequency and distribution from satellite observations. *Environ. Res. Lett.* **2018**, *13*, 075004. [CrossRef]

38. Du, J.; Kimball, J.S.; Azarderakhsh, M.; Dunbar, R.S.; Moghaddam, M.; McDonald, K.C. Classification of Alaska spring thaw characteristics using satellite L-band radar Remote Sens. *IEEE Trans. Geosci. Remote Sens.* **2014**, *53*, 542–556.

39. Montgomery, J.; Brisco, B.; Chasmer, L.; Devito, K.; Cobbaert, D.; Hopkinson, C. SAR and Lidar Temporal Data Fusion Approaches to Boreal Wetland Ecosystem Monitoring. *Remote Sens.* **2019**, *11*, 161. [CrossRef]

40. Davidson, S.; Santos, M.; Sloan, V.; Watts, J.; Phoenix, G.; Oechel, W.; Zona, D. Mapping Arctic tundra vegetation communities using field spectroscopy and multispectral satellite data in North Alaska, USA. *Remote Sens.* **2016**, *8*, 978. [CrossRef]

41. Lindenschmidt, K.E.; Li, Z. Radar Scatter Decomposition to Differentiate between Running Ice Accumulations and Intact Ice Covers along Rivers. *Remote Sens.* **2019**, *11*, 307. [CrossRef]

42. Veh, G.; Korup, O.; Roessner, S.; Walz, A. Detecting Himalayan glacial lake outburst floods from Landsat time series. *Remote Sens. Environ.* **2018**, *207*, 84–97. [CrossRef]

43. Rondeau-Genesse, G.; Trudel, M.; Leconte, R. Monitoring snow wetness in an Alpine Basin using combined C-band SAR and MODIS data. *Remote Sens. Environ.* **2016**, *183*, 304–317. [CrossRef]

44. Shi, J.; Dozier, J. Estimatino of snow water equivalence using SIR-C/X-SAR, Part I: Inferring snow density and subsurface properties. *IEEE Trans. Geosci. Remote Sens.* **2000**, *38*, 2465–2474.

45. Shi, J.; Dozier, J. Estimation of snow water equivalence using SIR-C/X-SAR, Part II: Inferring snow depth and particle size. *IEEE Trans. Geosci. Remote Sens.* **2000**, *38*, 2475–2488.

46. Rautiainen, K.; Parkkinen, T.; Lemmetyinen, J.; Schwank, M.; Wiesmann, A.; Ikonen, J.; Derksen, C.; Davydov, S.; Davydova, A.; Boike, J.; et al. SMOS prototype algorithm for detecting autumn soil freezing. *Remote Sens. Environ.* **2016**, *180*, 346–360. [CrossRef]

47. Baghdadi, N.; Bazzi, H.; El Hajj, M.; Zribi, M. Detection of frozen soil using Sentinel-1 SAR data. *Remote Sens.* **2018**, *10*, 1182. [CrossRef]

48. Chen, X.; Liu, L.; Bartsch, A. Detecting soil freeze/thaw onsets in Alaska using SMAP and ASCAT data. *Remote Sens. Environ.* **2019**, *220*, 59–70. [CrossRef]

49. Jansson, P.; Hock, R.; Schneider, T. The concept of glacier storage: A review. *J. Hydrol.* **2003**, *282*, 116–129. [CrossRef]

50. Williams, R.S. *Glaciers: Clues to Future Climate?* United States Geological Survey: Denver, CO, USA, 1983.

51. Sasgen, I.; Konrad, H.; Helm, V.; Grosfeld, K. High-Resolution Mass Trends of the Antarctic Ice Sheet through a Spectral Combination of Satellite Gravimetry and Radar Altimetry Observations. *Remote Sens.* **2019**, *11*, 144. [CrossRef]

52. Wahr, J.; Swenson, S.; Zlotnicki, V.; Velicogna, I. Time-variable gravity from GRACE: First results. *Geophys. Res. Lett.* **2004**, *31*. [CrossRef]

53. Wesche, C.; Jansen, D.; Dierking, W. Calving fronts of Antarctica: Mapping and classification. *Remote Sens.* **2013**, *5*, 6305–6322. [CrossRef]

54. Wang, X.; Cheng, X.; Gong, P.; Huang, H.; Li, Z.; Li, X. Earth science applications of ICESat/GLAS: A review. *Int. J. Remote Sens.* **2011**, *32*, 8837–8864. [CrossRef]

55. Markus, T.; Neumann, T.; Martino, A.; Abdalati, W.; Brunt, K.; Csatho, B.; Farrell, S.; Fricker, H.; Gardner, A.; Harding, D.; et al. The Ice, Cloud, and land Elevation Satellite-2 (ICESat-2): Science requirements, concept, and implementation. *Remote Sens. Environ.* **2017**, *190*, 260–273. [CrossRef]

56. Cook, A.J.; Murray, T.; Luckman, A.; Vaughan, D.G.; Barrand, N.E. A new 100-m Digital Elevation Model of the Antarctic Peninsula derived from ASTER Global DEM: Methods and accuracy assessment. *Earth Syst. Sci. Data* **2012**, *4*, 129–142. [CrossRef]

57. Toutin, T.; Schmitt, C.; Berthier, E.; Clavet, D. DEM generation over ice fields in the Canadian Arctic with along-track SPOT5 HRS stereo data. *Can. J. Remote Sens.* **2012**, *37*, 429–438. [CrossRef]

58. McMillan, M.; Shepherd, A.; Sundal, A.; Briggs, K.; Muir, A.; Ridout, A.; Hogg, A.; Wingham, D. Increased ice losses from Antarctica detected by CryoSat-2. *Geophys. Res. Lett.* **2014**, *41*, 3899–3905. [CrossRef]

59. Fahnestock, M.; Scambos, T.; Moon, T.; Gardner, A.; Haran, T.; Klinger, M. Rapid large-area mapping of ice flow using Landsat 8. *Remote Sens. Environ.* **2016**, *185*, 84–94. [CrossRef]

60. Li, T.; Liu, Y.; Li, T.; Hui, F.; Chen, Z.; Cheng, X. Antarctic Surface Ice Velocity Retrieval from MODIS-Based Mosaic of Antarctica (MOA). *Remote Sens.* **2018**, *10*, 1045. [CrossRef]

61. Strozzi, T.; Luckman, A.; Murray, T.; Wegmuller, U.; Werner, C.L. Glacier motion estimation using SAR offset-tracking procedures. *IEEE Trans. Geosci. Remote Sens.* **2002**, *40*, 2384–2391. [CrossRef]

62. Cheng, X.; Xu, G. The integration of JERS-1 and ERS SAR in differential interferometry for measurement of complex glacier motion. *J. Glaciol.* **2006**, *52*, 80–88. [CrossRef]

63. Gourmelen, N.; Kim, S.W.; Shepherd, A.; Park, J.W.; Sundal, A.V.; Björnsson, H.; Pálsson, F. Ice velocity determined using conventional and multiple-aperture InSAR. *Earth Planet. Sci. Lett.* **2011**, *307*, 156–160. [CrossRef]

64. Brown, L.C.; Duguay, C.R. The response and role of ice cover in lake-climate interactions. *Prog. Phys. Geogr.* **2010**, *34*, 671–704. [CrossRef]

65. Eerola, K.L.; Rontu, E.; Kourzeneva, H.; Kheyrollah, P.; Duguay, C.R. Impact of partly ice-free Lake Ladoga on temperature and cloudiness in an anticyclonic winter situation—A case study using HIRLAM model. *Tellus Ser. A Dyn. Meteorol. Oceanogr.* **2014**, *66*, 23929. [CrossRef]

66. Baijnath-Rodino, J.A.; Duguay, C.R. Historical spatiotemporal trends in snowfall extremes over the Canadian domain of the Great Lakes Basin. *Adv. Meteorol.* **2018**, *2018*. [CrossRef]

67. Baijnath-Rodino, J.A.; Duguay, C.R.; LeDrew, E.F. Climatological trends of snowfall over the Laurentian Great Lakes Basin. *Int. J. Climatol.* **2018**, *38*, 3942–3962. [CrossRef]

68. Baijnath-Rodino, J.A.; Duguay, C.R. Assessment of coupled CRCM5-FLake on the reproduction of wintertime lake-induced precipitation in the Great Lakes Basin. *Theor. Appl. Climatol.* **2019**. [CrossRef]

69. GCOS. *The Global Observing System for Climate: Implementation Needs, GCOS-200*; GCOS 2016 Implementation Plan; World Meteorological Organization: Geneva, Switzerland, 2016; p. 315.

70. Duguay, C.R.; Prowse, T.D.; Bonsal, B.R.; Brown, R.D.; Lacroix, M.P.; Ménard, P. Recent trends in Canadian lake ice cover. *Hydrol. Process.* **2006**, *20*, 781–801. [CrossRef]

71. Derksen, C.; Burgess, D.; Duguay, C.; Howell, S.; Mudryk, L.; Smith, S.; Thackeray, C.; Kirchmeier-Young, M. *Changes in Snow, Ice, and Permafrost across CANADA*; Chapter 5 in Canada's Changing Climate, Report; Bush, E., Lemmen, D.S., Eds.; Government of Canada: Ottawa, ON, USA, 2019; pp. 194–260.

72. Surdu, C.M.; Duguay, C.R.; Fernández Prieto, D. Evidence of recent changes in the ice regime of high arctic lakes from spaceborne satellite observations. *Cryosphere* **2016**, *10*, 941–960. [CrossRef]

73. Engram, M.; Arp, C.D.; Jones, B.M.; Ajadi, O.A.; Meyer, F.J. Analyzing floating and bedfast lake ice regimes across Arctic Alaska using 25 years of space-borne SAR imagery. *Remote Sens. Environ.* **2018**, *209*, 660–676. [CrossRef]

74. Duguay, C.; Brown, L. 2018: Lake Ice [in Arctic Report Card 2018. Available online: https://www.arctic.noaa.gov/Report-Card (accessed on 15 May 2019).

75. National Ice Center. *IMS Daily Northern Hemisphere Snow and Ice Analysis at 1 km, 4 km, and 24 km Resolutions, Version 1*; NASA National Snow and Ice Data Center Distributed Active Archive Center: Boulder, CO, USA, 2008; updated daily. [CrossRef]

76. Duguay, C.; Brown, L.; Kang, K.-K.; Pour, H.K. The Arctic Lake ice In State of the Climate in 2014. *Bull. Am. Meteorol. Soc.* **2015**, *96*, S144–S145.

77. Cai, Y.; Ke, C.Q.; Li, X.; Zhang, G.; Duan, Z.; Lee, H. Variations of lake ice phenology on the Tibetan Plateau From 2001 to 2017 based on MODIS Data. *J. Geophys. Res. Atmos.* **2019**, *124*, 1–19. [CrossRef]

78. Chen, J.; Wang, Y.; Cao, L.; Zheng, J. Variations in the ice phenology and water level of Ayakekumu Lake, Tibetan Plateau, derived from MODIS and satellite altimetry data. *J. Indian Soc. Remote Sens.* **2018**, *46*, 1689–1699. [CrossRef]

79. Gou, P.; Ye, Q.; Che, T.; Feng, Q.; Ding, B.; Lin, C.; Zong, J. Lake ice phenology of Nam Co, Central Tibetan Plateau, China, derived from multiple MODIS data products. *J. Great Lakes Res.* **2017**, *43*, 989–998. [CrossRef]

80. Murfitt, J.; Brown, L.C. Lake ice and temperature trends for Ontario and Manitoba: 2001 to 2014. *Hydrol. Process.* **2017**, *31*, 3596–3609. [CrossRef]

81. Qi, M.; Yao, X.; Li, X.; Duan, H.; Gao, Y.; Liu, J. Spatiotemporal characteristics of Qinghai Lake ice phenology between 2000 and 2016. *J. Geogr. Sci.* **2019**, *29*, 115–130. [CrossRef]

82. Šmejkalová, T.; Edwards, M.E.; Dash, J. Arctic lakes show strong decadal trend in earlier spring ice-out. *Sci. Rep.* **2016**, *6*, 1–8. [CrossRef]

83. Kang, K.-K.; Duguay, C.R.; Howell, S.E.L. Estimating ice phenology on large northern lakes from AMSR-E: Algorithm development and application to Great Bear Lake and Great Slave Lake, Canada. *Cryosphere* **2012**, *6*, 235–254. [CrossRef]

84. Kang, K.-K.; Duguay, C.R.; Lemmetyinen, J.; Gel, Y. Estimation of ice thickness on large northern lakes from AMSR-E brightness temperature measurements. *Remote Sens. Environ.* **2014**, *150*, 1–19. [CrossRef]

85. Wang, J.; Duguay, C.R.; Clausi, D.A.; Pinard, V.; Howell, S.E.L. Semi-automated classification of lake ice cover using dual polarization RADARSAT-2 imagery. *Remote Sens.* **2018**, *10*, 1727. [CrossRef]

86. Leigh, S.; Wang, Z.; Clausi, D.A. Automated ice-water classification using dual polarization SAR satellite imagery. *IEEE Trans. Geosci. Remote Sens.* **2014**, *52*, 5529–5539. [CrossRef]

87. Murfitt, J.; Brown, L.C.; Howell, S.E.L. Evaluating RADARSAT-2 for the automated monitoring of lake Ice phenology events in mid-latitudes. *Remote Sens.* **2018**, *10*, 1641. [CrossRef]

88. Surdu, C.M.; Duguay, C.R.; Pour, H.K.; Brown, L.C. Ice freeze-up and break-up detection of shallow lakes in Northern Alaska with spaceborne SAR. *Remote Sens.* **2015**, *7*, 6133–6159. [CrossRef]

89. Beckers., J.F.; Casey, J.A.; Haas, C. Retrievals of lake ice thickness from Great Slave Lake and Great Bear Lake using CryoSat-2. *IEEE Trans. Geosci. Remote Sens.* **2017**, *55*, 3708–3720. [CrossRef]

90. Murfitt, J.C.; Brown, L.C.; Howell, S.E.L. Estimating lake ice thickness in Central Ontario. *PLoS ONE* **2018**, *13*, e0208519. [CrossRef] [PubMed]

91. Pour, H.K.; Duguay, C.R.; Scott, A.; Kang, K.-K. Improvement of lake ice thickness retrieval from MODIS satellite data using a thermodynamic model. *IEEE Trans. Geosci. Remote Sens.* **2017**, *55*, 5956–5965. [CrossRef]

92. Duguay, C.R.; Flato, G.M.; Jeffries, M.O.; Ménard, P.; Morris, K.; Rouse, W.R. Ice cover variability on shallow lakes at high latitudes: Model simulations and observations. *Hydrol. Process.* **2003**, *17*, 3465–3483. [CrossRef]

93. Duguay, C.R.; Bernier, M.; Gauthier, Y.; Kouraev, A. Remote sensing of lake and river ice. In *Remote Sensing of the Cryosphere*; Tedesco, M., Ed.; Wiley-Blackwell: Oxford, UK, 2015; pp. 273–306.

94. Atwood, D.; Gunn, G.; Roussi, C.; Wu, J.; Duguay, C.; Sarabandi, K. Microwave backscatter from Arctic lake ice and polarimetric implications. *IEEE Trans. Geosci. Remote Sens.* **2015**, *53*, 5972–5982. [CrossRef]

95. Gunn, G.; Duguay, C.; Atwood, D.; King, J.; Toose, P. Observing scattering mechanisms of bubbled freshwater lake ice using polarimetric RADARSAT-2 (C-band) and UWScat (X-, Ku-band). *IEEE Trans. Geosci. Remote Sens.* **2018**, *56*, 2887–2903. [CrossRef]

96. Surdu, C.M.; Duguay, C.R.; Brown, L.C.; Fernández Prieto, D. Response of ice cover on shallow lakes of the North Slope of Alaska to contemporary climate conditions (1950–2011): Radar remote-sensing and numerical modeling data analysis. *Cryosphere* **2014**, *8*, 167–180. [CrossRef]

97. Antonova, S.; Duguay, C.; Kääb, A.; Heim, B.; Langer, M.; Westermann, S.; Boike, J. Monitoring bedfast ice and ice phenology in lakes of the Lena river delta using TerraSAR-X backscatter and coherence time series. *Remote Sens.* **2016**, *8*, 903. [CrossRef]

98. Barnett, T.P.; Adam, J.C.; Lettenmaier, D.P. Potential impacts of a warming climate on water availability in snow-dominated regions. *Nature* **2005**, *438*, 303–309. [CrossRef] [PubMed]

99. Tsai, Y.L.S.; Dietz, A.; Oppelt, N.; Kuenzer, C. Remote Sensing of Snow Cover Using Spaceborne SAR: A Review. *Remote Sens.* **2019**, *11*, 1456. [CrossRef]

100. Dozier, J. Spectral signature of alpine snow cover from the Landsat Thematic Mapper. *Remote Sens. Environ.* **1989**, *28*, 9–22. [CrossRef]

101. Hüsler, F.; Jonas, T.; Wunderle, S.; Albrecht, S. Validation of a modified snow cover retrieval algorithm from historical 1-km AVHRR data over the European Alps. *Remote Sens. Environ.* **2012**, *121*, 497–515. [CrossRef]

102. Hall, D.K.; Riggs, G.A.; Salomonson, V.V.; DiGirolamo, N.E.; Bayr, K.J. MODIS snow-cover products. *Remote Sens. Environ.* **2002**, *83*, 181–194. [CrossRef]

103. Wang, X.; Wang, J.; Che, T.; Huang, X.; Hao, X.; Li, H. Snow cover mapping for complex mountainous forested environments based on a multi-index technique. *IEEE J. Sel. Top. Appl. Earth Obs. Remote Sens.* **2018**, *11*, 1433–1441. [CrossRef]

104. Metsämäki, S.; Mattila, O.-P.; Pulliainen, J.; Niemi, K.; Luojus, K.; Böttcher, K. An optical reflectance model-based method for fractional snow cover mapping applicable to continental scale. *Remote Sens. Environ.* **2012**, *123*, 508–521. [CrossRef]

105. Hori, M.; Sugiura, K.; Kobayashi, K.; Aoki, T.; Tanikawa, T.; Kuchiki, K.; Niwano, M.; Enomoto, H. A 38-year (1978–2015) Northern Hemisphere daily snow cover extent product derived using consistent objective criteria from satellite-borne optical sensors. *Remote Sens. Environ.* **2017**, *191*, 402–418. [CrossRef]

106. Wayand, N.E.; Marsh, C.B.; Shea, J.M.; Pomeroy, J.W. Globally scalable alpine snow metrics. *Remote Sens. Environ.* **2018**, *213*, 61–72. [CrossRef]

107. Gascoin, S.; Grizonnet, M.; Bouchet, M.; Salgues, G.; Hagolle, O. Theia Snow collection: High-resolution operational snow cover maps from Sentinel-2 and Landsat-8 data. *Earth Syst. Sci. Data* **2019**, *11*, 493–514. [CrossRef]

108. Moosavi, V.; Malekinezhad, H.; Shirmohammadi, B. Fractional snow cover mapping from MODIS data using wavelet-artificial intelligence hybrid models. *J. Hydrol.* **2014**, *511*, 160–170. [CrossRef]

109. Czyzowska-Wisniewski, E.H.; van Leeuwen, W.J.D.; Hirschboeck, K.K.; Marsh, S.E.; Wisniewski, W.T. Fractional snow cover estimation in complex alpine-forested environments using an artificial neural network. *Remote Sens. Environ.* **2015**, *156*, 403–417. [CrossRef]

110. Roesch, A.; Wild, M.; Gilgen, H.; Ohmura, A.; Arugnell, N.C. A new snow cover fraction parameterization for the ECHAM4 GCM. *Clim. Dyn.* **2001**, *17*, 933–946. [CrossRef]

111. Salomonson, V.V.; Appel, I. Estimating fractional snow cover from MODIS using the normalized difference snow index. *Remote Sens. Environ.* **2004**, *89*, 351–360. [CrossRef]

112. Vikhamar, D.; Solberg, R. Snow-cover mapping in forests by constrained linear spectral unmixing of MODIS data. *Remote Sens. Environ.* **2003**, *88*, 309–323. [CrossRef]

113. Mishra, V.D.; Negi, H.S.; Rawat, A.K.; Chaturvedi, A.; Singh, R.P. Retrieval of sub-pixel snow cover information in the Himalayan region using medium and coarse resolution remote sensing data. *Int. J. Remote. Sens.* **2009**, *30*, 4707–4731. [CrossRef]

114. Painter, T.H.; Rittger, K.; McKenzie, C.; Slaughter, P.; Davis, R.E.; Dozier, J. Retrieval of subpixel snow covered area, grain size, and albedo from MODIS. *Remote Sens. Environ.* **2009**, *113*, 868–879. [CrossRef]

115. Sirguey, P.; Mathieu, R.; Arnaud, Y. Subpixel monitoring of the seasonal snow cover with MODIS at 250 m spatial resolution in the Southern Alps of New Zealand: Methodology and accuracy assessment. *Remote Sens. Environ.* **2009**, *113*, 160–181. [CrossRef]

116. Hao, S.; Jiang, L.; Shi, J.; Wang, G.; Liu, X. Assessment of MODIS-Based Fractional Snow Cover Products Over the Tibetan Plateau. *IEEE J. Sel. Top. Appl. Earth Obs. Remote Sens.* **2018**, *12*, 533–548. [CrossRef]

117. Romanov, P.; Tarpley, D. Enhanced algorithm for estimating snow depth from geostationary satellites. *Remote Sens. Environ.* **2007**, *108*, 97–110. [CrossRef]

118. De Ruyter de Wildt, M.; Seiz, G.; Gruen, A. Operational snow mapping using multitemporal Meteosat SEVIRI imagery. *Remote Sens. Environ.* **2007**, *109*, 29–41. [CrossRef]

119. Siljamo, N.; Hyvärinen, O. New geostationary satellite-based snow-cover algorithm. *J. Appl. Meteorol. Clim.* **2011**, *50*, 1275–1290. [CrossRef]

120. Yang, J.; Jiang, L.; Wu, F.; Sun, R. Monitoring snow cover over China with MTSAT-2 geostationary satellite. *J. Remot. Sens.* **2013**, *17*, 1264–1280.

121. Yang, J.; Jiang, L.; Shi, J.; Wu, S.; Sun, R.; Yang, H. Monitoring snow cover using Chinese meteorological satellite data over China. *Remote Sens. Environ.* **2014**, *143*, 192–203. [CrossRef]

122. Wang, G.; Jiang, L.; Wu, S.; Shi, J.; Hao, S.; Liu, X. Fractional Snow Cover Mapping from FY-2 VISSR Imagery of China. *Remote Sens.* **2017**, *9*, 983. [CrossRef]

123. Gao, Y.; Xie, H.; Lu, N.; Yao, T.; Liang, T. Toward advanced daily cloud-free snow cover and snow water equivalent products from Terra–Aqua MODIS and Aqua AMSR-E measurements. *J. Hydrol.* **2010**, *385*, 23–35. [CrossRef]

124. Liang, T.; Zhang, X.; Xie, H.; Wu, C.; Feng, Q.; Huang, X.; Chen, Q. Toward improved daily snow cover mapping with advanced combination of MODIS and AMSR-E measurements. *Remote Sens. Environ.* **2008**, *112*, 3750–3761. [CrossRef]

125. Yang, J.; Jiang, L.; Wu, S.; Wang, G.; Wang, J.; Liu, X. Development of a Snow Depth Estimation Algorithm over China for the FY-3D/MWRI. *Remote Sens.* **2019**, *11*, 977. [CrossRef]

126. Helfrich, S.R.; McNamara, D.; Ramsay, B.H.; Baldwin, T.; Kasheta, T. Enhancements to, and forthcoming developments in the Interactive Multisensor Snow and Ice Mapping System (IMS). *Hydrol. Process.* **2007**, *21*, 1576–1586. [CrossRef]

127. Qiu, Y.; Guo, H.; Chu, D.; Zhang, H.; Shi, J.; Shi, L.; Zheng, Z. MODIS daily cloud-free snow cover products over Tibetan Plateau. *Sci. Data Bank* **2016**. [CrossRef]

128. Hoang, T.; Phu, N.; Mohammed, O.; Kuo-lin, H.; Soroosh, S.; Xia, Q. A cloud-free MODIS snow cover dataset for the contiguous United States from 2000 to 2017. *Sci. Data* **2019**, *6*, 180300.

129. Du, J.; Shi, J.; Tjuatja, S.; Chen, K.S. A combined method to model microwave scattering from a forest medium. *IEEE Trans. Geosci. Remote Sens.* **2006**, *44*, 815–824.

130. Jiang, L.; Wang, P.; Zhang, L.; Yang, H.; Yang, J. Improvement of snow depth retrieval for FY3B-MWRI in China. *Sci. China Earth Sci.* **2014**, *57*, 1278–1292. [CrossRef]

131. Kelly, R. The AMSR-E Snow Depth Algorithm: Description and Initial Results. *Remote Sens. Soc. Jpn.* **2009**, *29*, 307–317.

132. Tedesco, M.; Reichle, R.; Low, A.; Markus, T.; Foster, J.L. Dynamic approaches for snow depth retrieval from spaceborne microwave brightness temperature. *IEEE Trans. Geosci. Remote Sens.* **2010**, *48*, 1955–1967. [CrossRef]

133. Pulliainen, J.; Hallikainen, M. Retrieval of regional snow water equivalent from space-borne passive microwave observations. *Remote Sens. Environ.* **2001**, *75*, 76–85. [CrossRef]

134. Jiang, L.; Shi, J.; Tjuatja, S.; Chen, K.S.; Du, J.; Zhang, L. Estimation of snow water equivalence using the polarimetric scanning radiometer from the cold land processes experiments (CLPX03). *IEEE Geosci. Remote Sens. Lett.* **2011**, *8*, 359–363. [CrossRef]

135. Pan, J.; Durand, M.T.; Vander Jagt, B.J.; Liu, D. Application of a Markov Chain Monte Carlo algorithm for snow water equivalent retrieval from passive microwave measurements. *Remote Sens. Environ.* **2017**, *192*, 150–165. [CrossRef]

136. Santi, E.; Pettinato, S.; Paloscia, S.; Pampaloni, P.; Macelloni, G.; Brogioni, M. An algorithm for generating soil moisture and snow depth maps from microwave spaceborne radiometers: HydroAlgo. *Hydrol. Earth Syst. Sci.* **2012**, *16*, 3659–3676. [CrossRef]

137. Bair, E.H.; Abreu Calfa, A.; Rittger, K.; Dozier, J. Using machine learning for real-time estimates of snow water equivalent in the watersheds of Afghanistan. *Cryosphere* **2018**, *12*, 1579–1594. [CrossRef]

138. Che, T.; Li, X.; Jin, R.; Huang, C. Assimilating passive microwave remote sensing data into a land surface model to improve the estimation of snow depth. *Remote Sens. Environ.* **2014**, *143*, 54–63. [CrossRef]

139. Larue, F.; Royer, A.; Sève, D.D.; Roy, A.; Cosme, E. Assimilation of passive microwave AMSR-2 satellite observations in a snowpack evolution model over northeastern Canada. *Hydrol. Earth Syst. Sci.* **2018**, *22*, 5711–5734. [CrossRef]

140. Rott, H.; Yueh, S.H.; Cline, D.W.; Duguay, C.; Essery, R.; Haas, C.; Heliere, F.; Kern, M.; MacElloni, G.; Malnes, E.; et al. Cold regions hydrology high-resolution observatory for snow and cold land processes. *Proc. IEEE* **2010**, *98*, 752–765. [CrossRef]

141. Du, J.; Shi, J.; Rott, H. Comparison between a multi-scattering and multi-layer snow scattering model and its parameterized snow backscattering model. *Remote Sens. Environ.* **2010**, *114*, 1089–1098. [CrossRef]

142. Zhu, J.; Tan, S.; King, J.; Derksen, C.; Lemmetyinen, J.; Tsang, L. Forward and inverse radar modeling of terrestrial snow using SnowSAR data. *IEEE Trans. Geosci. Remote Sens.* **2018**, *56*, 1–11. [CrossRef]

143. Thompson, A.; Kelly, R. Observations of a Coniferous Forest at 9.6 and 17.2 GHz: Implications for SWE Retrievals. *Remote Sens.* **2019**, *11*, 6. [CrossRef]

144. Ding, K.H.; Xu, X.; Tsang, L. Electromagnetic scattering by bicontinuous random microstructures with discrete permittivities. *IEEE Trans. Geosci. Remote Sens.* **2010**, *48*, 3139–3151. [CrossRef]

145. Tedesco, M.; Jeyaratnam, J. A new operational snow retrieval algorithm applied to historical AMSR-E brightness temperatures. *Remote Sens.* **2016**, *8*, 1037. [CrossRef]

146. Smyth, E.J.; Raleigh, M.S.; Small, E.E. Particle Filter Data Assimilation of Monthly Snow Depth Observations Improves Estimation of Snow Density and SWE. *Water Resour. Res.* **2019**, *55*, 1296–1311. [CrossRef]

147. Yueh, S.H.; Xu, X.; Shah, R.; Kim, Y.; Garrison, J.L.; Komanduru, A.; Elder, K. Remote Sensing of Snow Water Equivalent Using Coherent Reflection from Satellite Signals of Opportunity: Theoretical Modeling. *IEEE J. Sel. Top. Appl. Earth Obs. Remote Sens.* **2017**, *10*, 5529–5540. [CrossRef]

148. Conde, V.; Nico, G.; Mateus, P.; Catalão, J.; Kontu, A.; Gritsevich, M. On the estimation of temporal changes of snow water equivalent by spaceborne SAR interferometry: A new application for the Sentinel-1 mission. *J. Hydrol. Hydromech.* **2019**, *67*, 93–100. [CrossRef]

149. Kim, Y.; Kimball, J.S.; Zhang, K.; McDonald, K.C. Satellite detection of increasing Northern Hemisphere non-frozen seasons from 1979 to 2008: Implications for regional vegetation growth. *Remote Sens. Environ.* **2012**, *121*, 472–487. [CrossRef]

150. Parazoo, N.C.; Arneth, A.; Pugh, T.A.M.; Smith, B.; Steiner, N.; Luus, K.; Commance, R.; Benmergui, J.; Stofferahn, E.; Liu, J.; et al. Spring photosynthetic onset and net CO2 uptake in Alaska triggered by landscape thawing. *Glob. Chang. Biol.* **2018**, *24*, 3416–3435. [CrossRef] [PubMed]

151. Park, H.; Yoshikawa, Y.; Oshima, K.; Kim, Y.; Ngo-Duc, T.; Kimball, J.S.; Yang, D. Quantification of warming climate-induced changes in terrestrial arctic river ice thickness and phenology. *J. Clim.* **2016**, *29*, 1733–1754. [CrossRef]

152. Derksen, C.; Xu, X.; Dunbar, R.S.; Colliander, A.; Kim, Y.; Kimball, J.S.; Black, T.A.; Euskirchen, E.; Langlois, A.; Loranty, M.M.; et al. Retrieving landscape freeze/thaw state from Soil Moisture Active Passive (SMAP) radar and radiometer measurements. *Remote Sens. Environ.* **2017**, *194*, 48–62. [CrossRef]

153. Mortin, J.; Schröder, T.M.; Walløe Hansen, A.; Holt, B.; McDonald, K.C. Mapping of seasonal freeze-thaw transitions across the pan-Arctic land and sea ice domains with satellite radar. *J. Geophys. Res. Ocean.* **2012**, *117*. [CrossRef]

154. Hu, T.; Zhao, T.; Shi, J.; Wu, S.; Liu, D.; Qin, H.; Zhao, K. High-resolution mapping of freeze/thaw status in china via fusion of MODIS and AMSR2 data. *Remote Sens.* **2017**, *9*, 1339. [CrossRef]

155. Zhao, T.; Zhang, L.; Jiang, L.; Zhao, S.; Chai, L.; Jin, R. A new soil freeze/thaw discriminant algorithm using AMSR-E passive microwave imagery. *Hydrol. Process.* **2011**, *25*, 1704–1716. [CrossRef]

156. Jin, R.; Li, X.; Che, T. A decision tree algorithm for surface soil freeze/thaw classification over China using SSM/I brightness temperature. *Remote Sens. Environ.* **2009**, *113*, 2651–2660. [CrossRef]

157. Forman, B.A.; Reichle, R.H. Using a support vector machine and a land surface model to estimate large-scale passive microwave brightness temperatures over snow-covered land in North America. *IEEE J. Sel. Top. Appl. Earth Obs. Remote Sens.* **2014**, *8*, 4431–4441. [CrossRef]

158. Zwieback, S.; Bartsch, A.; Melzer, T.; Wagner, W. Probabilistic Fusion of Ku - and C-band Scatterometer Data for Determining the Freeze/Thaw State. *IEEE Trans. Geosci. Remote Sens.* **2011**, *50*, 2583–2594. [CrossRef]

159. McColl, K.A.; Roy, A.; Derksen, C.; Konings, A.G.; Alemohammed, S.H.; Entekhabi, D. Triple collocation for binary and categorical variables: Application to validating landscape freeze/thaw retrievals. *Remote Sens. Environ.* **2016**, *176*, 31–42. [CrossRef]

160. Kim, Y.; Kimball, J.S.; Glassy, J.; McDonald, K.C. *MEaSUREs Global Record of Daily Landscape Freeze/Thaw Status, Version 4. [Indicate Subset Used]*; NASA National Snow and Ice Data Center Distributed Active Archive Center: Boulder, CO, USA, 2017. [CrossRef]

161. Kim, Y.; Kimball, J.S.; Glassy, J.; McDonald, K.C. *MEaSUREs Northern Hemisphere Polar EASE-Grid 2.0 Daily 6 km Land Freeze/Thaw Status from AMSR-E and AMSR2, Version 1*; NASA National Snow and Ice Data Center Distributed Active Archive Center: Boulder, CO, USA, 2018. [CrossRef]

162. Roy, A.; Brucker, L.; Prince, M.; Royer, A.; Derksen, C. *Aquarius L3 Weekly Polar-Gridded Landscape Freeze/Thaw Data, Version 5. [Indicate Subset Used]*; NSIDC: National Snow and Ice Data Center: Boulder, CO, USA, 2018. [CrossRef]

163. Xu, X.; Dunbar, R.S.; Derksen, C.; Colliander, A.; Kim, Y.; Kimball, J.S. *SMAP L3 Radiometer Global and Northern Hemisphere Daily 36 km EASE-Grid Freeze/Thaw State, Version 2. [Indicate Subset Used]*; NASA National Snow and Ice Data Center Distributed Active Archive Center: Boulder, CO, USA, 2018. [CrossRef]

164. Xu, X.; Dunbar, R.S.; Derksen, C.; Colliander, A.; Kim, Y.; Kimball, J.S. *SMAP Enhanced L3 Radiometer Global and Northern Hemisphere Daily 9 km EASE-Grid Freeze/Thaw State, Version 2. [Indicate Subset Used]*; NASA National Snow and Ice Data Center Distributed Active Archive Center: Boulder, CO, USA, 2018. [CrossRef]

165. Steiner, N.; McDonald, K.C. *High Mountain Asia ASCAT Freeze/Thaw/Melt Status, Version 1. [Indicate Subset Used]*; NASA National Snow and Ice Data Center Distributed Active Archive Center: Boulder, CO, USA, 2018. [CrossRef]

166. Du, J.; Kimball, J.S.; Shi, J.; Jones, L.A.; Wu, S.; Sun, R.; Yang, H. Inter-Calibration of satellite passive microwave land observations from AMSR-E and AMSR2 using overlapping FY3B-MWRI sensor measurements. *Remote Sens.* **2014**, *6*, 8594–8616. [CrossRef]

167. Kim, Y.; Kimball, J.S.; Xu, X.; Dunbar, R.S.; Colliander, A.; Derksen, C. Global Assessment of the SMAP Freeze/Thaw Data Record and Regional Applications for Detecting Spring Onset and Frost Events. *Remote Sens.* **2019**, *11*, 1317. [CrossRef]

168. Tarolli, P. High-resolution topography for understanding Earth surface processes: Opportunities and challenges. *Geomorphology* **2014**, *216*, 295–312. [CrossRef]

169. Liu, L.; Schaefer, K.M.; Chen, A.C.; Gusmeroli, A.; Zebker, H.A.; Zhang, T. Remote sensing measurements of thermokarst subsidence using InSAR. *J. Geophys. Res. Earth Surf.* **2015**, *120*, 1935–1948. [CrossRef]

170. Luo, L.; Ma, W.; Zhang, Z.; Zhuang, Y.; Zhang, Y.; Yang, J.; Cao, X.; Liang, S.; Mu, Y. Freeze/Thaw-Induced Deformation Monitoring and Assessment of the Slope in Permafrost Based on Terrestrial Laser Scanner and GNSS. *Remote Sens.* **2017**, *9*, 198. [CrossRef]

171. Jorgenson, M.T.; Grosse, G. Remote Sensing of Landscape Change in Permafrost Regions. *Permafr. Periglac. Process.* **2016**, *27*, 324–338. [CrossRef]

172. Arenson, L.U.; Kääb, A.; O'Sullivan, A. Detection and Analysis of Ground Deformation in Permafrost Environments. *Permafr. Periglac. Process.* **2016**, *27*, 339–351. [CrossRef]

173. Eltner, A.; Kaiser, A.; Castillo, C.; Rock, G.; Neugirg, F.; Abellán, A. Image-based surface reconstruction in geomorphometry – merits, limits and developments. *Earth Surf. Dynam.* **2016**, *4*, 359–389. [CrossRef]

174. Meng, Y.; Lan, H.; Li, L.; Wu, Y.; Li, Q. Characteristics of Surface Deformation Detected by X-band SAR Interferometry over Sichuan-Tibet Grid Connection Project Area, China. *Remote Sens.* **2015**, *7*, 12265–12281. [CrossRef]

175. Stettner, S.; Beamish, A.L.; Bartsch, A.; Heim, B.; Grosse, G.; Roth, A.; Lantuit, H. Monitoring Inter- and Intra-Seasonal Dynamics of Rapidly Degrading Ice-Rich Permafrost Riverbanks in the Lena Delta with TerraSAR-X Time Series. *Remote Sens.* **2018**, *10*, 51. [CrossRef]

176. Chen, J.; Liu, L.; Zhang, T.; Cao, B.; Lin, H. Using persistent scatterer interferometry to map and quantify permafrost thaw subsidence: A case study of Eboling Mountain on the Qinghai-Tibet Plateau. *J. Geophys. Res. Earth Surf.* **2018**, *123*, 2663–2676. [CrossRef]

177. Pekel, J.F.; Cottam, A.; Gorelick, N.; Belward, A.S. High-resolution mapping of global surface water and its long-term changes. *Nature* **2016**, *540*, 418. [CrossRef] [PubMed]

178. Prigent, C.; Lettenmaier, D.P.; Aires, F.; Papa, F. Toward a high-resolution monitoring of continental surface water extent and dynamics, at global scale: From GIEMS (Global Inundation Extent from Multi-Satellites) to SWOT (Surface Water Ocean Topography). In *Remote Sensing and Water Resources*; Springer: Cham, Switzerland, 2016; pp. 149–165.

179. Du, J.; Kimball, J.S.; Velicogna, I.; Zhao, M.; Jones, L.A.; Watts, J.D.; Kim, Y. Multi-component satellite assessment of drought severity in the Contiguous United States from 2002 to 2017 using AMSR-E and AMSR2. *Water Resour. Res.* **2019**. [CrossRef]

180. Sakai, T.; Hatta, S.; Okumura, M.; Hiyama, T.; Yamaguchi, Y.; Inoue, G. Use of Landsat TM/ETM+ to monitor the spatial and temporal extent of spring breakup floods in the Lena River, Siberia. *Int. J. Remote Sens.* **2015**, *36*, 719–733. [CrossRef]

181. Cooley, S.W.; Smith, L.C.; Ryan, J.C.; Pitcher, L.H.; Pavelsky, T.M. Arctic-Boreal Lake Dynamics Revealed Using CubeSat Imagery. *Geophys. Res. Lett.* **2019**, *46*, 2111–2120. [CrossRef]

182. Nitze, I.; Grosse, G.; Jones, B.M.; Romanovsky, V.E.; Boike, J. Remote sensing quantifies widespread abundance of permafrost region disturbances across the Arctic and Subarctic. *Nat. Commun.* **2018**, *9*, 5423. [CrossRef]

183. Zhang, G.; Bolch, T.; Allen, S.; Linsbauer, A.; Chen, W.; Wang, W. Glacial lake evolution and glacier–lake interactions in the Poiqu River basin, central Himalaya, 1964–2017. *J. Glaciol.* **2019**, 1–19. [CrossRef]

184. Chand, M.B.; Watanabe, T. Development of Supraglacial Ponds in the Everest Region, Nepal, between 1989 and 2018. *Remote Sens.* **2019**, *11*, 1058. [CrossRef]

185. Carroll, M.L.; DiMiceli, C.M.; Townshend, J.R.G.; Sohlberg, R.A.; Elders, A.I.; Devadiga, S.; Sayer, A.M.; Levy, R.C. Development of an operational land water mask for MODIS Collection 6, and influence on downstream data products. *Int. J. Digit. Earth* **2017**, *10*, 207–218. [CrossRef]

186. Chapman, B.; McDonald, K.; Shimada, M.; Rosenqvist, A.; Schroeder, R.; Hess, L. Mapping regional inundation with spaceborne L-Band SAR. *Remote Sens.* **2015**, *7*, 5440–5470. [CrossRef]

187. Schroeder, R.; McDonald, K.; Chapman, B.; Jensen, K.; Podest, E.; Tessler, Z.; Bohn, T.; Zimmermann, R. Development and evaluation of a multi-year fractional surface water data set derived from active/passive microwave remote sensing data. *Remote Sens.* **2015**, *7*, 16688–16732. [CrossRef]

188. Du, J.; Kimball, J.S.; Jones, L.A.; Watts, J.D. Implementation of satellite based fractional water cover indices in the pan-Arctic region using AMSR-E and MODIS. *Remote Sens. Environ.* **2016**, *184*, 469–481. [CrossRef]

189. Pham-Duc, B.; Prigent, C.; Aires, F.; Papa, F. Comparisons of global terrestrial surface water datasets over 15 years. *J. Hydrometeorol.* **2017**, *18*, 993–1007. [CrossRef]

190. Du, J.; Kimball, J.S.; Galantowicz, J.; Kim, S.B.; Chan, S.K.; Reichle, R.; Jones, L.A.; Watts, J.D. Assessing global surface water inundation dynamics using combined satellite information from SMAP, AMSR2 and Landsat. *Remote Sens. Environ.* **2018**, *213*, 1–17. [CrossRef]

191. ARC, A.R.C.; Galantowicz, J.; Caruso-Marini, M.; Administrator, A.C., II; Picton, J.; Root, B. ARC Flood Extent Depiction Algorithm Description Document. *Contract.* **2017**, *4*, R00.

192. Aires, F.; Miolane, L.; Prigent, C.; Pham, B.; Fluet-Chouinard, E.; Lehner, B.; Papa, F. A global dynamic long-term inundation extent dataset at high spatial resolution derived through downscaling of satellite observations. *J. Hydrometeorol.* **2017**, *18*, 1305–1325. [CrossRef]

193. Fluet-Chouinard, E.; Lehner, B.; Rebelo, L.M.; Papa, F.; Hamilton, S.K. Development of a global inundation map at high spatial resolution from topographic downscaling of coarse-scale remote sensing data. *Remote Sens. Environ.* **2015**, *158*, 348–361. [CrossRef]

194. Wu, G.; Liu, Y. Downscaling surface water inundation from coarse data to fine-scale resolution: Methodology and accuracy assessment. *Remote Sens.* **2015**, *7*, 15989–16003. [CrossRef]

195. Bourgeau-Chavez, L.L.; Endres, S.; Powell, R.; Battaglia, M.J.; Benscoter, B.; Turetsky, M.; Kasischke, E.S.; Banda, E. Mapping boreal peatland ecosystem types from multitemporal radar and optical satellite imagery. *Can. J. For. Res.* **2016**, *47*, 545–559. [CrossRef]

196. Natali, S.M.; Schuur, E.A.G.; Rubin, R.L. Increased plant productivity in Alaskan tundra as a result of experimental warming of soil and permafrost. *J. Ecol.* **2012**, *100*, 488–498. [CrossRef]

197. Melaas, E.K.; Sulla-Menashe, D.; Friedl, M.A. Multidecadal Changes and Interannual Variation in Springtime Phenology of North American Temperate and Boreal Deciduous Forests. *Geophys. Res. Lett.* **2018**, *45*, 2679–2687. [CrossRef]

198. Friedl, M.A.; Sulla-Menashe, D.; Tan, B.; Schneider, A.; Ramankutty, N.; Sibley, A.; Huang, X. MODIS Collection 5 global land cover: Algorithm refinements and characterization of new datasets. *Remote Sens. Environ.* **2010**, *114*, 168–182. [CrossRef]

199. Zhu, Z.; Woodcock, C.E. Continuous change detection and classification of land cover using all available Landsat data. *Remote Sens. Environ.* **2014**, *144*, 152–171. [CrossRef]

200. Selkowitz, D.J.; Stehman, S.V. Thematic accuracy of the National Land Cover Database (NLCD) 2001 land cover for Alaska. *Remote Sens. Environ.* **2011**, *115*, 1401–1407. [CrossRef]

201. Wulder, M.A.; White, J.C.; Cranny, M.; Hall, R.J.; Luther, J.E.; Beaudoin, A.; Goodenough, D.G.; Dechka, J.A. Monitoring Canada's forests. Part 1: Completion of the EOSD land cover project. *Can. J. Remote Sens.* **2008**, *34*, 549–562. [CrossRef]

202. Karami, M.; Westergaard-Nielsen, A.; Normand, S.; Treier, U.A.; Elberling, B.; Hansen, B.U. A phenology-based approach to the classification of Arctic tundra ecosystems in Greenland. *ISPRS J. Photogramm. Remote Sens.* **2018**, *146*, 518–529. [CrossRef]

203. Stow, D.A.; Hope, A.; McGuire, D.; Verbyla, D.; Gamon, J.; Huemmrich, F.; Houston, S.; Racine, C.; Sturm, M.; Tape, K.; et al. Remote sensing of vegetation and land-cover change in Arctic Tundra Ecosystems. *Remote Sens. Environ.* **2004**, *89*, 281–308. [CrossRef]

204. Saatchi, S.S.; Rignot, E. Classification of boreal forest cover types using SAR images. *Remote Sens. Environ.* **1997**, *60*, 270–281. [CrossRef]

205. Ullmann, T.; Schmitt, A.; Roth, A.; Duffe, J.; Dech, S.; Hubberten, H.-W.; Baumhauer, R. Land Cover Characterization and Classification of Arctic Tundra Environments by Means of Polarized Synthetic Aperture X- and C-Band Radar (PolSAR) and Landsat 8 Multispectral Imagery—Richards Island, Canada. *Remote Sens.* **2014**, *6*, 8565–8593. [CrossRef]

206. Merchant, M.A.; Warren, R.K.; Edwards, R.; Kenyon, J.K. An Object-Based Assessment of Multi-Wavelength SAR, Optical Imagery and Topographical Datasets for Operational Wetland Mapping in Boreal Yukon, Canada. *Can. J. Remote Sens.* **2019**, 1–25. [CrossRef]

207. Whitcomb, J.; Moghaddam, M.; McDonald, K.; Kellndorfer, J.; Podest, E. Mapping vegetated wetlands of Alaska using L-band radar satellite imagery. *Can. J. Remote Sens.* **2009**, *35*, 54–72. [CrossRef]

208. Mohammadimanesh, F.; Salehi, B.; Mahdianpari, M.; Gill, E.; Molinier, M. A new fully convolutional neural network for semantic segmentation of polarimetric SAR imagery in complex land cover ecosystem. *ISPRS J. Photogramm. Remote Sens.* **2019**, *151*, 223–236. [CrossRef]

209. Xue, J.; Su, B. Significant remote sensing vegetation indices: A review of developments and applications. *J. Sens.* **2017**, *2017*, 17p. [CrossRef]

210. Chen, J.M.; Cihlar, J. Retrieving Leaf Area Index of Boreal Conifer Forests Using Landsat TM Images. *Remote Sens. Environ.* **1996**, *55*, 153–162. [CrossRef]

211. Hame, T.; Salli, A.; Andersson, K.; Lohi, A. A new methodology for the estimation of biomass of conifer dominated boreal forest using NOAA AVHRR data. *Int. J. Remote Sens.* **1997**, *18*, 3211–3243. [CrossRef]

212. Raynolds, M.K.; Walker, D.A.; Epstein, H.E.; Pinzon, J.E.; Tucker, C.J. A new estimate of tundra-biome phytomass from trans-Arctic field data and AVHRR NDVI. *Remote Sens. Lett.* **2011**, *3*, 403–411. [CrossRef]

213. Ueyama, M.; Ichii, K.; Iwata, H.; Euskirchen, E.S.; Zona, D.; Rocha, A.V.; Harazono, Y.; Iwama, C.; Nakai, T.; Oechel, W.C. Upscaling terrestrial carbon dioxide fluxes in Alaska with satellite remote sensing and support vector regression. *J. Geophys. Res. Biogeoscie.* **2013**, *118*, 1266–1281. [CrossRef]

214. Cui, Q.; Shi, J.; Du, J.; Zhao, T.; Xiong, C. An approach for monitoring global vegetation based on multiangular observations from SMOS. *IEEE J. Sel. Top. Appl. Earth Obs. Remote Sens.* **2015**, *8*, 604–616. [CrossRef]

215. Konings, A.G.; Piles, M.; Das, N.; Entekhabi, D. L-band vegetation optical depth and effective scattering albedo estimation from SMAP. *Remote Sens. Environ.* **2017**, *198*, 460–470. [CrossRef]

216. Du, J.; Kimball, J.S.; Jones, L.A.; Kim, Y.; Glassy, J.M.; Watts, J.D. A global satellite environmental data record derived from AMSR-E and AMSR2 microwave Earth observations. *Earth Syst. Sci. Data* **2017**, *9*, 791. [CrossRef]

217. Rao, K.; Anderegg, W.R.; Sala, A.; Martínez-Vilalta, J.; Konings, A.G. Satellite-based vegetation optical depth as an indicator of drought-driven tree mortality. *Remote Sens. Environ.* **2019**, *227*, 125–136. [CrossRef]

218. Tian, F.; Brandt, M.; Liu, Y.Y.; Verger, A.; Tagesson, T.; Diouf, A.A.; Rasmussen, K.; Mbow, C.; Wang, Y.; Fensholt, R. Remote sensing of vegetation dynamics in drylands: Evaluating vegetation optical depth (VOD) using AVHRR NDVI and in situ green biomass data over West African Sahel. *Remote Sens. Environ.* **2016**, *177*, 265–276. [CrossRef]

219. Jones, M.O.; Kimball, J.S.; Jones, L.A. Satellite microwave detection of boreal forest recovery from the extreme 2004 wildfires in Alaska and Canada. *Glob. Chang. Biol.* **2013**, *19*, 3111–3122. [CrossRef] [PubMed]

220. Sims, D.A.; Rahman, A.F.; Cordova, V.D.; El-Masri, B.Z.; Baldocchi, D.D.; Flanagan, L.B.; Goldstein, A.H.; Hollinger, D.Y.; Misson, L.; Monson, R.K.; et al. On the use of MODIS EVI to assess gross primary productivity of North American ecosystems. *J. Geophys. Res. Biogeosci.* **2006**, *111*. [CrossRef]

221. Yan, K.; Park, T.; Yan, G.; Chen, C.; Yang, B.; Liu, Z.; Nemani, R.; Knyazikhin, Y.; Myneni, R. Evaluation of MODIS LAI/FPAR Product Collection 6. Part 1: Consistency and Improvements. *Remote Sens.* **2016**, *8*, 359. [CrossRef]

222. Watts, J.D.; Kimball, J.S.; Parmentier, F.J.W.; Sachs, T.; Rinne, J.; Zona, D.; Oechel, W.; Tagesson, T.; Jackowicz-Korczyński, M.; Aurela, M. A satellite data driven biophysical modeling approach for estimating northern peatland and tundra CO_2 and CH4 fluxes. *Biogeosciences* **2014**, *11*, 1961–1980. [CrossRef]

223. Eitel, J.U.H.; Maguire, A.J.; Boelman, N.; Vierling, L.A.; Griffin, K.L.; Jensen, J.; Magney, T.S.; Mahoney, P.J.; Meddens, A.J.H.; Silva, C.; et al. Proximal remote sensing of tree physiology at northern treeline: Do late-season changes in the photochemical reflectance index (PRI) respond to climate or photoperiod? *Remote Sens. Environ.* **2019**, *221*, 340–350. [CrossRef]

224. Parazoo, N.C.; Bowman, K.; Fisher, J.B.; Frankenberg, C.; Jones, D.B.; Cescatti, A.; Perez-Priego, O.; Wohlfahrt, G.; Montagnani, L. Terrestrial gross primary production inferred from satellite fluorescence and vegetation models. *Glob. Chang. Biol.* **2014**, *20*, 3103–3121. [CrossRef] [PubMed]

225. Joiner, J.; Yoshida, Y.; Vasilkov, A.P.; Schaefer, K.; Jung, M.; Guanter, L.; Zhang, Y.; Garrity, S.; Middleton, E.M.; Huemmrich, K.F.; et al. The seasonal cycle of satellite chlorophyll fluorescence observations and its relationship to vegetation phenology and ecosystem atmosphere carbon exchange. *Remote Sens. Environ.* **2014**, *152*, 375–391. [CrossRef]

226. Sun, Y.; Frankenberg, C.; Jung, M.; Joiner, J.; Guanter, L.; Köhler, P.; Magney, T. Overview of Solar-Induced chlorophyll Fluorescence (SIF) from the Orbiting Carbon Observatory-2: Retrieval, cross-mission comparison, and global monitoring for GPP. *Remote Sens. Environ.* **2018**, *209*, 808–823. [CrossRef]

227. Gentine, P.; Alemohammad, S.H. Reconstructed solar-induced fluorescence: A machine learning vegetation product based on MODIS surface reflectance to reproduce GOME-2 solar-induced fluorescence. *Geophys. Res. Lett.* **2018**, *45*, 3136–3146. [CrossRef] [PubMed]

228. Drake, J.B.; Knox, R.G.; Dubayah, R.O.; Clark, D.B.; Condits, R.; Blair, J.B.; Hofton, M. Above-ground biomass estimation in closed canopy Neotropical forests using lidar remote sensing: Factors affecting the generality of relationships. *Glob. Ecol. Biogeogr.* **2003**, *12*, 147–159. [CrossRef]

229. Nelson, R.; Margolis, H.; Montesano, P.; Sun, G.; Cook, B.; Corp, L.; Andersen, H.E.; deJong, B.; Pellat, F.P.; Fickel, T.; et al. Lidar-based estimates of aboveground biomass in the continental US and Mexico using ground, airborne, and satellite observations. *Remote Sens. Environ.* **2017**, *188*, 127–140. [CrossRef]

230. Neigh, C.S.R.; Nelson, R.F.; Ranson, K.J.; Margolis, H.A.; Montesano, P.M.; Sun, G.; Kharuk, V.; Næsset, E.; Wulder, M.A.; Andersen, H.-E. Taking stock of circumboreal forest carbon with ground measurements, airborne and spaceborne LiDAR. *Remote Sens. Environ.* **2013**, *137*, 274–287. [CrossRef]

231. Margolis, H.A.; Nelson, R.F.; Montesano, P.M.; Beaudoin, A.; Sun, G.; Andersen, H.-E.; Wulder, M.A. Combining satellite lidar, airborne lidar, and ground plots to estimate the amount and distribution of aboveground biomass in the boreal forest of North America. *Can. J. For. Res.* **2015**, *45*, 838–855. [CrossRef]

232. Boudreau, J.; Nelson, R.; Margolis, H.; Beaudoin, A.; Guindon, L.; Kimes, D. Regional aboveground forest biomass using airborne and spaceborne LiDAR in Québec. *Remote Sens. Environ.* **2008**, *112*, 3876–3890. [CrossRef]

233. Neumann, M.; Saatchi, S.S.; Ulander, L.M.H.; Fransson, J.E.S. Assessing Performance of L- and P-Band Polarimetric Interferometric SAR Data in Estimating Boreal Forest Above-Ground Biomass. *IEEE Trans. Geosci. Remote Sens.* **2012**, *50*, 714–726. [CrossRef]

234. Sun, G.; Ranson, K.J.; Kharuk, V.I. Radiometric slope correction for forest biomass estimation from SAR data in the Western Sayani Mountains, Siberia. *Remote Sens. Environ.* **2002**, *79*, 279–287. [CrossRef]

235. Soja, M.J.; Sandberg, G.; Ulander, L.M.H. Regression-Based Retrieval of Boreal Forest Biomass in Sloping Terrain Using P-Band SAR Backscatter Intensity Data. *IEEE Trans. Geosci. Remote Sens.* **2013**, *51*, 2646–2665. [CrossRef]

236. Santoro, M.; Cartus, O. Research pathways of forest above-ground biomass estimation based on SAR backscatter and interferometric SAR observations. *Remote Sens.* **2018**, *10*, 608. [CrossRef]

237. Schlund, M.; Scipal, K.; Quegan, S. Assessment of a Power Law Relationship Between P-Band SAR Backscatter and Aboveground Biomass and Its Implications for BIOMASS Mission Performance. *IEEE J. Sel. Top. Appl. Earth Obs. Remote Sens.* **2018**, *99*, 1–10. [CrossRef]

238. Chen, X.; Liang, S.; Cao, Y.; He, T.; Wang, D. Observed contrast changes in snow cover phenology in northern middle and high latitudes from 2001–2014. *Sci. Rep.* **2015**, *5*, 16820. [CrossRef] [PubMed]

239. Boike, J.; Grau, T.; Heim, B.; Günther, F.; Langer, M.; Muster, S.; Gouttevin, I.; Lange, S. Satellite-derived changes in the permafrost landscape of central Yakutia, 2000–2011: Wetting, drying, and fires. *Glob. Planet. Chang.* **2016**, *139*, 116–127. [CrossRef]

240. Sulla-Menashe, D.; Woodcock, C.E.; Friedl, M.A. Canadian boreal forest greening and browning trends: An analysis of biogeographic patterns and the relative roles of disturbance versus climate drivers. *Environ. Res. Lett.* **2018**, *13*. [CrossRef]

241. Myneni, R.B.; Keeling, C.D.; Tucker, C.J.; Asrar, G.; Nemani, R.R. Increased plant growth in the northern high latitudes from 1981 to 1991. *Nature* **1997**, *386*, 698–702. [CrossRef]

242. Goetz, S.J.; Bunn, A.G.; Fiske, G.J.; Houghton, R.A. Satellite-observed photosynthetic trends across boreal North America associated with climate and fire disturbance. *Proc. Natl. Acad. Sci. USA* **2005**, *102*, 13521–13525. [CrossRef]

243. Fraser, R.H.; Olthof, I.; Carrière, M.; Deschamps, A.; Pouliot, D. Detecting long-term changes to vegetation in northern Canada using the Landsat satellite image archive. *Environ. Res. Lett.* **2011**, *6*. [CrossRef]

244. Didan, K.; Munoz, A.B.; Solano, R.; Huete, A. *MODIS Vegetation Index User's Guide (MOD13 Series)*; Vegetation Index and Phenology Lab, University of Arizona: Tucson, AZ, USA, 2015; pp. 1–38.

245. Rignot, E.; Thomas, R.H. Mass balance of polar ice sheets. *Science* **2002**, *297*, 1502–1506. [CrossRef] [PubMed]

246. Trusel, L.D.; Das, S.B.; Osman, M.B.; Evans, M.J.; Smith, B.E.; Fettweis, X.; van den Broeke, M.R. Nonlinear rise in Greenland runoff in response to post-industrial Arctic warming. *Nature* **2018**, *564*, 104. [CrossRef] [PubMed]

247. Bevis, M.; Harig, C.; Khan, S.A.; Brown, A.; Simons, F.J.; Willis, M.; Fettweis, X.; van den Broeke, M.R.; Madsen, F.B.; Kendrick, E.; et al. Accelerating changes in ice mass within Greenland, and the ice sheet's sensitivity to atmospheric forcing. *Proc. Natl. Acad. Sci. USA* **2019**, *116*, 1934–1939. [CrossRef] [PubMed]

248. Graeter, K.; Osterberg, E.C.; Lewis, G.; Hawley, R.; Marshall, H.P.; Meehan, T.; Overly, T.; Birkel, S. Ice core records of West Greenland melt and climate forcing. *Geophys. Res. Lett.* **2018**, *45*, 3164–3172. [CrossRef]

249. The IMBIE Team. Mass balance of the Antarctic Ice Sheet from 1992 to 2017. *Nature* **2018**, *558*, 219–222. [CrossRef] [PubMed]

250. Seo, M.; Kim, H.C.; Huh, M.; Yeom, J.M.; Lee, C.; Lee, K.S.; Choi, S.; Han, K.S. Long-term variability of surface albedo and its correlation with climatic variables over antarctica. *Remote Sens.* **2016**, *8*, 981. [CrossRef]

251. Meyer, H.; Katurji, M.; Appelhans, T.; Müller, M.; Nauss, T.; Roudier, P.; Zawar-Reza, P. Mapping daily air temperature for Antarctica based on MODIS LST. *Remote Sens.* **2016**, *8*, 732. [CrossRef]

252. Hall, D.; Cullather, R.; DiGirolamo, N.; Comiso, J.; Medley, B.; Nowicki, S. A Multilayer Surface Temperature, Surface Albedo, and Water Vapor Product of Greenland from MODIS. *Remote Sens.* **2018**, *10*, 555. [CrossRef]

253. Nicolas, J.P.; Vogelmann, A.M.; Scott, R.C.; Wilson, A.B.; Cadeddu, M.P.; Bromwich, D.H.; Verlinde, J.; Lubin, D.; Russell, L.M.; Jenkinson, C.; et al. January 2016 extensive summer melt in West Antarctica favoured by strong El Niño. *Nat. Commun.* **2017**, *8*, 15799. [CrossRef]

254. Tedesco, M.; Fettweis, X.; Mote, T.; Wahr, J.; Alexander, P.; Box, J.E.; Wouters, B. Evidence and Analysis of 2012 Greenland Records from Spaceborne Observations, a Regional Climate Model and Reanalysis Data. *Cryosphere* **2013**, *7*, 615–630. [CrossRef]

255. Li, X.; Zhang, Y.; Liang, L. Snowmelt detection on the Greenland ice sheet using microwave scatterometer measurements. *Int. J. Remote Sens.* **2017**, *38*, 796–807. [CrossRef]

256. Rignot, E.; Mouginot, J.; Scheuchl, B. Ice flow of the Antarctic ice sheet. *Science* **2011**, *333*, 1427–1430. [CrossRef]

257. Kaab, A.; Berthier, E.; Nuth, C.; Gardelle, J.; Arnaud, Y. Contrasting patterns of early twenty-first-century glacier mass change in the Himalayas. *Nature* **2012**, *488*, 495–498. [CrossRef]

258. Jacob, T.; Wahr, J.; Pfeffer, W.T.; Swenson, S. Recent contributions of glaciers and ice caps to sea level rise. *Nature* **2012**, *482*, 514–518. [CrossRef]

259. Neckel, N.; Kropacek, J.; Bolch, T.; Hochschild, V. Glacier mass changes on the Tibetan Plateau 2003–2009 derived from ICESat laser altimetry measurements. *Environ. Res. Lett.* **2014**, *9*, 014009. [CrossRef]

260. Gardner, A.S.; Moholdt, G.; Cogley, J.G.; Wouters, B.; Arendt, A.A.; Wahr, J.; Berthier, E.; Hock, R.; Pfeffer, W.T.; Kaser, G.; et al. A Reconciled Estimate of Glacier Contributions to Sea Level Rise: 2003 to 2009. *Science* **2013**, *340*, 852–857. [CrossRef]

261. Xiao, Z.; Duan, A. Impacts of Tibetan Plateau snow cover on the interannual variability of the East Asian summer monsoon. *J. Clim.* **2016**, *29*, 8495–8514. [CrossRef]

262. Basang, D.; Barthel, K.; Olseth, J.A. Satellite and Ground Observations of Snow Cover in Tibet during 2001–2015. *Remote Sens.* **2017**, *9*, 1201. [CrossRef]

263. Tang, Z.G.; Wang, J.; Li, H.Y.; Yan, L.L. Spatiotemporal changes of snow cover over the Tibetan plateau based on cloud-removed moderate resolution imaging spectroradiometer fractional snow cover product from 2001 to 2011. *Appl. Remote Sens.* **2013**, *7*, 073582. [CrossRef]

264. Chen, X.N.; Long, D.; Liang, S.L.; He, L.; Zeng, C.; Hao, X.H.; Hong, Y. Developing a composite daily snow cover extent record over the Tibetan Plateau from 1981 to 2016 using multisource data. *Remote Sens. Environ.* **2018**, *215*, 284–299. [CrossRef]

265. Kou, X.K.; Jiang, L.M.; Yan, S.; Zhao, T.J.; Lu, H.; Cui, H.Z. Detection of land surface freeze-thaw status on the Tibetan Plateau using passive microwave and thermal infrared remote sensing data. *Remote Sens. Environ.* **2017**, *199*, 291–301. [CrossRef]

266. Li, X.; Jin, R.; Pan, X.D.; Zhang, T.J.; Guo, J.W. Changes in the near-surface soil freeze-thaw cycle on the Qinghai-Tibetan Plateau. *Int. J. Appl. Earth Obs.* **2012**, *17*, 33–42. [CrossRef]

267. Ma, R.H.; Yang, G.S.; Duan, H.T.; Jiang, J.H.; Wang, S.M.; Feng, X.Z.; Li, A.N.; Kong, F.X.; Xue, B.; Wu, J.L.; et al. China's lakes at present: Number, area and spatial distribution. *Sci. China Earth Sci.* **2011**, *54*, 283–289. [CrossRef]

268. Zhang, G.Q.; Zheng, G.X.; Gao, Y.; Xiang, Y.; Lei, Y.B.; Li, J.L. Automated Water Classification in the Tibetan Plateau Using Chinese GF-1 WFV Data. *Photogramm. Eng. Remote Sens.* **2017**, *83*, 509–519. [CrossRef]

269. Wang, L.H.; Lu, A.X.; Yao, T.D.; Wang, N.L. The study of typical glaciers and lakes fluctuations using remote sensing in Qinghai-Tibetan Plateau. In Proceedings of the 2007 IEEE International Geoscience and Remote Sensing Symposium, Barcelona, Spain, 23–27 July 2007; pp. 4526–4529.

270. Wan, W.; Xiao, P.; Feng, X.; Li, H.; Ma, R.; Duan, H. Remote sensing analysis for changes of lakes in the southeast of Qiangtang area, Qinghai-Tibet Plateau in recent 30 years. *J. Lake Sci.* **2010**, *22*, 874–881. (In Chinese)

271. Zhang, G.Q.; Yao, T.D.; Xie, H.J.; Zhang, K.X.; Zhu, F.J. Lakes' state and abundance across the Tibetan Plateau. *Chin. Sci. Bull.* **2014**, *59*, 3010–3021. [CrossRef]

272. Yang, X.; Lu, X.; Park, E.; Tarolli, P. Impacts of Climate Change on Lake Fluctuations in the Hindu Kush-Himalaya-Tibetan Plateau. *Remote Sens.* **2019**, *11*, 1082. [CrossRef]

273. Hwang, C.W.; Peng, M.F.; Ning, J.S.; Luo, J.; Sui, C.H. Lake level variations in China from TOPEX/Poseidon altimetry: Data quality assessment and links to precipitation and ENSO. *Geophys. J. Int.* **2005**, *161*, 1–11. [CrossRef]

274. Zhang, G.Q.; Xie, H.J.; Kang, S.C.; Yi, D.H.; Ackley, S.F. Monitoring lake level changes on the Tibetan Plateau using ICESat altimetry data (2003–2009). *Remote Sens. Environ.* **2011**, *115*, 1733–1742. [CrossRef]

275. Phan, V.H.; Lindenbergh, R.; Menenti, M. ICESat derived elevation changes of Tibetan lakes between 2003 and 2009. *Int. J. Appl. Earth Obs.* **2012**, *17*, 12–22. [CrossRef]

276. Wang, X.W.; Gong, P.; Zhao, Y.Y.; Xu, Y.; Cheng, X.; Niu, Z.G.; Luo, Z.C.; Huang, H.B.; Sun, F.D.; Li, X.W. Water-level changes in China's large lakes determined from ICESat/GLAS data. *Remote Sens. Environ.* **2013**, *132*, 131–144. [CrossRef]

277. Gao, L.; Liao, J.J.; Shen, G.Z. Monitoring lake-level changes in the Qinghai-Tibetan Plateau using radar altimeter data (2002–2012). *Remote Sens.* **2013**, *7*, 073470. [CrossRef]

278. Zhang, Z.X.; Chang, J.; Xu, C.Y.; Zhou, Y.; Wu, Y.H.; Chen, X.; Jiang, S.S.; Duan, Z. The response of lake area and vegetation cover variations to climate change over the Qinghai-Tibetan Plateau during the past 30 years. *Sci. Total Environ.* **2018**, *635*, 443–451. [CrossRef]

279. Pang, G.J.; Wang, X.J.; Yang, M.X. Using the NDVI to identify variations in, and responses of, vegetation to climate change on the Tibetan Plateau from 1982 to 2012. *Quatern. Int.* **2017**, *444*, 87–96. [CrossRef]

280. Shen, X.J.; An, R.; Feng, L.; Ye, N.; Zhu, L.J.; Li, M.H. Vegetation changes in the Three-River Headwaters Region of the Tibetan Plateau of China. *Ecol. Indic.* **2018**, *93*, 804–812. [CrossRef]

281. Yu, H.Y.; Luedeling, E.; Xu, J.C. Winter and spring warming result in delayed spring phenology on the Tibetan Plateau. *Proc. Natl. Acad. Sci. USA* **2010**, *107*, 22151–22156. [CrossRef]

282. Shen, M.G. Spring phenology was not consistently related to winter warming on the Tibetan Plateau. *Proc. Natl. Acad. Sci. USA* **2011**, *108*, E91–E92. [CrossRef] [PubMed]

283. Chen, H.; Zhu, Q.A.; Wu, N.; Wang, Y.F.; Peng, C.H. Delayed spring phenology on the Tibetan Plateau may also be attributable to other factors than winter and spring warming. *Proc. Natl. Acad. Sci. USA* **2011**, *108*, E93. [CrossRef] [PubMed]

284. Li, M.; Wu, J.; Song, C.C.; He, Y.; Niu, B.; Fu, G.; Tarolli, P.; Tietjen, B.; Zhang, X. Temporal Variability of Precipitation and Biomass of Alpine Grasslands on the Northern Tibetan Plateau. *Remote Sens.* **2019**, *11*, 360. [CrossRef]

285. Wang, H.S.; Liu, D.S.; Lin, H.; Montenegro, A.; Zhu, X.L. NDVI and vegetation phenology dynamics under the influence of sunshine duration on the Tibetan plateau. *Int. J. Climatol.* **2015**, *35*, 687–698. [CrossRef]

286. Wang, S.Y.; Wang, X.Y.; Chen, G.S.; Yang, Q.C.; Wang, B.; Ma, Y.X.; Shen, M. Complex responses of spring alpine vegetation phenology to snow cover dynamics over the Tibetan Plateau, China. *Sci. Total Environ.* **2017**, *593*, 449–461. [CrossRef] [PubMed]

287. Yi, S.H.; Zhou, Z.Y. Increasing contamination might have delayed spring phenology on the Tibetan Plateau. *Proc. Natl. Acad. Sci. USA* **2011**, *108*, E94. [CrossRef] [PubMed]

288. Myers-Smith, I.H.; Forbes, B.C.; Wilmking, M.; Hallinger, M.; Lantz, T.; Blok, D.; Tape, K.D.; Macias-Fauria, M.; Sass-Klaassen, U.; Lévesque, E.; et al. Shrub expansion in tundra ecosystems: Dynamics, impacts and research priorities. *Environ. Res. Lett.* **2011**, *6*, 045509. [CrossRef]

289. Piermattei, L.; Carturan, L.; de Blasi, F.; Tarolli, P.; Dalla Fontana, G.; Vettore, A.; Pfeifer, N. Suitability of ground-based SfM–MVS for monitoring glacial and periglacial processes. *Earth Surf. Dyn.* **2016**, *4*, 425–443. [CrossRef]

290. Abdalati, W.; Zwally, H.J.; Bindschadler, R.; Csatho, B.; Farrell, S.L.; Fricker, H.A.; Harding, D.; Kwok, R.; Lefsky, M.; Markus, T.; et al. The ICESat-2 laser altimetry mission. *Proc. IEEE* **2010**, *98*, 735–751. [CrossRef]

291. Durand, M.; Fu, L.L.; Lettenmaier, D.P.; Alsdorf, D.E.; Rodriguez, E.; Esteban-Fernandez, D. The surface water and ocean topography mission: Observing terrestrial surface water and oceanic submesoscale eddies. *Proc. IEEE* **2010**, *98*, 766–779. [CrossRef]

292. Li, S.; Tan, H.; Liu, Z.; Zhou, Z.; Liu, Y.; Zhang, W.; Liu, K.; Qin, B. Mapping high mountain lakes using space-borne near-nadir SAR observations. *Remote Sens.* **2018**, *10*, 1418. [CrossRef]

293. van der Sanden, J.J.; Geldsetzer, T. Compact polarimetry in support of lake ice breakup monitoring: Anticipating the RADARSAT Constellation Mission. *Can. J. Remote Sens.* **2015**, *41*, 440–457. [CrossRef]

294. Alvarez-Salazar, O.; Hatch, S.; Rocca, J.; Rosen, P.; Shaffer, S.; Shen, Y.; Sweetser, T.; Xaypraseuth, P. November. Mission design for NISAR repeat-pass Interferometric SAR. In Sensors, Systems, and Next-Generation Satellites XVIII. *Int. Soc. Opt. Photonics* **2014**, *9241*, 92410C.

295. Du, J.; Kimball, J.S.; Moghaddam, M. Theoretical modeling and analysis of l-and p-band radar backscatter sensitivity to soil active layer dielectric variations. *Remote Sens.* **2015**, *7*, 9450–9472. [CrossRef]

296. George, A.D.; Wilson, C.M. Onboard processing with hybrid and reconfigurable computing on small satellites. *Proc. IEEE* **2018**, *106*, 458–470. [CrossRef]

297. Gorelick, N.; Hancher, M.; Dixon, M.; Ilyushchenko, S.; Thau, D.; Moore, R. Google Earth Engine: Planetary-scale geospatial analysis for everyone. *Remote Sens. Environ.* **2017**, *202*, 18–27. [CrossRef]

298. Thompson, A.A. Overview of the RADARSAT constellation mission. *Can. J. Remote Sens.* **2015**, *41*, 401–407. [CrossRef]

299. Le Toan, T.; Quegan, S.; Davidson, M.W.J.; Balzter, H.; Paillou, P.; Papathanassiou, K.; Plummer, S.; Rocca, F.; Saatchi, S.; Shugart, H.; et al. The BIOMASS mission: Mapping global forest biomass to better understand the terrestrial carbon cycle. *Remote Sens. Environ.* **2011**, *115*, 2850–2860. [CrossRef]

300. Spencer, M.; Ulaby, F. Spectrum Issues Faced by Active Remote Sensing: Radio frequency interference and operational restrictions Technical Committees. *IEEE Geosci. Remote Sens. Mag.* **2016**, *4*, 40–45. [CrossRef]

301. Altena, B.; Kääb, A. Weekly glacier flow estimation from dense satellite time series using adapted optical flow technology. *Front. Earth Sci.* **2017**, *5*, 53. [CrossRef]

302. Yang, Z.; Kang, Z.; Cheng, X.; Yang, J. Improved multi-scale image matching approach for monitoring Amery ice shelf velocity using Landsat 8. *Eur. J. Remote Sens.* **2019**, *52*, 56–72. [CrossRef]

303. Rittger, K.; Bair, E.H.; Kahl, A.; Dozier, J. Spatial estimates of snow water equivalent from reconstruction. *Adv. Water Resour.* **2016**, *94*, 345–363. [CrossRef]

304. Bair, E.H.; Rittger, K.; Davis, R.E.; Painter, T.H.; Dozier, J. Validating reconstruction of snow water equivalent in California's Sierra Nevada using measurements from the NASA Airborne Snow Observatory. *Water Resour. Res.* **2016**, *52*. [CrossRef]

305. Deems, J.S.; Painter, T.H.; Finnegan, D.C. Lidar measurement of snow depth: A review. *J. Glaciol.* **2013**, *59*, 467–479. [CrossRef]

306. Painter, T.H.; Berisford, D.F.; Boardman, J.W.; Bormann, K.J.; Deems, J.S.; Gehrke, F.; Hedrick, A.; Joyce, M.; Laidlaw, R.; Marks, D.; et al. The Airborne Snow Observatory: Fusion of scanning lidar, imaging spectrometer, and physically-based modeling for mapping snow water equivalent and snow albedo. *Remote Sens. Environ.* **2016**, *184*, 139–152. [CrossRef]

307. Leinss, S.; Löwe, H.; Proksch, M.; Lemmetyinen, J.; Wiesmann, A.; Hajnsek, I. Anisotropy of seasonal snow measured by polarimetric phase differences in radar time series. *Cryosphere* **2016**, *10*, 1771–1797. [CrossRef]

308. Henkel, P.; Koch, F.; Appel, F.; Bach, H.; Prasch, M.; Schmid, L.; Schweizer, J.; Mauser, W. Snow water equivalent of dry snow derived from GNSS carrier phases. *IEEE Trans. Geosci. Remote Sens.* **2018**, *56*, 3561–3572. [CrossRef]

309. Qiu, Y.; Xie, P.; Leppäranta, M.; Wang, X.; Lemmetyinen, J.; Lin, H.; Shi, L. MODIS-based Daily Lake Ice Extent and Coverage dataset for Tibetan Plateau. *Big Earth Data* **2019**, *3*, 170–185. [CrossRef]

310. Pulliainen, J.; Aurela, M.; Laurila, T.; Aalto, T.; Takala, M.; Salminen, M.; Kulmala, M.; Barr, A.; Heimann, M.; Lindroth, A.; et al. Early snowmelt significantly enhances boreal springtime carbon uptake. *Proc. Natl. Acad. Sci. USA* **2017**, *114*, 11081–11086. [CrossRef]

311. Bokhorst, S.; Pedersen, S.H.; Brucker, L.; Anisimov, O.; Bjerke, J.W.; Brown, R.D.; Ehrich, D.; Essery, R.L.; Heilig, A.; Ingvander, S.; et al. Changing Arctic snow cover: A review of recent developments and assessment of future needs for observations, modelling, and impacts. *Ambio* **2016**, *45*, 516–537. [CrossRef]

312. Qiu, Y.; Massimo, M.; Li, X.; Birendra, B.; Joni, K.; Narantuya, D.; Liu, S.; Gao, Y.; Cheng, B.; Wu, T.; et al. Observing and understanding high mountain and cold regions using big earth data. *Bull. Chin. Acad. Sci.* **2017**, *32*, 82–94.

313. Arendt, A.A.; Houser, P.; Kapnick, S.B.; Kargel, J.S.; Kirschbaum, D.; Kumar, S.; Margulis, S.A.; McDonald, K.C.; Osmanoglu, B.; Painter, T.H.; et al. NASA's High Mountain Asia Team (HiMAT): Collaborative research to study changes of the High Asia region. In Proceedings of the 2017 AGU Fall Meeting Abstracts, New Orleans, LA, USA, 11–15 December 2017.

MDPI

St. Alban-Anlage 66

4052 Basel

Switzerland

Tel. +41 61 683 77 34

Fax +41 61 302 89 18

www.mdpi.com

Remote Sensing Editorial Office

E-mail: remotesensing@mdpi.com

www.mdpi.com/journal/remotesensing

www.ingramcontent.com/pod-product-compliance
Lightning Source LLC
Chambersburg PA
CBHW051848210326
41597CB00033B/5816